Basics of CNC Programming

RIVER PUBLISHERS SERIES IN MATHEMATICAL AND ENGINEERING SCIENCES

Indexing: All books published in this series are submitted to the Web of Science Book Citation Index (BkCI), to SCOPUS, to CrossRef and to Google Scholar for evaluation and indexing.

Mathematics is the basis of all disciplines in science and engineering. Especially applied mathematics has become complementary to every branch of engineering sciences. The purpose of this book series is to present novel results in emerging research topics on engineering sciences, as well as to summarize existing research. It engrosses mathematicians, statisticians, scientists and engineers in a comprehensive range of research fields with different objectives and skills, such as differential equations, finite element method, algorithms, discrete mathematics, numerical simulation, machine leaning, probability and statistics, fuzzy theory, etc.

Books published in the series include professional research monographs, edited volumes, conference proceedings, handbooks and textbooks, which provide new insights for researchers, specialists in industry, and graduate students.

Topics covered in the series include, but are not limited to:

- Advanced mechatronics and robotics
- Artificial intelligence
- Automotive systems
- Discrete mathematics and computation
- Fault diagnosis and fault tolerance
- Finite element methods
- Fuzzy and possibility theory
- Industrial automation, process control and networked control systems
- Intelligent control systems
- Neural computing and machine learning
- Operations research and management science
- Optimization and algorithms
- Queueing systems
- Reliability, maintenance and safety for complex systems
- Resilience
- Stochastic modelling and statistical inference
- Supply chain management
- System engineering, control and monitoring
- Tele robotics, human computer interaction, human-robot interaction

For a list of other books in this series, visit www.riverpublishers.com

Basics of CNC Programming

Pawan Kumar Negi

Graphic Era Deemed to be University, Dehradun
India

Mangey Ram

Graphic Era Deemed to be University, Dehradun
India

Om Prakash Yadav

North Dakota State University, North Dakota
USA

LONDON AND NEW YORK

Published 2019 by River Publishers
River Publishers
Alsbjergvej 10, 9260 Gistrup, Denmark
www.riverpublishers.com

Distributed exclusively by Routledge
4 Park Square, Milton Park, Abingdon, Oxon OX14 4RN
605 Third Avenue, New York, NY 10017, USA

First issued in paperback 2023

Basics of CNC Programming / by Pawan Kumar Negi, Mangey Ram, Om Prakash Yadav.

Routledge is an imprint of the Taylor & Francis Group, an informa business

ISBN 13: 978-87-7022-965-4 (pbk)
ISBN 13: 978-87-7022-043-9 (hbk)
ISBN 13: 978-1-003-33731-7 (ebk)

Contents

Preface

Before the invention of automatic machines or automation, industrial manufacturing of machines and machine parts, for the automobile industry, aviation industry, defense industry, etc. were manufactured using manual machines. Because of this, manufacturers could not make complex profile or shapes with high accuracy. Apart from this, the manufacturing (production) time cost was very slow, the production cost was very high, the rejection rate was high and the manufacturer was not completing the buyer demands at a given time.

Semi-automatic manufacturing machine known as the **NC** machine gave a new direction and boost to the manufacturing industry. This machine was introduced in 1950s by the MIT institute in United States. After the invention of **NC** machine, the actual problem of machining of typical profile or complex shapes was solved. After solving this major issue, production rate was increased along with high accuracy.

But it is a unique quality of human beings to differentiate between humans and other species. This quality makes us transform dreams into reality. Because of this human quality, one big revolutionary change came in the 1970s and that change was the introduction of *Computer Numerical Control* **(CNC)** machine. This is the biggest boon at that time to the manufacturing industry.

Currently, **CNC** has become a very huge name in the field of manufacturing industry. Because of the extensive use of **CNC** machine and machine technology it is used in every industry directly or indirectly like the automobile, industry, aviation industry, defense industry, oil field industry, medical industry, electronics industry, optical industry etc.

Now-a-days **CNC** is working in every field of our life. It means, it is a concept of modern manufacturing technology, which replaced the conventional machine (manual machine) from all fields, we cannot imagine our modern life without CNC technology. CNC technology means fully computerized and automatically controlled manufacturing machine. Today's need is mass production with high accuracy, best quality, very less rejection, cost effective product, save more time during production and CNC machine fulfills all these qualities. Due to these qualities and best result, manufacturers are adopting this modern technology worldwide.

This book tells us about how to make CNC program and what cutting parameters are required to make a good CNC program. We explained about cutting parameters of CNC machine like cutting feed, depth of cut, rpm, cutting speed etc. and explained about **G codes & M codes**. We explained the simple method of CNC program writing and how to cut the material in different operations like straight turning, step turning, taper turning, drilling, chamfering, radius profile, profile turning etc. We try to develop the level of CNC programming from basics to industrial format. Also we tried to provide basic knowledge of cutting tool and its nomenclature. Here in this book, we have given some drawings and CNC programs for your practice.

Important things about this book:

- This book provides the readers programming support on how to make CNC program in simple language.
- We hope the readers will follow all the CNC programming instructions which are given in this book.
- It is important to understand that each CNC program is different from another CNC program, due to its drawing and coordinates.
- It is notable that, few starting and ending CNC programming lines (blocks) of the CNC program will be common. It depends on CNC control software.

So here, the learning process is being so easy for beginners, because in this book, a method of learning is simple, different with a new idea. We need only machining operations in the form of G & M codes with X, Z coordinates of turning machine.

So, we advise all beginners to read the definitions & meanings of the given G-codes & M-codes, types of operations, some important conditions and important notes, which are applied in some special condition(s) in the **CNC** program.

Authors

Acknowledgement

We wish to thank the people who helped us write this book. We extend our special thanks to Mr. Durgeshwar Pratap Singh, of Graphic Era (Deemed to be University), India, for advice throughout the project.

Authors are grateful to the authorities of the Graphic Era (Deemed to be University), India, who provided excellent environment, opportunities and facilities to undertake this task. In particular, the author Mr. Pawan Negi would like to express his sincere thanks to the Mechanical engineering department of Graphic Era (Deemed to be University). Also we are very thankful to the companies, Rawat Engg. Tech Pvt Ltd., India and Sara Sae India Pvt Ltd, India and other companies who provided us information, images and illustrations directly or indirectly for use in the book.

We are also thankful to our students, for their probing questions and comments which contributed to the subject.

Last, but not the least, a very special thanks to our family who supported us and gave us love, patience and motivation in making this project possible.

One of the authors of this book Mr. Pawan Negi dedicates this book to his entire family and his son Arihant Negi.

Er. Cum CNC Programmer Pawan Kumar Negi

Prof. (Dr.) Mangey Ram

Prof. (Dr.) Om Prakash Yadav

List of Figures

List of Abbreviations

μ	Micron
Ø	Diameter
ATC	Automatic Tool Changer
AWC	Auto Work Changer/Automatic Pallet changer
BHC	Bolt Hole Circle
CAM	Computer Aided Manufacturing
CCW	Counter Clockwise Direction
CMM	Coordinate Measuring Machine
CNC	Computer Numerical Control
CPU	Central processing unit
CW	Clockwise Direction
D	Diameter
d	Diameter of the machined surface (diameter of the work piece after machining)
DOC	Depth of cut
EOB	End of Block
EPR	Ending Point of Radius
EPRC	Ending Point of Rough Chamfer
EPRT	Ending Point of Rough Taper
EPT	End Point of Taper
F	Feed
FPM	Feet per minute
G	Preparatory function (code)
G00	Rapid Traverse
G01	Linear Motion
G02	Circular motion in clock wise direction
G03	Circular motion in Counter clock wise direction
IPM	Inch Per Minute
M	Miscellaneous code
MCU	Machine control unit
MS	Mild Steel

N	Spindle speed in rpm
NC	Numerical Control
O	Program number/Program name
R	Radius
RPM	Revolution Per Minute
S	Spindle Speed
SPM	Special Purpose Machine
SPR	Starting Point of Radius
SPRC	Starting Point of Rough Chamfer
SPRT	Starting Point of Rough Taper
SPT	Starting Point of Taper
SPTT	Single Point Turning Tool
t	Cutting time
T	Tool
TIP	Cutting Tool tip position/Tool type
TPI	Threads Per Inch
V_c	Cutting speed
X	Main axis in X-axis direction
Y	Main axis in Y-axis direction
Z	Main axis in Z-axis direction

1

CNC Machine and Its Importance

1.1 History of CNC Machine

Computer numerical control (CNC) machine is the advanced version of the conventional/old manual machine. The first manual metal working lathe was invented in the year 1800 by Henry Maudslay, a British machine tool innovator and inventor. In the beginning, the lathe machine was a simple machine which held the work piece between two rigid and strong supports.

After the time period of the first metal cutting machine, manual machines became little advanced but still needed the operator for its operation. Apart from this, these machines have some problems such as inability to make complex shapes, causes variation in dimensions, high scrap rate, fitting problem, wastage of raw material, high production cost and machine run time. Despite this, we are not saying that conventional/manual machine was not good. During that time, these machines were the best ones.

Due to the above-mentioned drawbacks, during World War I and II the American Air force faced some problems in equipments like the helicopter rotor blade and other machine parts because the manually operated machine was unable to perform three-dimensional complex operations as intended. In 1952, the Massachusetts Institute of Technology (MIT) introduced the first semiautomatic NC machine with the support of Mr. John T Parson and IBM. It was an American government funded project for the American Air force. This initial machine solved the initial requirements of the American Air force and gave revolutionary results to the manufacturing industry. **Numerical control (NC)** machine was the design of the mechanical and electronics technology. This NC machine performed its work perfectly and resolved maximum machine-related problems. These machines technology were the best machine technology of that time. We can say these machines were the back bone of that time industry like aviation industry etc. But after development in machine technologies, in the 1970s, Mr. John T Parson and the MIT together developed the first CNC machine tools with the help of IBM. This CNC machine was fully automatic and owns a computation system. After the development of CNC machine technology, the aircraft and manufacturing industries got the facility of contouring operations, automatic computation, typical profile machining solution through modern CNC machine technology and aircraft industry got the solution of aerodynamic parts. CNC machine technology gave a high degree of accuracy required by the aircraft industry [1].

1.2 What is CNC?

CNC stands for **Computer Numerical Control**. A CNC machine is a sophisticated metal (material) removing computerized machine on which we can make any complex shapes, which was not possible before, with high accuracy, repeatability, better quality, less rejection and mass production. Before NC and CNC machine, engineers cannot machining same product again and again with all the above qualities. CNC machine can perform works as lathe machine, milling machine, gear cutting machine, laser cutting, water jet CNC machine, stone cutting machine and many other different areas (Figure 1.1) [2].

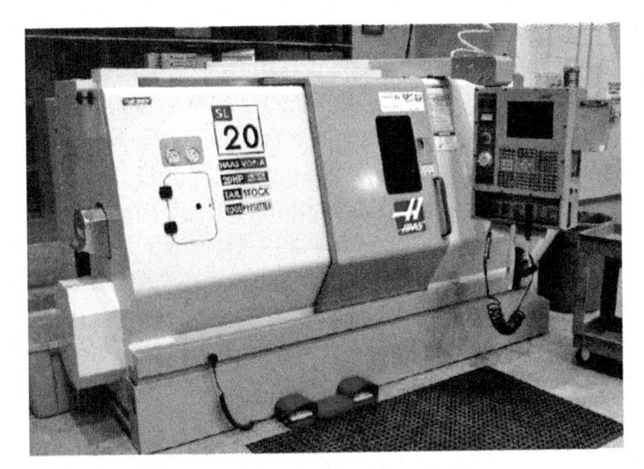

Figure 1.1 CNC turning machine.

(Courtesy: Haas SL 20, Sara Sae India Pvt. Ltd, India)

1.3 Working of CNC Machine

Computer numerical control (CNC) machines are automatic machines; they work automatically during manufacturing and fabrication of parts & components. CNC machines are used Alpha numeric instructions with using of drawing coordinates. These alpha numeric instructions are called G & M codes.

1.4 Language of CNC Machine

CNC machines are using for the purpose of manufacturing in every field of life. CNC machines are very accurate and powerful industrial robots. Most CNC machine tools use a language set by the Electronics Industry Association (EIA) in the 1960's. The official name of this **language is RS-274D**, but everyone says it "G-code" or "**G & M Code**" because mostly CNC program developed with the help of letters G or M.

The G-code language was developed when machine controls had very little memory. It was therefore designed to be as compact as possible. The modern machine tool language is the safest and most efficient way yet.

1.5 Benefits: After Coming to the CNC Machine

CNC machine gives many benefits over manual machines. CNC machine works rapidly & better than a manual machine. From the period of the manual machine, complex & complicated parts that were not previously possible but now being manufactured in CNC machine. We can design the complex shape and convert to G & M codes. That G & M codes program can transfer to CNC machine and immediately we can machine. The result, we can save time and money without any errors.

After coming of CNC machine technology, industrial automation picked the top gear of automation. And the result is operator's physical work reduced too much. Many CNC machines can work without manpower during their entire machining cycle, it gives the opportunity of several benefits to do operator to perform another task like it reduces operator fatigue, removed human error and consistent and predictable machining time for each work piece. After coming of CNC machine technology, no need of highly skilled operator compares to the conventional machine operator.

The major benefit of CNC machine technology is consistent and high quality work piece. Today's CNC machine gives unbelievable accuracy and repeatability of the work piece. It means if the CNC program is ready as per drawing, the operator can start bulk production.

1.6 Fast Change Over

Fast change-over, it means suppose one lot is machining on the CNC machine, we need urgent to machine another different lot. We can change the program and tool according to next job order very quickly and start the next production.

1.7 What is the Importance of CNC Machine in the Modern World?

Before 1950's, engineers were not able to make complex shapes or profile (as per drawing like helicopter's rotor shaft, blades etc.) on manual machines with high accuracy and best quality, because the manual machine was not able to machine that complex/complicated shapes or profile.

But in the 1950's, one semi-automatic machine introduced by **MIT** (Massachusetts Institute of Technology), which is based on computer technology, that machine name was NC (Numerical Control) machine. After came that NC machine technology, manufacturing industry was totally changed. Before NC machine, that products/parts were impossible for manufacturing? After coming of this CNC machine technology, machine parts or products were easy to manufacturing.

CNC machine technology is using in every field of human life or in every industry. CNC machines are used in the aerospace, automobile, machine parts manufacturing, optical, and defense and many other industries. It is said to better that these machines are full fill our daily needs in every field of life. They are playing his roll directly or indirectly very well. **Without CNC machine technology, we cannot imagine our high-tech life**.

1.8 Advantages and Disadvantages of CNC Machine

We cannot imagine modern manufacturing industry without CNC machine technology. CNC machines are widely used in the manufacturing industry. Traditional machines such as lathe, milling, shaper machine machines etc.

Operate by trained engineer/operator but for CNC machine, we need a semiskilled operator to operate the CNC machine.

1.8.1 Advantages

- CNC machine can be used continuously for 365 days a year. It is switched off, only when required for maintenance.
- CNC machine can manufacture one or thousands of product pieces continuously with the help of CNC programming which is programmed as specified in the drawing. It is important that each manufactured product to be the same.
- For the operation of CNC machine, semi-skilled/less-trained workers are required, on the other hand, manual machines (lathe, milling and other manual machines) need skilled engineers/operators.
- CNC machine software can be improved or updated.
- CNC machine trainee can be trained through the virtual CNC simulator and CNC program can be made through the CNC machine simulator. Nowadays, this is easily available.
- CNC machine can be programmed by advance design software such as Master CAM/Del CAM/solid CAM etc.
- One person can supervise many CNC machines as once machined are programmed.

1.8.2 Some Most Important Advantages of CNC Machine

- High accuracy and repeatability
- Reduce inspection
- Ease of assembly and interchangeability
- Better productivity
- Less rejection/rework
- Reduced tooling
- Increased overall efficiency
- Mass production
- Space saving
- Low product cost
- Customer satisfaction
- Less cycle time
- Less material handling
- Less paper work

- Less inventory cost
- High flexibility for design changes
- Design freedom for complex shapes
- Increased operator efficiency
- Semi-skilled operator/engineer required

1.8.3 Disadvantages

- CNC machines are very expensive compared to a manual machine.
- Less manpower (operators) is required to operate the CNC machine.

Figures 1.2 and 1.3 show heavy duty and extremely heavy duty CNC Lathe machine.

Figure 1.2 A heavy duty CNC machine.

(Courtesy: Tuscan 5 Bed Way Heavy Duty CNC Lathe, United Kingdom)

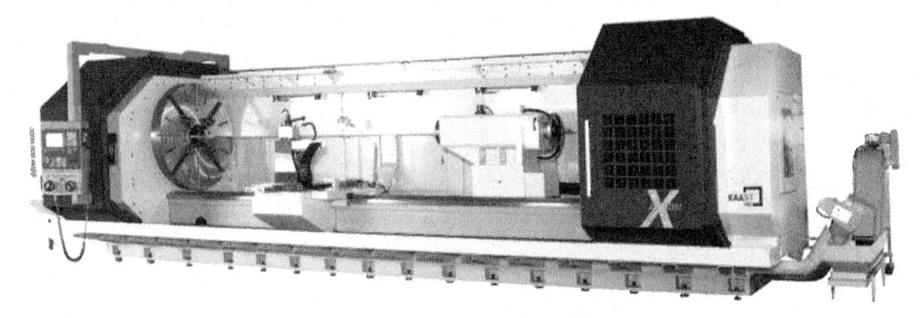

Figure 1.3 Extremely heavy duty CNC lathe machine.

(Courtesy: Extremely Heavy Duty CNC machine, KAAST Machine Tools, Inc. USA)

1.9 Types of CNC Machines

CNC machines are available in many types as follows [3]:

- Horizontal Turning Machine
- Vertical Turning Machine
- Horizontal Milling Machine
- Vertical Milling Machine
- Indexing Head Machining Center
- Universal milling and drilling machine
- Multi-Axis Machining Center
- Cylindrical Grinding Machine
- Surface Grinding Machine
- Wire Cut Machine
- Router machine
- CNC punch press
- Water Jet Machine
- Gear-Cutting Machine
- Special Purpose machine (SPM) like Oil Field Turning Machine
- Coordinate Measuring Machine (CMM), etc.

1.10 Some Constructional Features of CNC Machine [3]

Figure 1.4 shows light and medium work CNC turning machine.

Figure 1.5 shows a Flat Bed CNC Lathe machine.

Spindle Drive, Feed Drive, Turret/ATC, Bed (30°/45°/60°), Chuck, Servo Motor, Slide Ways (Bed), Ball Screw, Conveyor, Coolant Motor, Encoder, Transducer, AWC, Lubrication System, etc.

Figure 1.6 shows tool turret of CNC turning machine.

CNC machine Turret: tool turrets are used as a cutting tool station, where cutting tool mounted.

If you want to write the CNC program, you will use the CNC control Panel. See (Figure 1.7).

Figure 1.8 shows a Slant Bed CNC turning machine. Slant bed CNC turning machine is available in 30°/45°/60° angle.

Figure 1.9 shows 3 jaws chuck.

Figure 1.10 shows foot pedals and foot pedals are used for clamp and unclamp of the work piece in CNC machines.

Figure 1.4 Slant-Bed CNC turning machine.
(Courtesy: CNC Horizontal Turning machine, JYOTI, Rawat Engg. Tech Pvt Ltd., India)

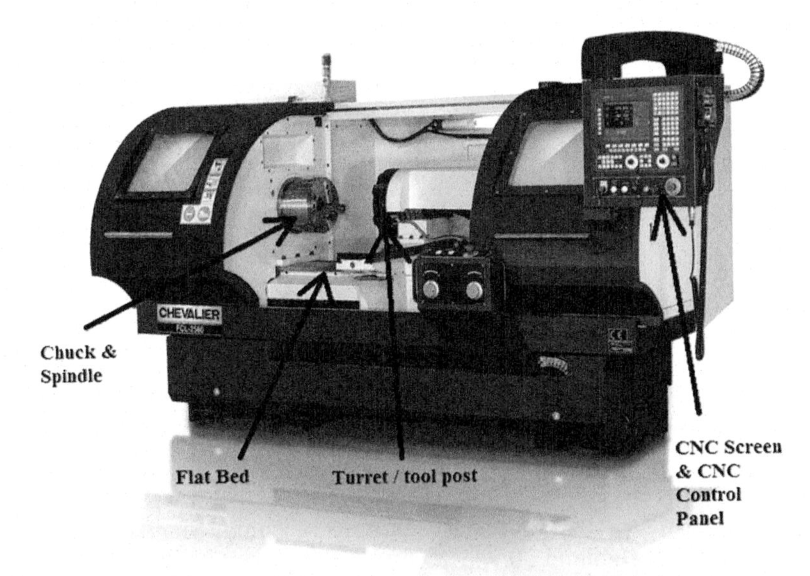

Figure 1.5 Flat Bed CNC turning machine with main parts.
(Courtesy: Flat Bed CNC Lathe machine, Chevalier FCL-2560, Chevalier_Fagor Automation, Spain)

Figure 1.6 CNC machine turret.
(Courtesy: Rawat Engg. Tech Pvt Ltd., Dehradun, India)

Figure 1.7 CNC control Panel.
(Courtesy: Haas SL 20, CNC control Panel, Sara Sae India Pvt Ltd, India)

Figure 1.8 Slant Bed of CNC turning machine.

(Courtesy: Rawat Engg. Tech Pvt Ltd., Dehradun, India)

Figure 1.9 3 Jaw Chuck.

(Courtesy: 3 Jaws Chuck, Sara Sae India Pvt Ltd, India)

Figure 1.10 Foot pedals of CNC turning machine.
(Courtesy: Haas SL 20, CNC control Panel, Sara Sae India Pvt Ltd, India)

1.11 Important Parts of CNC Machine

We can divide CNC machine into five important parts:

i. Input media
ii. Machine control unit (MCU)
iii. Servo drive unit
iv. Feedback transducer
v. Cutting tool

The first part is **input media**—a bunch of instructions, which gives the instructions to the machine and cutting tools as per the CNC program. The position and movement of the cutting tool relative to the work piece given by the input media. In the beginning of the NC technology, NC machine program was making through punch tape and magnetic tape. Nowadays, most CNC machines use G & M codes for making CNC program which is transferred in the CNC machine via electronic devices like data cable, pen drive, memory card etc.

The second part is the **Machine control unit** (MCU), which is the mind of the CNC machine. All machine instructions and data are stored in the MCU unit. When CNC machine program runs that time MCU gives the instructions to the cutting tool or all necessary machine parts for executing of machining and smooth functioning of CNC machine.

The third part is the **Servo Drive Unit**. It controls the spindle drive and feeds drive mechanism. This unit controls the all servo machine tool mechanism including a servo amplifier.

The numerical command converted into signal voltage by the MCU unit and this converted voltage send to servo amplifiers, the amplifier amplifies these low voltage signals into high voltage signal and send to servo drive motors and spindle drive motor for the motion of axis and spindle.

The fourth part of the NC machine is the **feedback transducer**. When servo motors move, transducers and sensors detect and measure its value, actual movement or position. The difference between the required position and the actual position detected by the comparison circuit and the action is taken. Servomotor full fills these required differences.

The last or final important part of the NC machine is the **Cutting Tool**. All machine operations, which are performed by the cutting tool are commanded or directed by the MCU (machine control unit).

1.12 What Do You Mean By NC And CNC Machine?

Figure 1.11 shows Perforated Paper tape reader on a CNC machine, it is a medium of data storage in NC and CNC control unit.

Figure 1.11 Tape Reader unit on a CNC machine, the medium of data storage.

(Courtesy: Punch Tape, Wikipedia)

Numerical control (NC) machine is a semiautomatic machine. The first generation of NC machine was controlled via electronic circuits and its working was based on NC. It had no memory and central processing unit (CPU). NC machine cannot save any data. It was programmed through punch tape and magnetic tape. Therefore, it is called the NC machine.

NC machine performed any machine operation that was impossible to perform on a conventional machine. Apart from this, NC machine could perform complex shapes, contouring operation and high precision machining of aircraft parts. So we can say, the NC machine fulfilled all requirements of complex machining of that time.

For making the program in NC machine, engineers must know about the standard programming format and machining sequence also should know in which sequence cutting tool, spindle speed and feed will come. Because of all these specialties the NC machine worked through the numerical input which is called a part program.

The most popular method of part programming is typing (writing) the codes as a CNC program (G & M codes) as per drawing on CNC control panel. When CNC machine program runs, the CNC control software sends the program codes to MCU unit to convert all codes and transfer it to drive/servo drive, servo drive(s) send the signal to the machine tool. Now cutting tool takes the action and remove the material and manufacture the product as per given CNC program.

1.13 Difference Between the NC Machine and CNC Machine

1.13.1 NC Machine

In the numerical control (NC) machine, the program is feed to the machine through magnetic tape or punch tape therefore it was not familiar in programming. It has no self-memory and it was a semiautomatic machine. (Figures 1.12 and 1.13).

1.13.2 CNC Machine

But in a CNC machine, it has self-memory. An engineer can write and save the CNC programs in his memory drive. It is a fully automatic machine. It is a high productivity machine compare to NC machine and manual machine.

Figure 1.12 Magnetic tape reader, which is used in NC machine.

(Courtesy: Tape Reader Unit, FANUC LTD., JAPAN)

Figure 1.13 Punched paper tape.

(Courtesy: Five-hole and eight-hole punched paper tape, Wikipedia)

1.14 What is CNC Control Software?

CNC control software is one of the most important part of the CNC machine. CNC control software operates by the CNC control panel. Control panel looks like a computer where you will get hard keys and soft keys for operating this control panel. Through the control panel you can control all machining functions of CNC machine-like operations, spindle speed, coolant motor, conveyor movements etc. In this software, programmer can see or enable/disable machine parameter setting as per requirement like cutting tool offset, work zero offset, cutting tool geometry, required parameters and settings etc. and also can see machining simulation, programmed cutting tool path on the graph and can make compact and better error free CNC program as per the required drawing.

1.14.1 Some Popular CNC Control Software Names are Given Below

i. Fanuc
ii. Siemens
iii. Haas
iv. Mazak
v. Heidenhain
vi. Bosch
vii. Fagor
viii. Denford etc.

References

[1] Fundamentals of Metal Cutting and Machine Tools, B.L. Juneja, G.S. Sekhon and Nitin Seth, Revised Second Edition (2005), New Age International Publisher, India.
[2] Production Technology (Manufacturing Process, Technology and Automation), 17th Edition, 2009, R.K. Jain, Khanna Publishers, India.
[3] CNC Technology & Programming, Tilak Taj, 2016, Dhanapat Rai Publishing Company, India.

2

Turning Process and Its Cutting Parameters

2.1 What is Turning?/What is Turning Operation?

Turning is a machine operation, which is performed in lathe machine or CNC turning machine. During turning operation, the material is removed by cutting tool in cylindrical shapes.

The most common machine which we see in the workshop is called horizontal lathe machine/horizontal CNC turning machine.

2.2 What is Adjustable Cutting Factor in Turning Machine?

The three basic primary factors in any turning operation are spindle speed (rpm), feed and depth of cut. These three are called adjustable cutting factors. Other factors are material types and types of tools. These two types are very important during the cutting operation. But the above three factors can be changed/adjusted by the operator at any time.

2.3 What is the Feed, Cutting Speed, Spindle Speed and Depth of Cut?

2.3.1 Spindle Speed

In turning machine, the angular velocity of the spindle is called the "spindle speed". The unit of spindle speed is revolution per minute (rpm).

2.3.2 Feed

Feed rate is defined as the distance the tool travels during one revolution of the part. Cutting feed and cutting speed determines the surface finish.

We can define the **unit of feed into** two types: mm/revolution and mm/minute.

Note:
Mostly, we use mm/revolution as a feed unit in turning machine, whereas mm/minute is used as a feed unit in milling machine.

2.3.3 Cutting Speed

1. Cutting speed is defined as the speed at which the work piece moves (rotates) with respect to the tool [1].

$$\text{Cutting speed in metric system} = Vc = \frac{\pi \times D \times N}{1000}$$

$$\text{Cutting speed in inch system} = Vc = \frac{\pi \times D \times N}{12}$$

Where,

V_c = Cutting speed
π = 3.1416
D = diameter, where from the cutting tool will remove the material
N = spindle speed in rpm

The cutting speed denotes feet per minute (FPM). To obtain uniform speed during machining on lathe, lathe spindle must revolve faster for a small diameter and slower for a larger diameter. The proper cutting speed depends on the hardness of the material being machined as shown in Figure 2.1.

Note:
Spindle speed shown in RPM and Cutting speed is shown in surface feet per minute (sfpm).

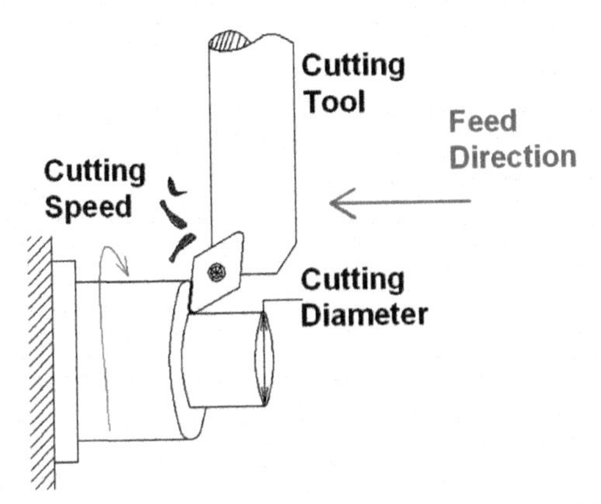

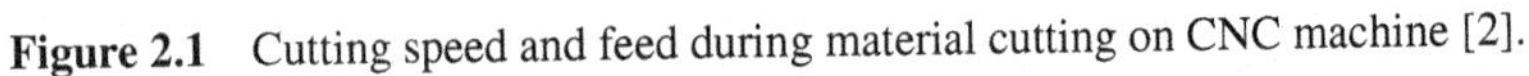
Figure 2.1 Cutting speed and feed during material cutting on CNC machine [2].

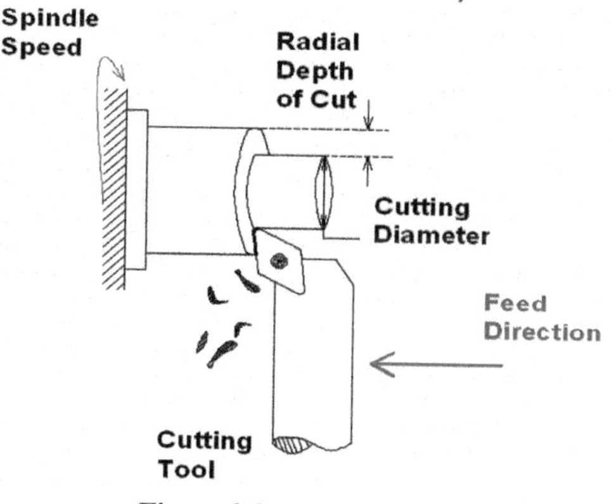

Figure 2.2 Depth of cut.

2.3.4 Depth of Cut

Depth of cut is the thickness of the layer (generally, we say material chips) of the removed material or distance from the uncut surface to the cut surface is called depth of cut. Or we can say, depth of cut is the amount of metal which is removed by the cutting tool in per pass as shown in Figure 2.2.

Note:

In turning machine, the material is removed two times (diametrically) at each depth of cut and that cut is called diametrical cut.

$$Depth\ of\ cut\ (DOC) = \frac{D - d}{2}$$

D = diameter of the work piece before machining
d = diameter of the machined surface (diameter of the work piece after machining).

2.4 What are the Turning, Facing, Straight Turning, Step Turning, Drilling, Boring and Threading Operations?

2.4.1 Turning Operation

Turning is the machining operation, which produces cylindrical parts in its basic form. The lathe/turning machine reduces the diameter of a work piece to a desired dimension. See Figure 2.3.

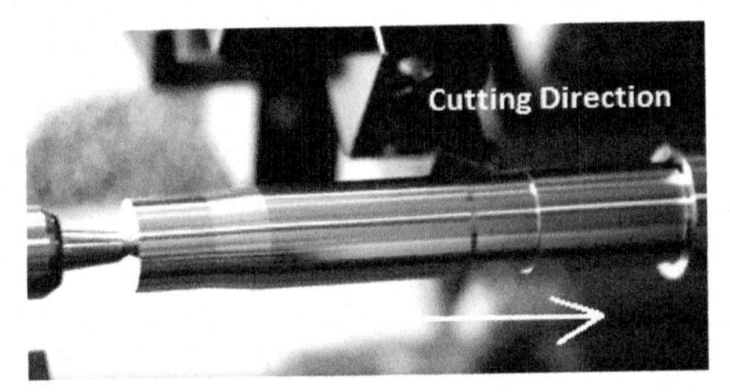

Figure 2.3 Turning operation is performing by turning tool.

(Courtesy: CNC Horizontal Turning machine, JYOTI, Rawat Engg. Tech Pvt Ltd., India)

2.4.2 Important Tips

- For rough turning, always use cutting tool with a big nose radius. You can use round, square or 80° shaped cutting bit/insert for rough turning.
- For finish turning, always use sharp tool (small nose radius bit/insert). You can use such as a triangle, 55°, 80° or 35° shaped cutting bit/insert.

2.4.3 Facing Operation

1. Mostly a facing operation is used for removing extra material from the length or we can say it is used for maintaining the length as per drawing. Generally, facing is the first machining operation. After facing operation, we can set zero offset for all other tools in the CNC machine. Also facing operation is used for removing the unwanted surface from the face of the work piece [3].

During facing operation, cutting tool can take one or more than one rough cut as per drawing requirement but the last cut should be the finishing cut with less removal of material. Use always a rigid cutting tool for the facing operation. Example: you can use square, 55° or 80° etc. cutting bits (inserts) in facing operation (Figures 2.4 and 2.5).

2.4.4 Plain Turning/Straight Turning Operation

Figure 2.6 is showing a plain turning operation but in this operation cutting tool goes to one direction during cutting time respectively parallel to the central axis of the spindle and produce only one diameter from start to end.

Figure 2.4 Cutting tool is performing facing operation.

(Courtesy: Flat Bed CNC Lathe machine, HAAS, Rawat Engg. Tech Pvt Ltd., India)

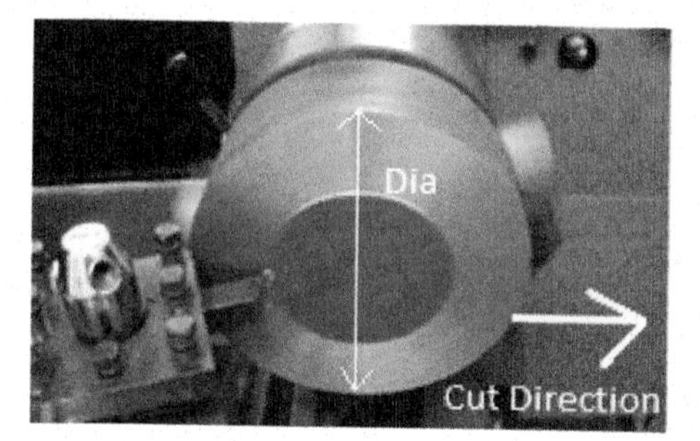

Figure 2.5 Cutting tool is going towards the center of the work piece from the bigger diameter during the facing operation.

(Courtesy: Flat Bed CNC Lathe machine, HAAS, Rawat Engg. Tech Pvt Ltd., India)

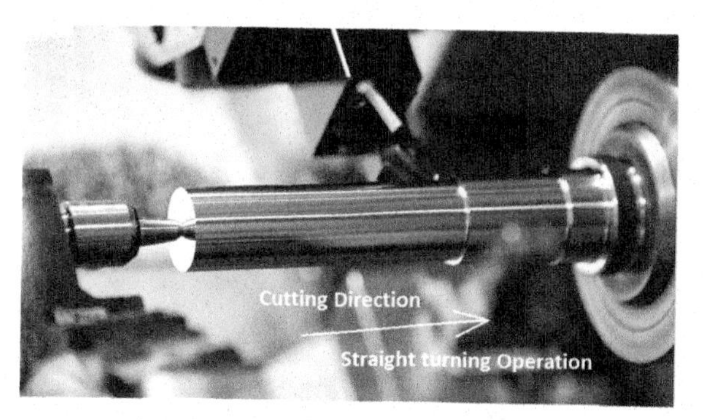

Figure 2.6 Straight turning operation.

(Courtesy: CNC Horizontal Turning machine, JYOTI, Rawat Engg. Tech Pvt Ltd., India)

2.4.5 Step Turning Operation

In this operation (Figure 2.7) cutting tool makes the different diameter of different length on the other hand we can say after material removal we get the different diameter of different length in cylindrical form.

2.4.6 Drilling

A turning machine is used also for the drilling operation. Drill tool can drill a hole at the center of the work piece. We can say when drilling tool drills the hole; the hole will be parallel to the central axis of the spindle (Figure 2.8) [4].

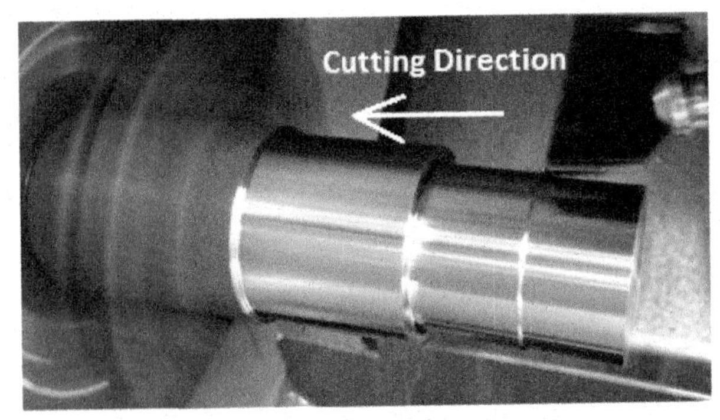

Figure 2.7 Step turning operation.

(Courtesy: CNC Horizontal turning machine, JYOTI, Rawat Engg. Tech Pvt Ltd., India)

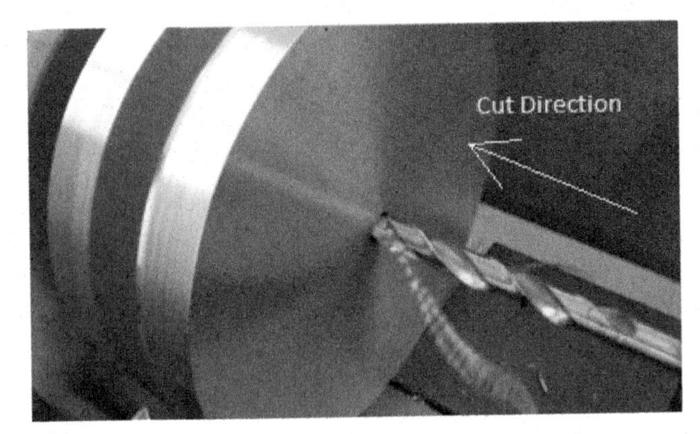

Figure 2.8 Drilling operation.

(Courtesy: CNC Horizontal Turning machine, JYOTI, Rawat Engg. Tech Pvt Ltd., India)

Note:
Always use drill tool length carefully (drill tool length must be little more longer than drill hole depth) for drilling if possible because it reduces the chatter and gives more accurate hole. The more accurate hole means, good drill tool length gives more a concentric hole with respect to the axis of rotation.

2.4.7 Taper Turning Operation

Taper turning is an operation for producing a conical surface by gradual reduction in diameter of a cylindrical work piece as shown in Figure 2.9.

2.4.8 Boring Operation

Boring operation is the same as turning operation only difference is during boring operation material is removed inside the hollow work piece. The work piece can be previously hollow or drilled by the drilling tool. Boring is the process of turning and enlarging the hole by removing material from an internal surface with a single point cutting tool (Figure 2.10).

2.4.9 Threading Operation

Taping and die threading are machining operations in whicn internal and external threads are produced by the helical cutting motion. For these threading operations, we use single point or multi point cutting tools [5]. (Figure 2.11).

Figure 2.9 Taper turning operation.

(Courtesy: Flat Bed CNC Lathe machine, HAAS, Rawat Engg. Tech Pvt Ltd., India)

Figure 2.10 Boring operation.

(Courtesy: CNC Horizontal Turning machine, JYOTI, Rawat Engg. Tech Pvt Ltd., India)

Figure 2.11 Threading operation.

(Courtesy: CNC Turning machine, JYOTI, Rawat Engg. Tech Pvt Ltd., India)

2.5 Why We Choose the CNC Machine for Manufacturing the Products?

Need of the day is that delivery should be in the right time frame with cost effective and quality products—This requirement is fulfilled by the CNC machines.

If we manufacture the products in the CNC machine, we get more advantages like we get quality products, high accuracy of work piece profiles, less rejection, mass production, cost-effective products, etc. We can manufacture the products within a time frame and deliver without any interruption.

2.6 What Should We Need for Manufacturing the Product in CNC Machine?

We need the following facilities before manufaturing the product in the CNC machine.

- Drawing study
- Selection of the right material
- Machining process (turning, milling, hardening, etc.) including premachining (if necessary).
- Selection of the machine
- Cutting tool selection
- Work holding device (fixture, vice or jaw)
- Cutting tool parameters (feed, rpm, depth of cut, etc.)
- CNC program write and enter in to compatible CNC machine.
- Documentation
- Simulate the CNC program and check
- Estimate the cycle time.
- Machining operation start
- Quality measurement of the first sample piece as per drawing before starting the production.

Note:
If you follow the above steps then you can avoid the machine problems like machine accident, product rejection or any confusion regarding production etc.

References

[1] Production Technology (Manufacturing Process, Technology and Automation), 17th Edition, R.K. Jain, Khanna Publishers, India.
[2] Drawings and sketches has been generated from Auto CAD 2014.
[3] http://www.engineeringarticles.org/lathe-lathe-operations-types-and-cutting-tools/
[4] software.https://en.wikipedia.org/wiki/Turning
[5] Fundamentals of Metal Cutting and Machine Tools, B.L. Juneja, G.S. Sekhon and Nitin Seth, Revised Second Edition (2005), New Age International Publisher, India.

3

Importance of Alphabets in CNC Programming

3.1 Alphabets, Which are Used in CNC Programming?

The following alphabets (letters) are used in CNC programming as code. Every code has different work during the execution of CNC programming.

3.1.1 A Rotary axis around the X-axis (unit in degrees)

3.1.2 B Rotary axis around Y-axis (unit in degrees)

3.1.3 C Rotary axis around Z-axis (unit in degrees) [1]

3.1.4 D Depth of cut

3.1.4.1 Formula to Find Out Unknown Diameter on Turning Machine

It is the total amount of material which is cut from the work piece surface by the cutting tool.

See Figure 3.1 for Depth of cut.

Both side depth of cut is called diametrical depth of cut = 2 × Depth of cut = --?-- mm

Minor Diameter (M. Ø) = Major Diameter − 2 × Depth of cut = Ø?

Major Diameter (M. Ø) = Minor Diameter + 2 × Depth of cut = Ø?

Example

Minor Diameter (M. Ø) = Ø27 mm (Major Diameter) − 2 × 2.5 mm = Ø22.0 mm

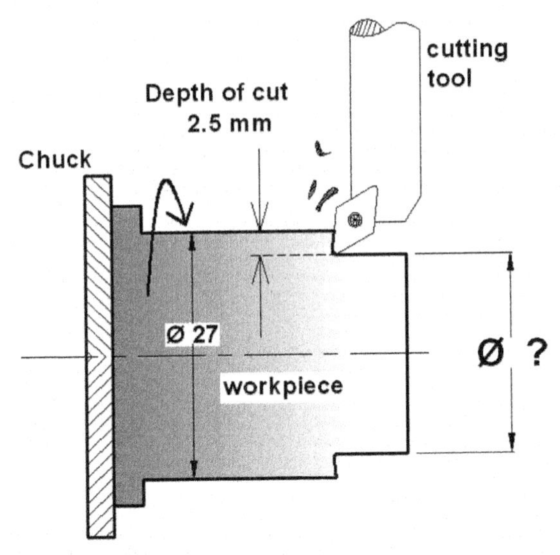

Figure 3.1 Depth of cut [2].

3.1.5 E For accuracy control during machining operation/engraving feed rate or contouring accuracy

Note: Alphabet E is used, with G187, to control accuracy during high-speed machining [3].

3.1.6 F Feed is like a cutting force of the tool. This is applied during material removing.

Unit of feed	millimeter/revolution
	millimeter/minute

3.1.7 G Preparatory function (G-code/job operation code)
G codes are used for machining like facing/drilling/threading/radius profile etc. [4].

G codes are two types

i. Model command
ii. Non-Model command

G codes are job operation codes; it means any type of operation in CNC machine like straight turning, step turning, chamfering, radius profile, grooving, threading, countering or complex profile operations etc. can be perform using G-codes. Also G codes are used as CNC machine parameters (Figure 3.2).

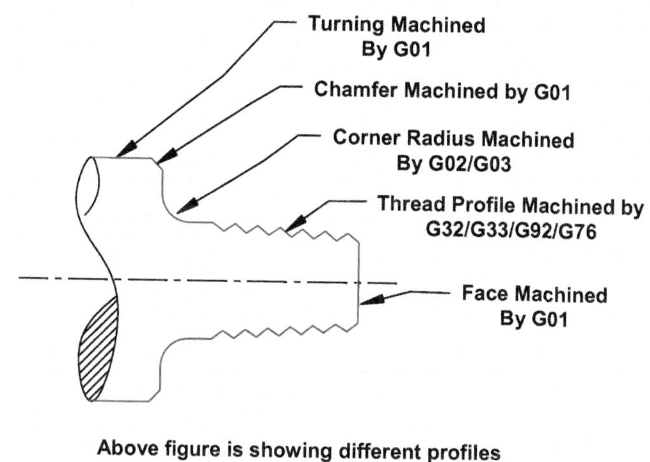

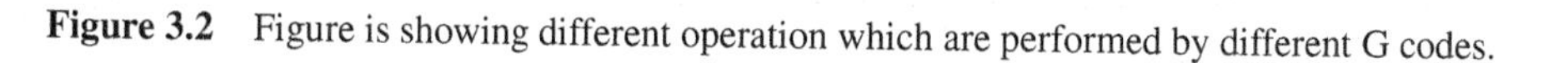

Figure 3.2 Figure is showing different operation which are performed by different G codes.

3.1.8 H Tool height offset/data

It is known as a tool height or tool length offset. Actually it is a tool data that is define cutting tool height position in program. Always tool length data saved in tool geometry. It is a very necessary setting before using the cutting tool in programming. Generally it is used in milling machine operation.

3.1.9 I Auxiliary axis in X-axis direction. It works in X axis direction same like X-axis but within some special conditions. We can say it is an angular distance in X-axis direction.

Example: It is used as a incremental axis in X-axis for arc coordinates with G02/G03 command or taper amount in threading operation in X-axis and in CNC milling programming for depth of cut in X-axis or used as a parameter in some fixed canned cycles. Figure 3.3 shows I & K location.

3.1.10 J is an auxiliary axis in Y-axis. It is used in CNC milling programming. **J** is used as a incremental axis in Y-axis for radius profile with G02 or G03 command. It is also used as a depth of cut in Y-axis in milling machine or used as a parameter in some fixed canned cycle. Figure 3.4 shows I & J location on milling machine's job. I & J coordinates are used in the Milling machine programming for X & Y-axis during arc profile.

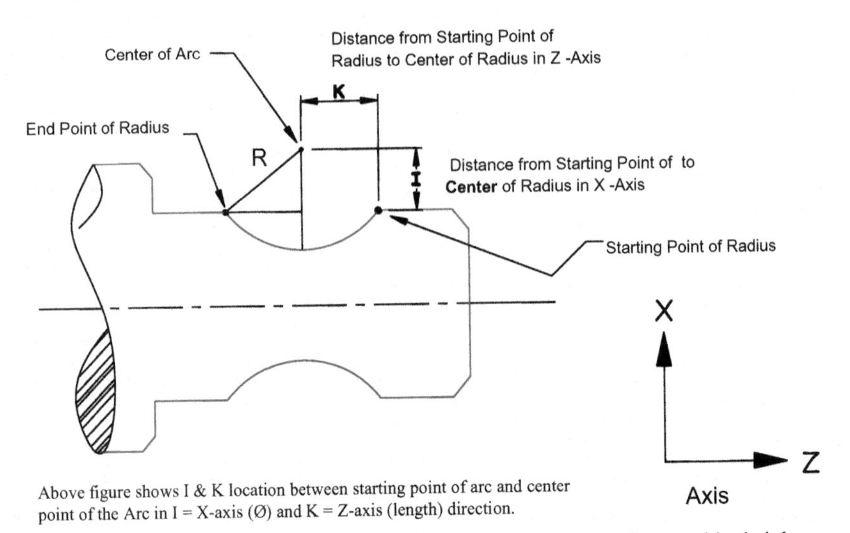

Figure 3.3 Angular increment commands (I and K) on turning machine's job.

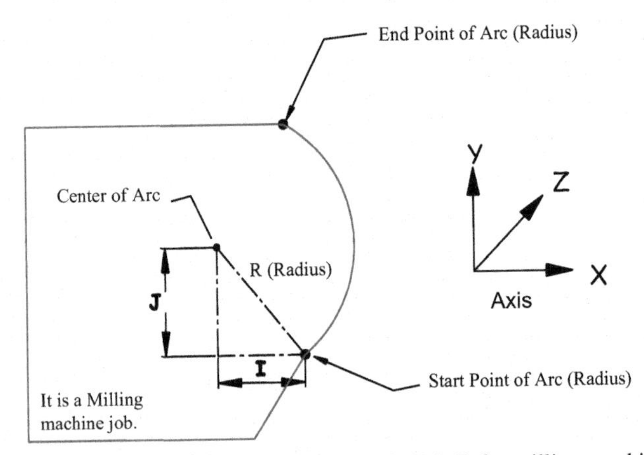

Figure 3.4 Angular increment commands (I & J) for milling machine.

3.1.11 K is an auxiliary axis in the Z-axis. It is used as a incremental axis in Z-axis for arc coordinates with G02/G03 command or taper amount in threading operation in Z-axis. You can see Figure 3.3.

Example It is used an angular distance in Z-axis for radius profile with G02 or G03 command and is used as a depth of cut in Z-axis or parameter settings in some fixed canned cycles.

3.1.12 L is used as a subprogram number or **L** is used as a number of repetition (loops) in canned cycle. In some CNC softwares, **L** is used as a cutting tool length offset/data.

3.1.13 M Miscellaneous code (M-code/Machine operating code)
M-codes are used mostly for machine operating.

Example M-codes are used for spindle rotation (M03/M04), tool change (M06), coolant on/off (M08/M09), spindle stop, spindle orientation (M19), gear selections (G41/G 42), CNC machine door open/close, etc. [4].

3.1.14 N Block number/Sequence number/Line number in CNC program
N is used as a line number in the CNC program. In some CNC control softwares, during CNC programming, N is used as a line number in every block of CNC programming [5].

```
O14006;
N1 G54;
N2 G21 G90 G95;
N3 G50 S2400;
N4 T0202;
N5 G90 G00 X100 Z100;
N6 G00 X60 Z10;
N7 G96 M03 S210;
N8 G00 X60 Z2;
N9 M08;
N10 G00 Z0;
N11 G01 X0.0 F0.1;
N12 G00 Z2;
N13 M09;
N14 G00 X100 Z100;
N15 M05;
N16 M30;
```

3.1.15 O Program number/Program name
When we start to write CNC program, first of all we write the program name and the program name starts with letter **O**. After letter **O**, we write the drawing number which comes from the drawing title box. Including of both (alphabet **O** and drawing numbers) are called program number [5].

Example
O14007: Here, **O** is used as a program number and digits belong to the drawing number of the work piece.)

G54;
G21 G90 G95;
G91 G28 U0 W0;
G50 S1200;

3.1.16 P is used as a block sequence number/Line number in some canned cycles in turning machine. It is used with G04 code and also is used for calling the sub programs (M98 **P**126).

Example: G72 U1.0 R1.5;
G72 **P10** Q11 F0.22;
N10 G00 Z0;
N11 G01 X0;

3.1.17 Q is used as a Line number in some canned cycles in turning machine. Some CNC softwares are used as a depth of cut during canned cycles.

Example: G72 U1.0 R1.5;
G72 P10 **Q11** F0.22;
N10 G00 Z0;
N11 G01 X0;

3.1.18 R denotes radius and R is used for radius profile in CNC programming. It is also used for tool retraction.

Example: G01 X64 Z0 F0.12;
G02 X70 Z-3 **R3**.0 F0.12;

3.1.19 S Spindle speed is the angular velocity of the work piece is called the spindle speed. The unit of spindle speed is **RPM** (revolution/minute). During programming, **S** is used for spindle speed.

Example: G97 M03 **S700**;

3.1.20 T Tool T is used as a cutting tool in CNC programming.

Example: T0505;

In the above example, first two digits (05) are shown tool station number where the required tool is mounted if operator calls T05 it means the tool will come in cutting position and second two digits (05) are tool geometry, where tool data stored. Tool geometry is the brain of the cutting tool. It is stored coordinates of the cutting tool respect of the work piece and machine.

3.1.21 U Incremental axis in X-axis direction

U is an incremental command in X-axis direction. It means when cutting tool takes the little/small movement in X-axis direction; in that case **U** can be applied. Generally increment commands are used for little movement (approximately movement 0.1 mm to 5.0 millimeter) in X-axis. In some CNC control (FANUC) softwares are used **U** as a depth of cut in canned cycles in X-axis direction and is also used as a finishing allowance in some control softwares.

Example: G00 X65 Z2 M08; G01 X65 Z-74 F0.12;
G01 **U3.0** (X68.0) F0.22; G00 Z2 M09;

3.1.22 V Incremental axis in Y-axis direction

V is an incremental command in Y-axis direction. It means when cutting tool takes the little/small movement in Y-axis direction; in that case **V** can be applied. Generally increment commands are used for little movement (approximately movement 0.1 mm to 5.0 millimeter) in Y-axis.

3.1.23 W Incremental axis in Z-axis direction

W is an incremental command in X-axis direction. It means when cutting tool takes the little/small movement in X-axis direction; in that case **W** can be applied. Generally increment commands are used for little movement (approximately movement 0.1 mm to 5.0 millimeter) in W-axis. Some CNC control (FANUC) software are used was a depth of cut in canned cycles in Z-axis direction and also is used as a finishing allowance in some control softwares.

Example: G00 X98 Z2 M08; G01 Z-74 F0.12;
G01 X98.0 **W2.0** (Z-72) F0.22; G00 Z100 M09;

3.1.24 X Main axis in X-axis direction

It is an axis, which applies when CNC machine bed/cutting tool travels towards to cross feed (across the diameter) on turning machine. We can say X-axis is used for diameter (Ø) on CNC turning machine and in milling machine, X-axis is used for movement of milling machine's bed towards to left/right direction when you stand in front of the milling machine.

3.1.25 Y Main axis in Y-axis direction

It is an axis, which applies when CNC machine bed/cutting tool travels towards to cross of the bed (when milling machine bed comes towards to operator/retracts from towards to operator. You can understand this movement when you stand in front of the milling machine or on milling machine). We can say Y-axis is used for work piece width on milling machine.

3.1.26 Z Main axis in Z-axis direction

It is an axis, which applies when CNC machine bed/cutting tool travels longitudinally on turning machine. We can say Z-axis is used for work piece length on CNC turning machine and in milling machine, Z-axis is used for movement of milling machine's spindle towards to up/down direction respect of milling machine's bed.

3.2 How to Write a CNC Program?/Write the Procedure of CNC Program?

CNC program writing is not one or two days learning procedure. It takes time. In the market you see many control systems and mostly are little similar to other, but always start with simple and user-friendly CNC control system. Here, we will start to write CNC program with FANUC control system. CNC program executes line by line in sequence from the top to bottom. CNC programming lines/blocks should be arranged in sequence.

Making Procedure of Simple CNC Programming, Step by Step:

i. **First, the programmer will write the program name.**

   Example: 041603;

   Where, alphabet O is the symbol of program name and numeric value is the drawing number of the work piece. Numeric value should be 4 or 5 digits, its depends on machine to machine.

ii. **Now programmer will call/write required cutting tool.**

   Example: T0404;

   Cutting tool will be assigned from the turret, where the tool is already mounted. We can say the required cutting tool will take its cutting position.

iii. **Cutting tool will move rapidly towards the work piece but keep some safe distance.**

   Example: G00 X80 Z10;

iv. **Spindle will start and rotate with given spindle speed (rpm) in required direction.**

   Example: G97 M03 S2100;

v. **Finally cutting tool will come nearest the desired cutting location (coordinates) of the work piece as per drawing.**

   Example: G00 X64 Z1.0;

vi. **Coolant motor will start.**

   Example: M08;

vii. **Now cutting tool will start the cutting and make the required profile as per drawing.**

Example: G01 X34 Z-15.5 F0.25; G01 X59 Z-33 F0.25;
G02 X75 Z-56 R34.09 F0.25; G01 X78 Z-62 F0.25;
G00 X78 Z1.0;

viii. **After removing the material by cutting tool(s) coolant motor will stop.**
Example: M09;

ix. **Now Spindle motor will stop.**
Example: M05;

x. **Cutting tool will go far from the work piece and take safe position.**
Example: G00 X100 Z200;

xi. **Finally CNC program will stop & reset. Reset means, program will rewind and ready for new production.**
Example: M30;

References

[1] CNC programming Manual of MTAB company: (Certificate course on CNC Turning, MTAB Technology Centre, MTAB, Chennai, India).
[2] Drawings and sketches has been generated from Auto CAD 2014.
[3] Haas CNC software.
[4] Production Technology (Manufacturing Process, Technology and Automation), 17th Edition, 2009, R.K. Jain, Khanna Publishers, India.
[5] CNC Technology & Programming, Tilak Taj, 2016, Dhanapat Rai Publishing Company, India.

4

Cutting Tool Geometry Settings in CNC Software

4.1 What is the Cutting Tool Geometry in CNC Software?

Tool geometry: When the cutting tool travels from one place to another in the CNC machine during machining/non-machining operation the movements depend on tool data and the tool data is stored in tool geometry. Tool data is the coordinates value of the cutting tools respectively to work piece zero (0, 0). Figure 4.1 shows tool data in tool geometry.

When cutting tool cuts the material from the work piece surface as per drawing at that time cutting tool geometry plays a most important role. Tool geometry is one of the most important tool data settings before machining (operation). Tool geometry is the collection of different tools data which are used by cutting tools during different cutting operation as per program to know coordinates location respectively to work piece zero (0, 0).

Tool geometry is the brain of all cutting tools. Without it we cannot imagine the CNC program and auto machining. On the other hand we can say, cutting tool geometry is fully responsible for all cutting tool movements.

Figure 4.1 shows cutting tool geometry setting page on CNC screen.

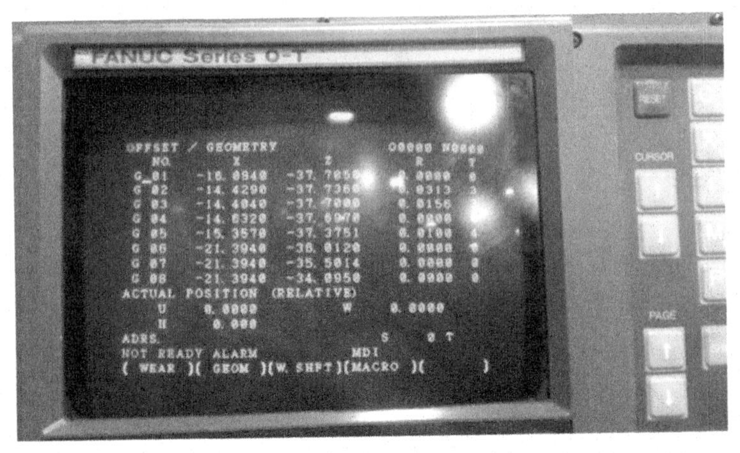

Figure 4.1 Tool offset/geometry.

(Courtesy: FANUC Series O-T, Sara Sae India Pvt Ltd., India)

4.2 Another Example of Tool Offset/Geometry

Figure 4.2 shows an example of the cutting Tool Geometry [1] and Figure 4.3 shows the actual cutting tool geometry image.

Tool Geometry				
Tool No.	**X-offset**	**Z-offset**	**Tool Nose Radius**	**Tool Tip Location**
1	-32.549	0.785	0.80	**3** (for turning tool)
2	-54.321	51.022	0.80	**2** (for boring tool)
3	-38.010	1.054	0.40	**3** (for turning tool)
4	-39.588	2.546	0.04	**8** (for external grooving)
5	-21.589	1.025	0.80	**7** (face grooving)
....	....	....	....	....
....	....	....	....	....
12	....	....	....	....

Figure 4.2 Tool geometry.

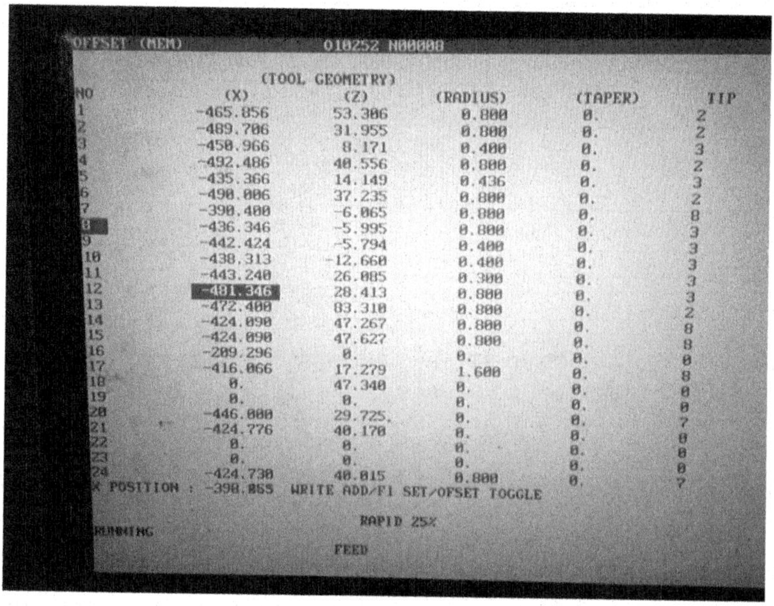

OFFSET (MEM) 010252 N00000

(TOOL GEOMETRY)

NO	(X)	(Z)	(RADIUS)	(TAPER)	TIP
1	-465.856	53.306	0.800	0.	2
2	-489.706	31.955	0.800	0.	2
3	-450.966	8.171	0.400	0.	3
4	-492.486	40.556	0.800	0.	2
5	-435.366	14.149	0.436	0.	3
6	-490.006	37.235	0.800	0.	2
7	-390.400	-6.065	0.800	0.	8
8	-436.346	-5.995	0.800	0.	3
9	-442.424	-5.794	0.400	0.	3
10	-438.313	-12.660	0.400	0.	3
11	-443.240	26.085	0.300	0.	3
12	-481.346	28.413	0.800	0.	3
13	-472.400	83.310	0.800	0.	2
14	-424.090	47.267	0.800	0.	8
15	-424.090	47.627	0.800	0.	8
16	-209.296	0.	0.	0.	0
17	-416.066	17.279	1.600	0.	8
18	0.	47.340	0.	0.	0
19	0.	0.	0.	0.	0
20	-446.000	29.725	0.	0.	7
21	-424.776	40.170	0.	0.	0
22	0.	0.	0.	0.	0
23	0.	0.	0.	0.	0
24	-424.730	40.815	0.800	0.	7

X POSITION : -390.865 WRITE ADD/F1 SET/OFSET TOGGLE

RAPID 25%

RUNNING

FEED

Figure 4.3 Tool offset geometry.

(Courtesy: Haas SL 20, CNC control Panel, Sara Sae India Pvt Ltd, India)

4.3 What is Cutting Tool Offset (Wear)?

Cutting tool offset (wear) is used for minor dimension adjustment (increasing and decreasing) when the work piece is undersized or oversized after machining. In the tool offset (wear), we adjust the dimensions between tolerances as per drawing.

Example

Let us assume, by using the tool number 01 we machined the work piece diameter but after machining when we measure the machined diameter we found the diameter oversized (dimension including tolerance) with 0.024 mm over the tolerance limit as per drawing.

For the maintenance in the size of the oversized diameter as per drawing now we **open the tool offset (wear) page on the control panel and we give the** oversized value with a negative sign (−0.024) to the tool number 01 in X-axis. Now again, when we run the same CNC program with tool

Tool No.	X-axis	Z-axis	Radius
1	0.567	0.04	0.80
2	0.214	0.100	0.80
3	1.489	0	0.40
4	0.230	0.01	0.04
....	....	....	....
....	....	....	....
32	....	....	....

Figure 4.4 Tool offset wear.

number 01. The tool will cut the oversized material from the work piece surface and we get the correct diameter as per drawing. See Figure 4.4.

4.4 An Example of the Tool Offset (Wear)

Figure 4.5 shows actual cutting Tool Offset Wear image.

OFFSET (MEM) O10252 N00112

(TOOL WEAR)

NO	(X)	(Z)	(RADIUS)
1	-0.060	0.	0.
2	0.100	0.	0.
3	-0.220	0.	0.
4	1.000	0.100	0.
5	0.550	0.100	0.
6	-0.030	0.	0.
7	12.050	0.	0.
8	0.050	0.030	0.
9	0.150	0.	0.
10	-0.300	0.	0.
11	-0.200	-0.060	0.
12	0.150	0.	0.
13	0.	0.	0.
14	0.	0.	0.
15	0.	0.	0.
16	0.	0.	0.
17	0.	0.	0.
18	0.	0.	0.
19	0.	0.	0.
20	0.	0.	0.
21	0.	0.	0.
22	-1.450	0.	0.
23	0.	0.	0.
24	0.	0.	0.

X POSITION : -231.196 WRITE ADD/F1 SET/OFSET TOGGLE

RAPID 25%

Figure 4.5 Tool offset wear.

(Courtesy: Haas SL 20, CNC control Panel, Sara Sae India Pvt Ltd, India)

4.5 What are Tool Offset Geometry and Tool Offset Wear in CNC Turning Machine?

"FANUC and other control manufacturers provide two sets of offsets for their turning center controls. One set, the geometry offsets, is used to help with program zero assignment. The other set, the wear offsets, is used to help with sizing adjustments. It is important to fully understand these two offset types since they provide a benefit that many CNC users overlook.

Again, geometry offsets are used to assign program zero. By one means or another, the setup person determines the distances in X and Z from the tool tip at the zero return position to the program zero point. The actual procedure varies depending on whether or not the work shift function is being used, but generally speaking, geometry offsets contain rather large negative values.

Wear offsets, on the other hand, are used to deal with adjustments that cause cutting tools to machine surfaces within their tolerance bands. If, for instance, a finish turning tool is machining a turned diameter 0.002 inch oversize on the initial work piece, most setup people will make the (−0.002-inch) adjustment in the X axis register of the tool's wear offset" [2].

4.6 What is Tool Nose Radius in CNC Machine?

Nose radius is used on the cutting edge surface of the cutting bits. It is called the nose radius of the cutting bits. You can see Figure 4.6 with nose radius. Cutting edge with nose radius, where from the cutting bit cuts/removes the material from the work piece surface. The following picture is showing cutting tool nose radius. Generally in machining operation, we observe nose radius with different values like 0.0 mm (very sharp edge for light/finishing cut), 0.4 mm (little sharp edge for light/finishing cut), 0.8 mm (medium radius edge for heavy cut), 1.2 mm (big radius edge for very heavy cut) etc.

Figure 4.6 shows cutting insert (bit).

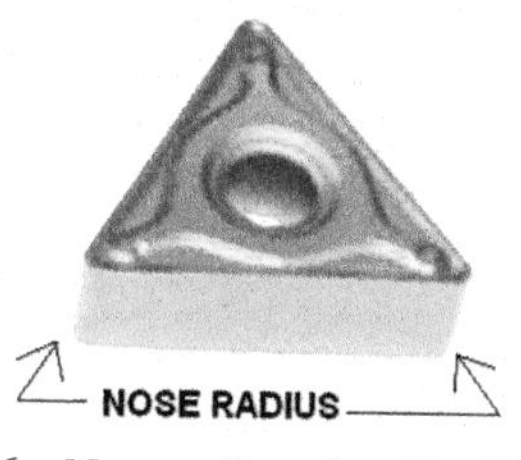

Figure 4.6 Nose radius of cutting bit (insert).

(Courtesy: TNMG Insert, Rawat Engg. Tech Pvt Ltd., India)

4.7 What is Tool Tip Location in CNC Machine? Is it Same for All CNC Turning Machines?

In CNC turning machine, the value of TIP location is used in Tool geometry page. Tool tip shows the location of the cutting tool in the CNC machine. It is very helpful during machining the job.

We can say tool tip is used for identifying the location of the cutting tool **so that cutting tool can produce the correct tool path as per the given drawing**.

When we apply the cutting tool nose radius compensation (G41 & G42) in CNC programming, we must use tip location in tool geometry.

Tool **TIP location** is used in different ways for different types of machining. Figure 4.5 shows slant bed CNC turning machine's TIP location.

Example

In **slant bed** CNC turning machine, if the turning tool is removing the material from the outer surface (OD) then location no. 3 will be used for turning tool as a tip location (see Figure 4.2). If the tool is removing the internal material (ID) then location no. 2 will be used for the boring tool as a tip location (see Figure 4.2), same like in face grooving operation, location no.7 will be used for face grooving tool as a tip location (see Figure 4.2). In external grooving operation, location no. 8 will be used as an external grooving tool (see Figure 4.2), etc.

Figure 4.7 is showing cutting tool tip location for slant bed CNC turning machine [Slant bed means: the cutting tool is situated far from the operator & spindle. On the other hand we can say, in the CNC machine first we see operator than we see work piece and last we see turret (tool post) with slant bed].

Figure 4.8 shows cutting tool tip location for Flat Bed CNC turning machine (when tool post situated between the operator and work piece same like lathe machine).

Note:

In CNC turning machine, the tip location can vary from machine to machine.

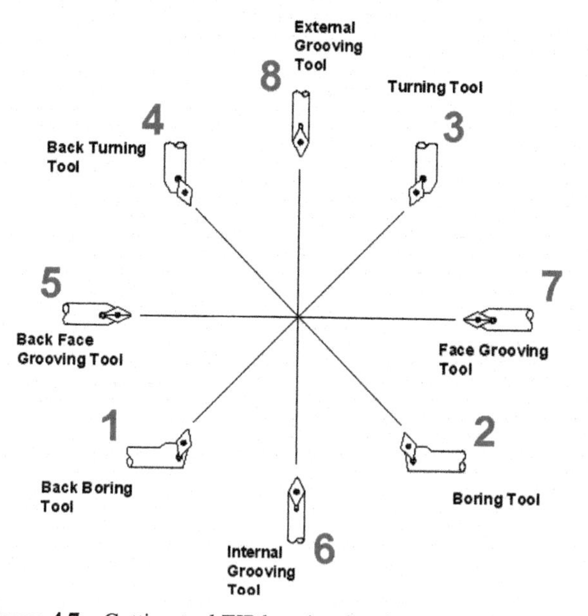

Figure 4.7 Cutting tool TIP location for slant bed condition [3].

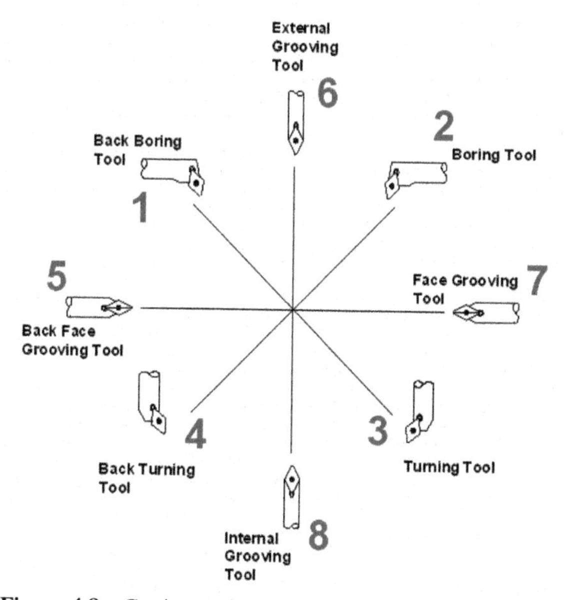

Figure 4.8 Cutting tool TIP location for flat bed condition.

4.8 Where You Enter Tool Nose Radius and Tool Tip Location?

We enter the value of tool nose radius and tool tip in cutting tool geometry page on CNC machine screen, which we can get in CNC control software.

References

[1] CNC programming Manual of MTAB company: (Certificate course on CNC Turning, MTAB Technology Centre, MTAB, Chennai, India).
[2] Columns Post: 10/15/2013, Mike Lynch, Founder and President, *CNC Concepts Inc., https://www.mmsonline.com/columns/taking-full-advantage-of-geometry-and-wear-offsets*
[3] Drawings and sketches has been generated from Auto CAD 2014.

5

Dimension Methods, Machine Zero, Work Zero and Machine Axis

5.1 What is Absolute Dimension and Increment Dimension?

Absolute dimension method (G90) is used during CNC programming. In the drawing where we want one reference point to measure all dimensions, we use incremental dimension method (G91) in CNC programming.

Note:

i. In few conditions, absolute and increment method are used during sub programming.
ii. Mostly, absolute (G90) dimensioning method is used in CNC programming.

5.1.1 Absolute Dimension (G90)

In this method, all dimensions must be measured from one point and this point is called origin/reference point (0, 0). *With respect to turning machine this point is called work zero (0, 0).* We use G90 command for absolute dimension in CNC programming [1].

Figure 5.1 shows absolute dimension method (G90).

5.1.2 Incremental Dimension (G91)

It is totally different to absolute dimension method. In absolute dimension method, every dimension start from the fixed point and this fixed point is called origin point.

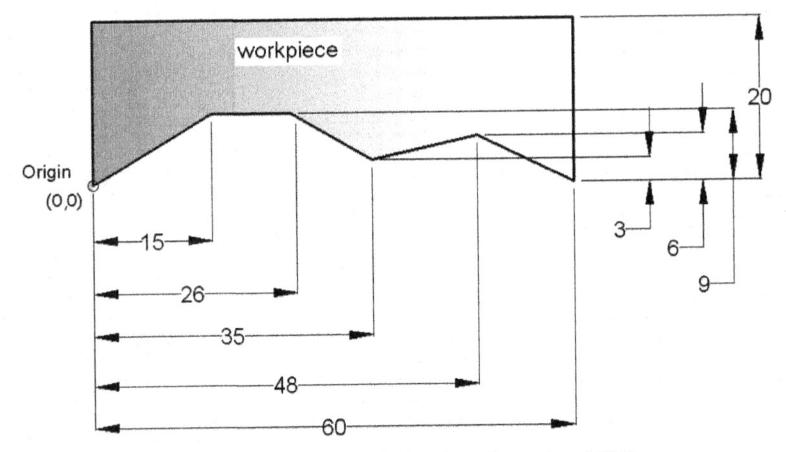

Figure 5.1 Example of Absolute dimension (G90).

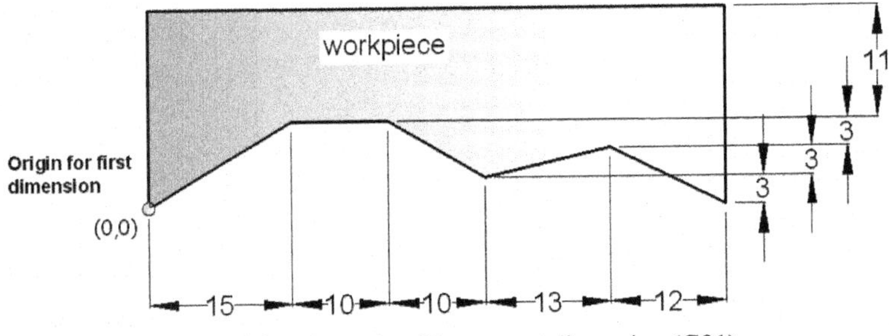

Figure 5.2 Example of Increment dimension (G91).

But in this case, every dimension has own reference point (0, 0). It means when programmer takes the dimensions during CNC programming, every dimension measure from the end point (reference point for the next dimension) of the previous dimension. Figure 5.2 shows incremental dimension method (G91).

5.2 What is Diametrical Method and Radius Method?

5.2.1 Diametrical Method

The diametrical method is one of the dimension setting methods for round-shaped work piece in CNC turning machine. Generally, diametrical setting activates in turning machine. After activation of diameter setting, in CNC

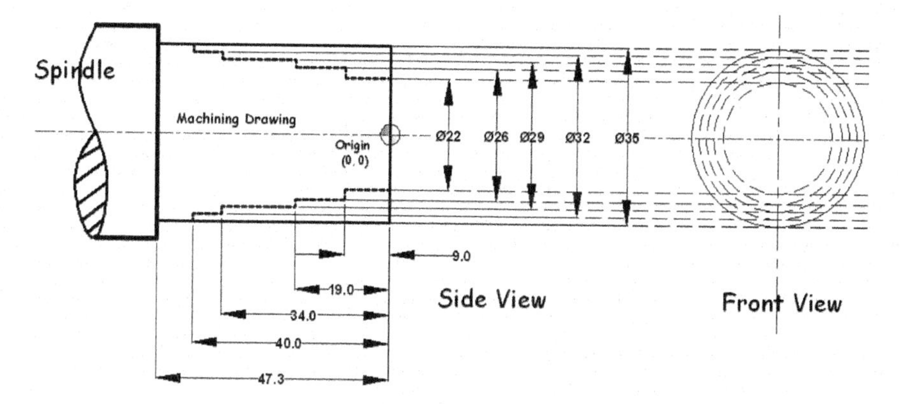

Figure 5.3 Diametrical dimension of the cylindrical work piece.

program value of round shape will be read as a diameter. In CNC programming, **X-axis** represents the round shape of the work piece.

Example X46.0 [X46.0 is showing diameter (Ø46) of work piece].

Note: Symbol Ø represents the diametrical dimension.

Figure 5.3 shows diametrical method.

5.2.2 Radius Method

Radius method is another machine parameter setting for round shape dimension. If this setting activates, it means half value of round shapes will take from the drawing during CNC programming. Generally this parameter setting does not use during programming.

5.3 What is the Machine Zero?

Figure 5.4 shows machine zero position.

In CNC lathe/turning machine, machine zero is the default position, which we cannot change. It is not a changeable position. On the other hand, we can say machine zero position set by the machine manufacturer during machine manufacturing [2].

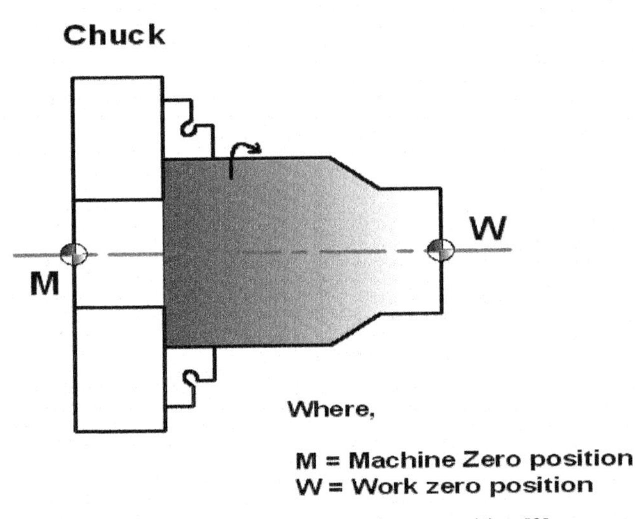

Figure 5.4 Machine and work zero position [3].

5.4 What is the Work Zero Offset/Origin (0, 0) in CNC Turning Machine and What is the Importance of the Work Zero in CNC Machine?

Work zero offset is the origin (0, 0) of the work piece, where the origin is X = 0 and Z = 0, in turning machine.

Any CNC machine program cannot perform the operation without work zero offset (0, 0). It is also called program zero. Work zero is always taken on the center of the work piece on turning machine. It means when cutting tools move toward the coordinates, it happens only due to the respect of work zero (X = 0, Z = 0). Without taking work zero cutting tools cannot move according to the CNC program.

After taking work zero (0, 0) offset, then CNC machine program runs. Cutting tools know his zero position and that zero position (0, 0) is situated on the center of the front face of the work piece. See Figure 5.5.

Note: *In the horizontal turning machine, the value of X will be zero on the center of the work piece. It is a universal truth for this type of machine. This center is the longitudinal axis or horizontal axis of the horizontal CNC turning machine.*

WORK ZERO OFFSET

G CODE	(X)	(Z)	(B)	
G 52	0.	0.	0.	
G 54	0.	-888.797	0.	
G 55	0.	0.	0.	
G 56	0.	0.	0.	
G 57	0.	0.	0.	
G 58	0.	0.	0.	
G 59	0.	0.	0.	
G154 P1	0.	0.	0.	(G110)
G154 P2	0.	0.	0.	(G111)
G154 P3	0.	0.	0.	(G114)
G154 P4	0.	0.	0.	(G115)
G154 P5	0.	0.	0.	(G116)
G154 P6	0.	0.	0.	(G117)
G154 P7	0.	0.	0.	(G118)
G154 P8	0.	0.	0.	(G119)
G154 P9	0.	0.	0.	(G120)
G154 P10	0.	0.	0.	(G121)
G154 P11	0.	0.	0.	(G122)
G154 P12	0.	0.	0.	(G123)
G154 P13	0.	0.	0.	(G124)
G154 P14	0.	0.	0.	(G125)
G154 P15	0.	0.	0.	(G126)
G154 P16	0.	0.	0.	(G127)
G154 P17	0.	0.	0.	(G128)

X POSITION : -448.036 WRITE ADD/F1 SET/OFSET TOGGLE

RUNNING: RAPID 25%

Figure 5.5 Work zero offset.
(Courtesy: Haas SL 20 CNC turning machine, Sara Sae India Pvt Ltd, India)

5.5 What is the Importance of the Central Axis of the Spindle in CNC Turning Machine?

"The value of the X-axis must be zero on the central axis of the spindle in CNC turning machine (CNC turning machine can be horizontal or vertical)". When we hold the work piece and rotates it between the chuck's jaws, it rotates parallel to CNC machine spindle. We can say spindle and chuck rotate on the same axis (central axis). The value of X-axis must be zero (X = 0) for all cutting tools on the central axis of the work piece/spindle, it is a universal truth for turning machine". So we do not need to take X = 0 value in CNC turning machine. But cutting tools data of X-axis and Z-axis must be taken before machining.

5.6 How Many Axis in CNC Turning Machine?

Generally CNC turning machine has **two main axis**.

i. **X-axis**
ii. **Z-axis**

In turning machine, when cutting tool travels across (cross feed) the work piece diameter, that cutting tool movement is called **X**-axis travel and the axis are called X-axis. When the cutting tool travels longitudinally

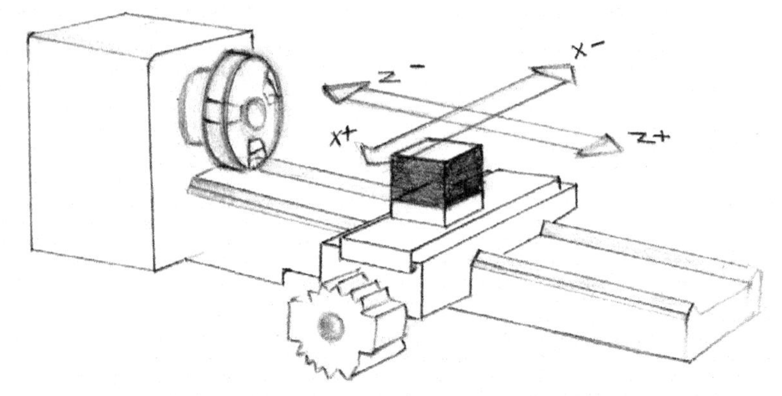

Figure 5.6 Flat Bed turning machine sketch and it is showing Axis directions.

towards to spindle or chuck, this cutting tool movement is called **Z-**axis travel and the axis is called Z-axis.

Figure 5.6 shows axis position on flat bed CNC machine.

X-axis (when cutting tool moves one diameter to another diameter which is called X-axis).

Figure 5.7 shows work piece diameter and X-axis position.

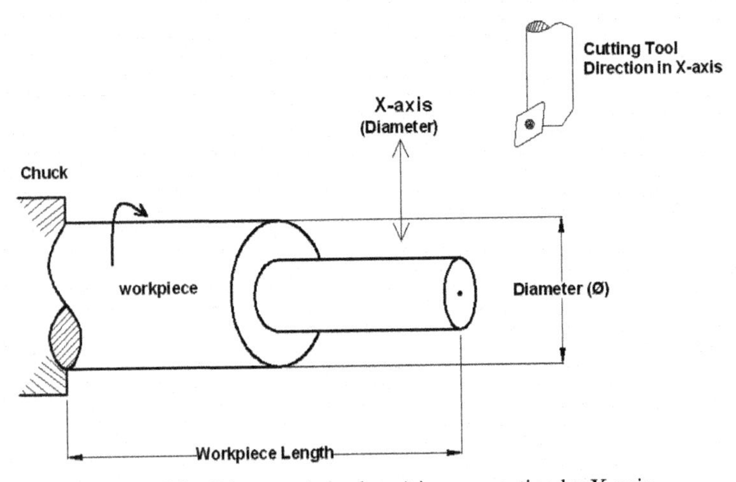

Figure 5.7 Diameter (roundness) is representing by X-axis.

5.6.1 Z-Axis (When Cutting Tool Moves Towards to Work Piece Length, Which is Called Z-Axis)

Figure 5.8 shows length direction and Z-axis position.

5.6.2 Some Important Facts About X = 0 & Z = 0

Figure 5.9 shows work zero (0, 0) position.

Figure 5.10 shows work zero (0, 0) position and axis position (X & Z-axis).

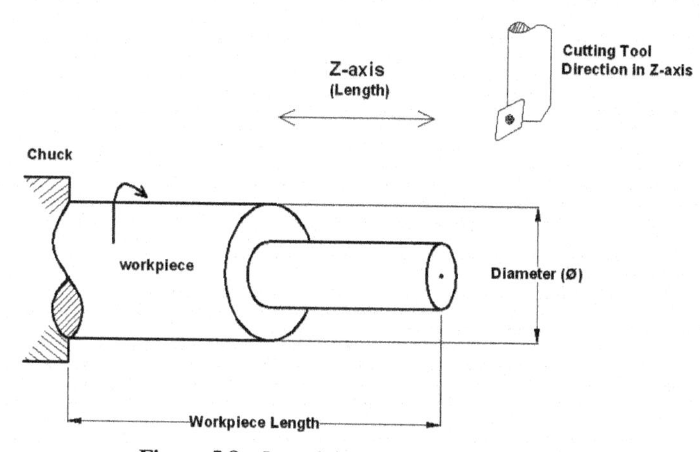

Figure 5.8 Length is representing by Z-axis.

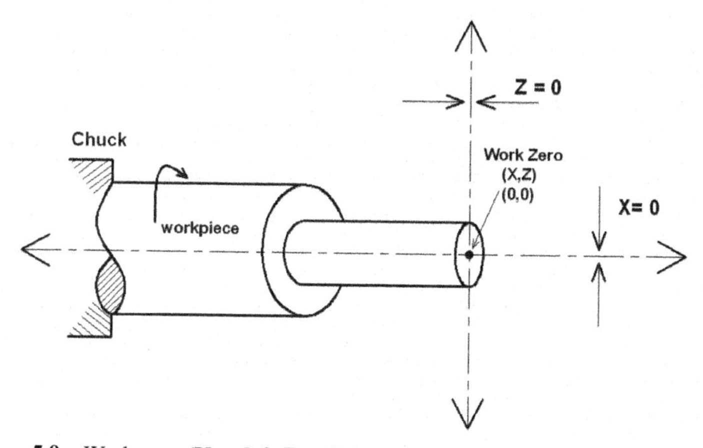

Figure 5.9 Work zero (X = 0 & Z = 0) is representing on the work piece face [4].

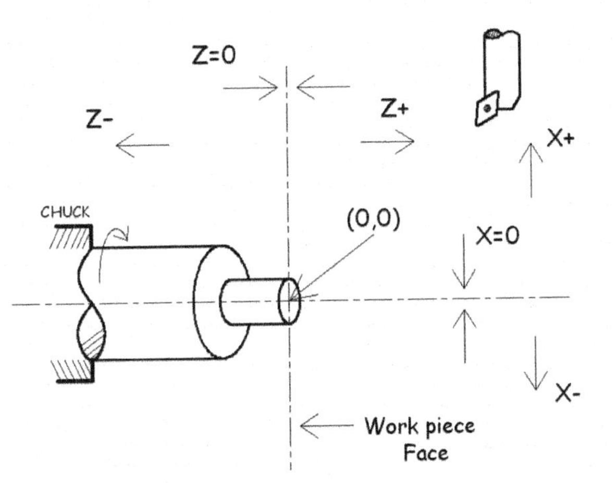

Figure 5.10 X & Z-axis value for slant bed CNC turning machine.

5.6.2.1 Before CNC programming and running the CNC turning machine, the programmer must know about the X = 0 & Z = 0. It means, programmer should know where is X = 0 (zero) and Z = 0 (zero) situated

When the cutting tool moves towards the work piece length (longitudinal direction) it is called Z direction. But when the cutting tool is touched on the front face of the work piece at anywhere, it is called Z = 0.

When we apply the above condition and let the cutting tool go from the face of the work piece (Z = 0) toward spindle, this tool movement is called Z– (negative) and when cutting tool retracts from the face of work piece (Z = 0) toward tailstock. This tool movement is called Z+ (positive).

5.6.2.2 Always in CNC turning machine, the value of X-axis will be zero (X = 0) in the center of the rotating work piece. It is a universal truth for turning machine

When the cutting tool stays above the center line of the rotating work piece, it is called X+ (positive) and when the cutting tool goes under the central line of the rotating work piece it is called X– (negative).

5.7 In Some Special Cases, We Can Take Z = 0 Anywhere on the Work Piece Surface. It Depends on the Machining Condition

Figure 5.11 shows axis zero X = 0 & Z = 0 position.

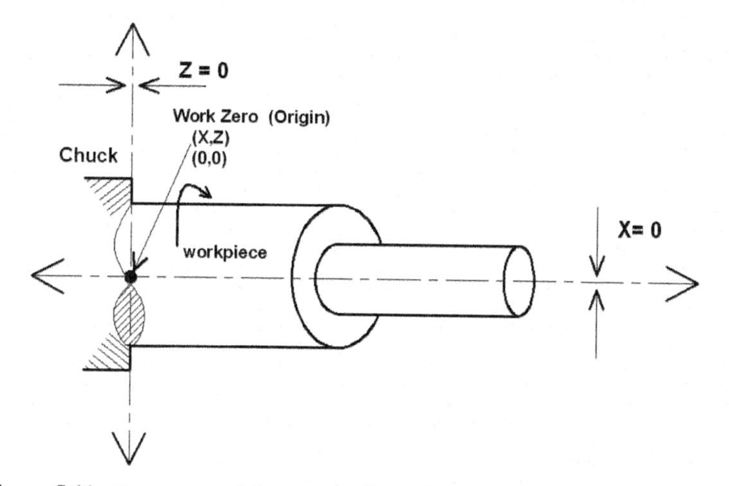

Figure 5.11 In some special cases, the figure is showing work zero position [2].

In some special cases, the programmer can take **Z** = **0** (zero) anywhere on the work piece either middle of the work piece or **front face of the jaws**.

Mostly CNC programmer sets the work zero offset (0, 0) on the **front face of the work piece**.

References

[1] Fundamentals of Metal Cutting and Machine Tools, B.L. Juneja, G.S. Sekhon and Nitin Seth, Revised Second Edition (2005), New Age International Publisher, India.
[2] CNC programming Manual of MTAB company: (Certificate course on CNC Turning, MTAB Technology Centre, MTAB, Chennai, India).
[3] CAD/CAM Principles and Applications, Second Edition, P.N. Rao, Fifth reprint 2006, Tata McGraw-Hill Company Limited.
[4] Drawings and sketches has been generated from Auto CAD 2014.

6

Turning Machine and Its Coordinates

6.1 What is the Importance of the Coordinate System in CNC Machine?

All movements of the CNC machine tools (cutting tools) depend on coordinates system like cutting tool movements, depth of cut, etc. we can say coordinates are responsible for cutting tool movements and tell the story about the work piece shape.

CNC machining operation depends on the CNC program and CNC program depends on drawing and dimensions. If we use drawing and dimensions with respect to the axis, those dimensions are called coordinates. If CNC programmer wants to make program he will use drawing coordinates during programming.

6.2 Find the Coordinates of Given Figure in Absolute Method (G90 Code)

Following drawing is the simplest coordinates drawing. Before we used this type of coordinates in our academic course(s). We use this type of drawing in milling machine programming. Figure 6.1 shows the simple coordinate system in X and Y axis direction [1].

Following coordinate example has been given for better understanding of students.

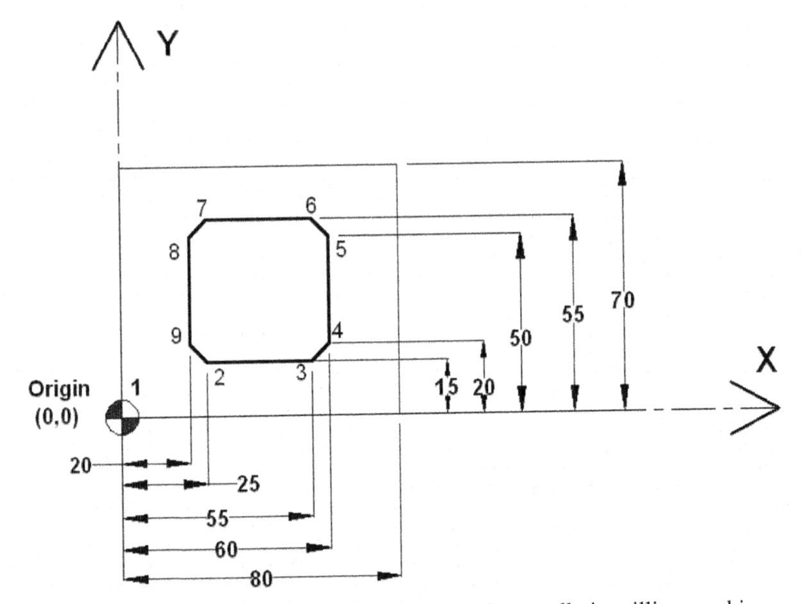

Figure 6.1 Coordinates system which are used generally in milling machine.

Coordinates of Figure 6.1.

(X, Y)	(X, Y)
1. (0, 0)	**6.** (55, 55)
2. (25, 15)	**7.** (25, 55)
3. (55, 15)	**8.** (20, 50)
4. (60, 20)	**9.** (20, 20)
5. (60, 50)	

6.3 Find the Absolute (G90) Coordinates of Given Figure in Diametrical (Ø) Method

Figure 6.2 shows step turning operation.

Figure 6.3 shows isometric dimensional view.

Figure 6.4 shows two-dimensional (2D) view.

Figure 6.2 Three-dimensional image of Step Turning Operation.
(Courtesy: CNC Horizontal Turning machine, JYOTI, Rawat Engg. Tech Pvt Ltd., India)

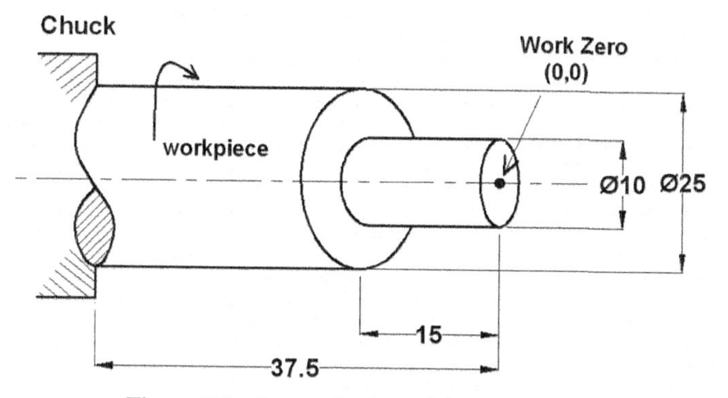

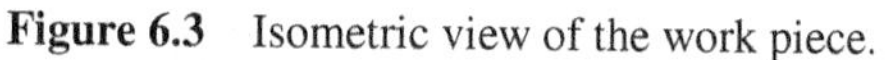

Figure 6.3 Isometric view of the work piece.

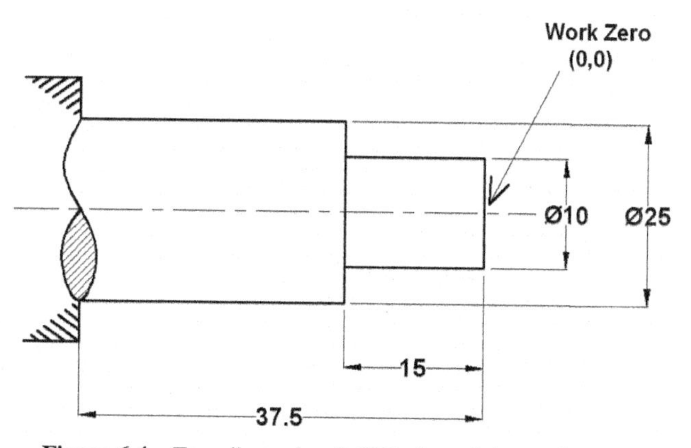

Figure 6.4 Two-dimensional (2D) view of the work piece.

6.3.1 In Turning Machine X-Axis is a Diametrical Axis, During Programming and Coordinates X Will Use with Diameter Value

Example X10 & X25 (these two examples are showing the value of diameter).

6.3.2 In Turning Machine Z-Axis Works in Length, During Programming and Coordinates Z Will Use with Length Value

Example Z-15 & Z-37.5 (these two examples are showing length value, the value of Z is negative because Z = 0 is on the face of work piece therefore all length dimensions will negative inside the work piece or towards the chuck.)

Figure 6.4 has cylindrical (round) surface, it means, its round surface is called diameter so it is a diametrical figure and these (Ø10, Ø25) are diametrical surface and machining operation will call **Step Turning** operation.

Figure 6.5 shows X-axis (Ø) and Z-axis (length) with dimensions.

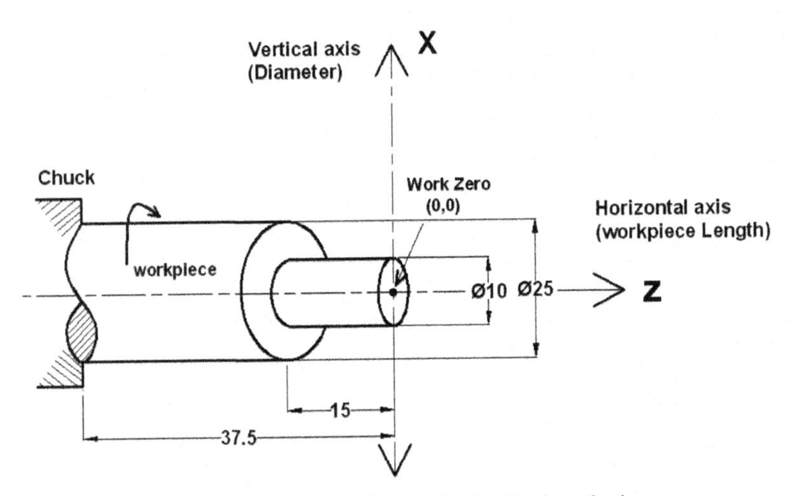

Figure 6.5 Dimensional axis of cylindrical work piece.

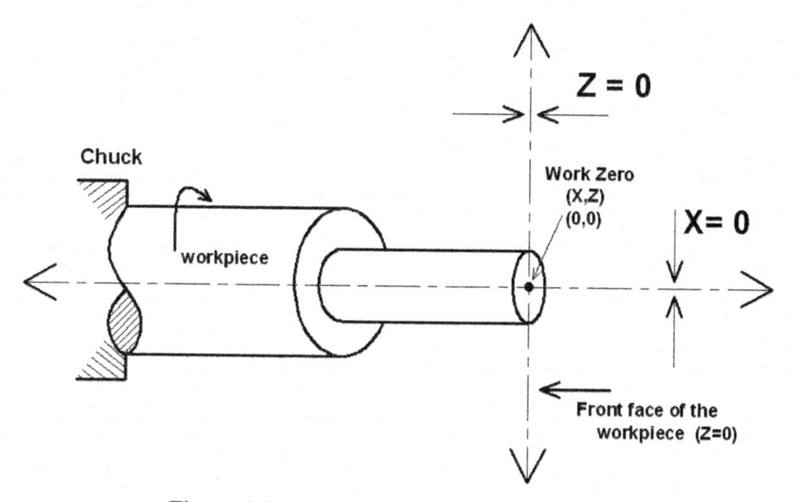

Figure 6.6 X zero and Z zero value of the axis.

Figure 6.6 shows Z = 0 on the front face of the work piece, it means the value of the front face of the work piece is **ZERO** in Z-axis and X = 0 is showing on the central axis of the rotating work piece.

In the Figure 6.6, the value of X = 0 will be in X-axis and value of Z = 0 will be in Z- axis. These two axis X & Z are met on the center of the work piece on the front face. Therefore the value of the center of the work piece is X = 0 & Z = 0. So coordinates of the center of the work piece will be (0, 0) [2].

6.3.2.1 In Figure 6.7, *all length dimensions* will be count in Z-axis with negative value because whole work piece is coming inside the negative zone (left side from work zero) of the Z axis. So length dimension will write with negative sign

Example Z-15, Z-37.5

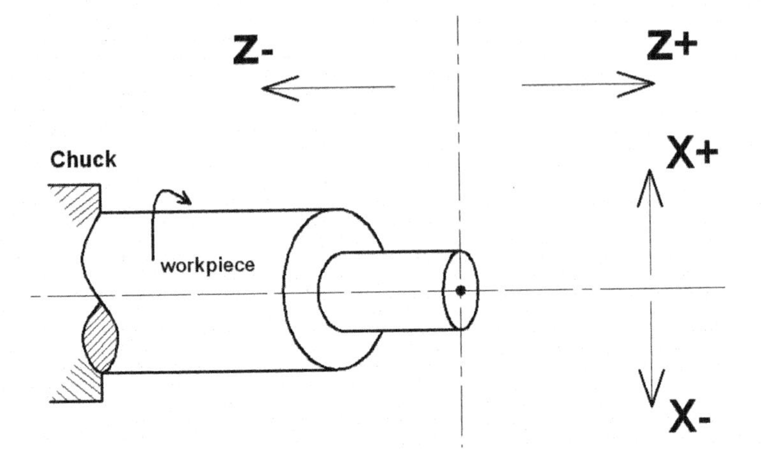

Figure 6.7 Negative and positive direction of the work piece in X & Z-axis [3].

6.3.2.2 All *diametrical dimensions* (Ø10, Ø25) are in X- axis, so it will use with X-axis with positive value (+) because cutting tool cuts the material from above the center line and above the center line value of the X is always positive (+). Therefore all X-axis dimensions (diameter) will take with positive value

Example X10, X25

6.3.2.3 Before Taking the Coordinates from the Drawing You Should Know Some Facts

It is very clear that, by default all cutting tools come for material cutting at particular cutting position (axis position of the tool) and this cutting position depends on the machine bed (machine bed can be slant or flat).

This cutting position (axis position of the tool) followed by the cutting tool, in the respect of machine bed. For an example, in turning operation, the turning tool has an own cutting position (axis position of the tool) and this position will be the same for all turning tools during different turning operations. For better understanding please see the above turning operation's in Figure 6.9.

In Figure 6.8, during rotation of the work piece cutting tool travels (moves) one by one towards to different coordinates and remove the extra material as per drawing. The cutting tool starts the movement from the

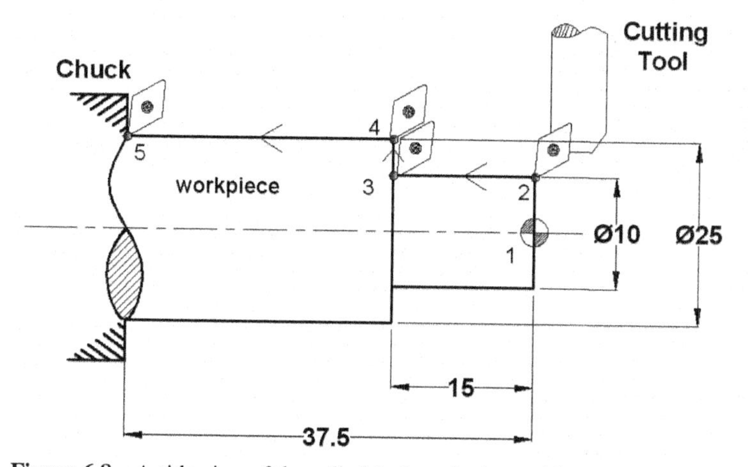

Figure 6.8 A side view of the cylindrical work piece with cutting tool path.

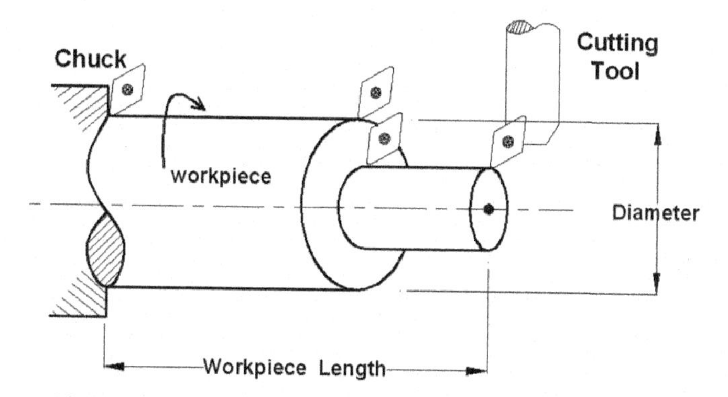

Figure 6.9 Isometric view of the cylindrical work piece with cutting tool path.

coordinates point (10, 0) respect of origin point (0, 0). Cutting tool moves towards to 3rd coordinates point (10, −15) and makes straight turning profile. When tool moves towards to 4th coordinates point (25, −15) this time tool performs step turning operation. Again tool moves towards to 5th coordinates point (25, −37.5) and performs straight turning operation. Whenever cutting tool travels from one point (coordinates) to another point (coordinates), the cutting tool removes the material in a linear motion (G01) with feed (F).

Following coordinates positions are showing with their location 1, 2, 3, 4 and 5 which is given in the Figure 6.8.

Coordinates

(X, Z)	X is diametrical axis & Z represents the length in CNC turning machine.
i. **(0, 0)**	coordinates of the center of the work piece in term of X & Z (X = 0 & Z = 0).
ii. **(10, 0)**	**First Starting straight turning diameter Ø10** (X10) & **starting length value** Z = 0 on the work piece face, where the tool will take position for material cutting.
iii. **(10, −15)**	at this coordinates, due to straight diameter. cutting tool will cut the material and make straight turning diameter Ø10 (X10) on Z-15 length.
iv. **(25, −15)**	At the forth co-ordinates, the diameter of the step is Ø25 (X = 25) and length value of the step will be same like previous value Z-15 from the work piece face.
v. **(25, −37.5)**	Second Straight turning diameter Ø25 (X25) & length value at the end of the fifth coordinates are Z -37.5 from the work piece face.

6.4 Find the Coordinates in the Absolute and Diametrical Method of Given Figure

Following work piece images are showing side view (Figure 6.10) and isometric view (Figure 6.11) with dimension and also showing 2D cutting tool path view (Figure 6.12) and isometric view (Figure 6.13).

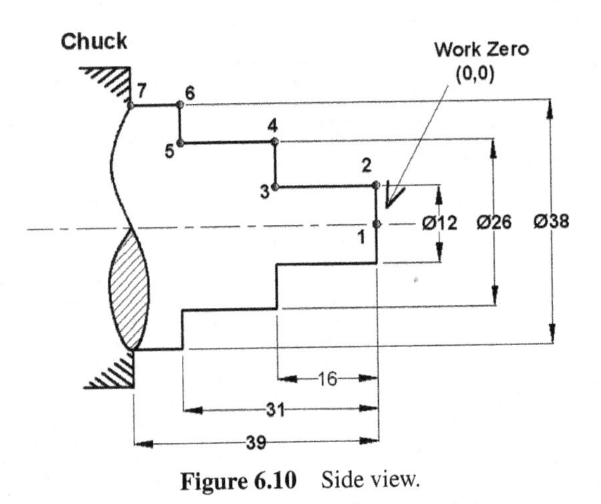

Figure 6.10 Side view.

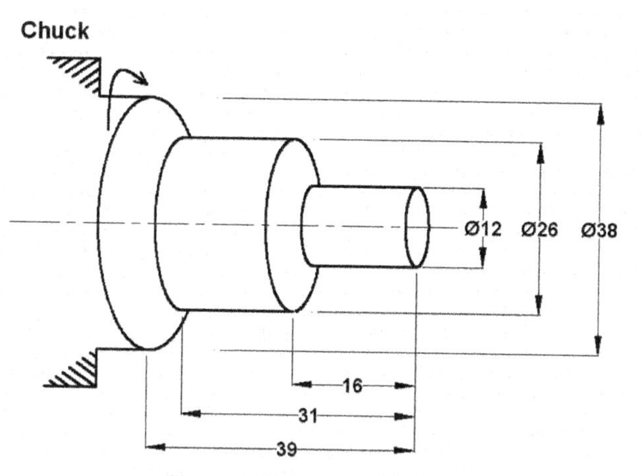

Figure 6.11 Isometric view.

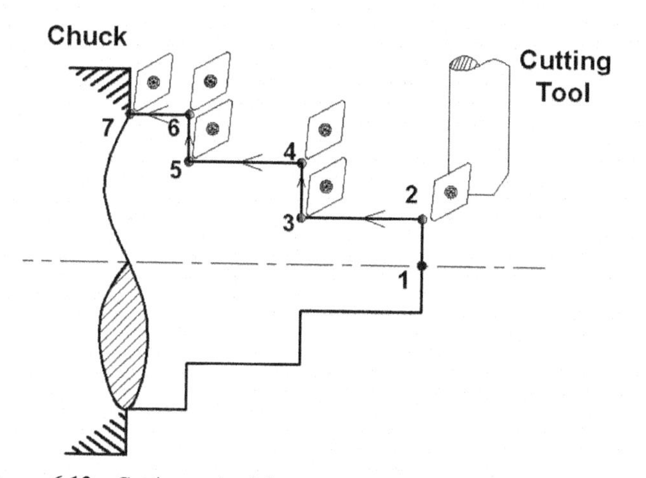

Figure 6.12 Cutting path of the step turning operation with side view.

Figure 6.13 shows 3D view. Here, the tool is traveling and cutting the material from a smaller diameter to bigger diameter. In previous pages, we told about cutting tool position. Cutting tool will always cutting the material above the central line of the rotating work piece. And the value of **X** will be positive in turning machine during cutting.

So following coordinates are taken from Figure 6.10.

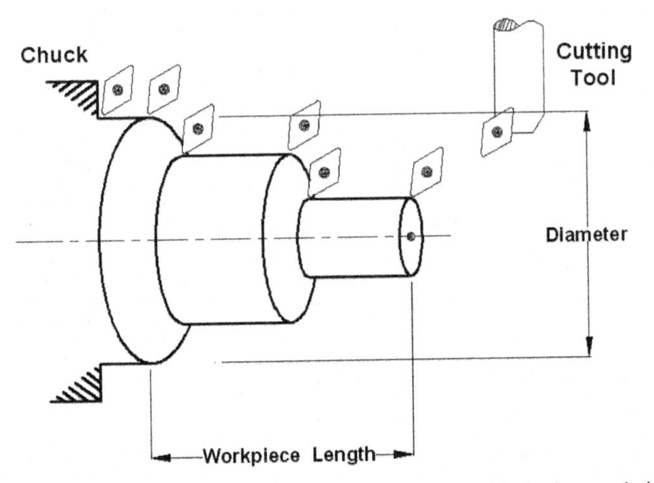

Figure 6.13 Cutting path of the step turning operation with the isometric image.

6.4.1 Cutting Tool Will Follow Figure 6.10 Coordinates Points 1, 2, 3, 4, 5, 6 & 7 and Remove the Extra Material as Per Drawing

(X, Z)	(X, Z)
1. (0, 0)	**5.** (26, −31)
2. (12, 0)	**6.** (38, −31)
3. (12, −16)	**7.** (38, −39)
4. (26, −16)	

6.5 Find the Coordinates in Absolute and Diametrical Method of the Given Figure

Following work piece images are showing side view (Figure 6.14) and also showing 2D cutting tool path view (Figure 6.15) and isometric view with cutting tool path (Figure 6.16).

6.5.1 During in Final Cut, Cutting Tool Will Follow Figure 6.14 Coordinates Points

(X, Z)	(X, Z)
1. (0, 0)	**5.** (31, −18)
2. (16, 0)	**6.** (35, −21)
3. (22, −3)	**7.** (35, −55)
4. (22, −18)	

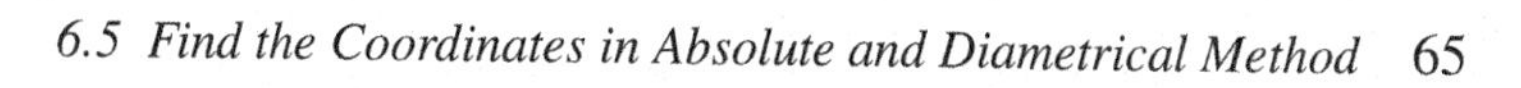

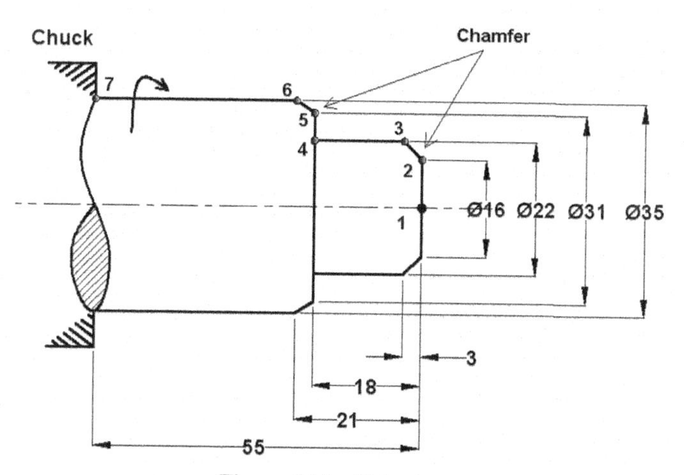

Figure 6.14 Side view.

Figure 6.15 3D view.

Figure 6.16 Cutting tool path with 3D view.

6.6 Find the Coordinates in Absolute and Diametrical Method of the Given Figure

Following work piece images are showing side view (Figure 6.17) and also showing 3D view (Figure 6.18) and isometric view with cutting tool path (Figure 6.19).

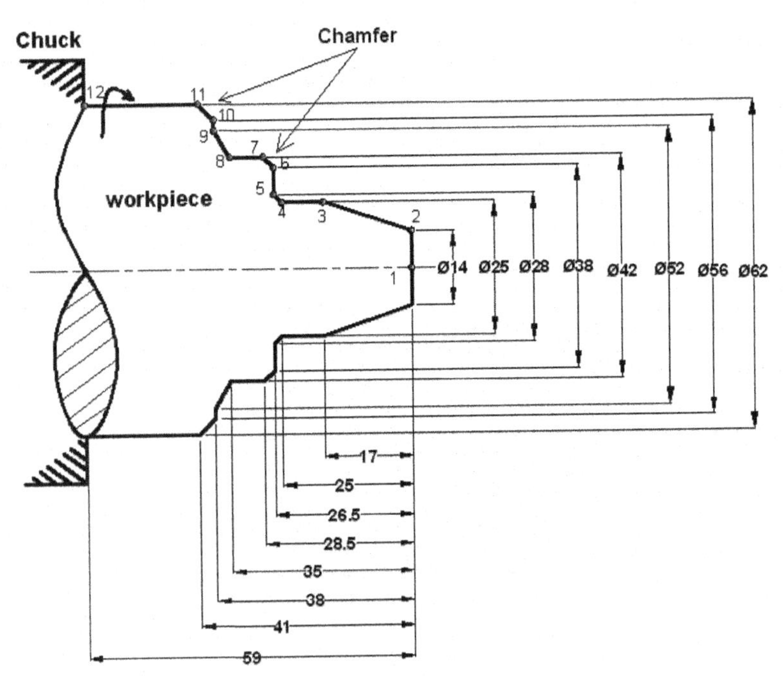

Figure 6.17 2D view of the cylindrical work piece.

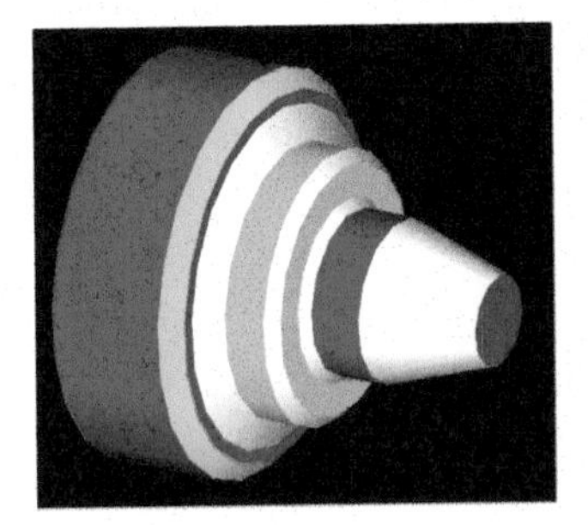

Figure 6.18 3D view.

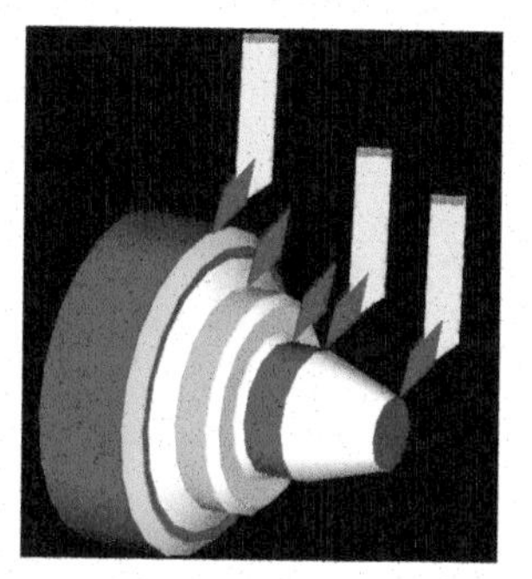

Figure 6.19 Cutting tool path with 3D view.

6.6.1 During in Final Cut, Cutting Tool Will Follow Figure 6.17 Coordinates Points

(X, Z)	(X, Z)
1. (0, 0)	**7.** (42, −28.5)
2. (14, 0)	**8.** (42, −35)
3. (25, −17)	**9.** (52, −38)
4. (25, −25)	**10.** (56, −38)
5. (28, −26.5)	**11.** (62, −41)
6. (38, −26.5)	**12.** (62, −59)

6.7 Find the Coordinates in Absolute and Diametrical Method of the Given Figure

The following image is showing side view (Figure 6.20).

6.7.1 During in Final Cut, Cutting Tool Will Follow Figure 6.20 Coordinates Points

(X, Z)	(X, Z)
1. (0, 0)	**7.** (25, −42)
2. (10, 0)	**8.** (32, −47)
3. (10, −15)	**9.** (32, −52)
4. (20, −25)	**10.** (35, −52)
5. (20, −30)	**11.** (35, −70)
6. (25, −37)	

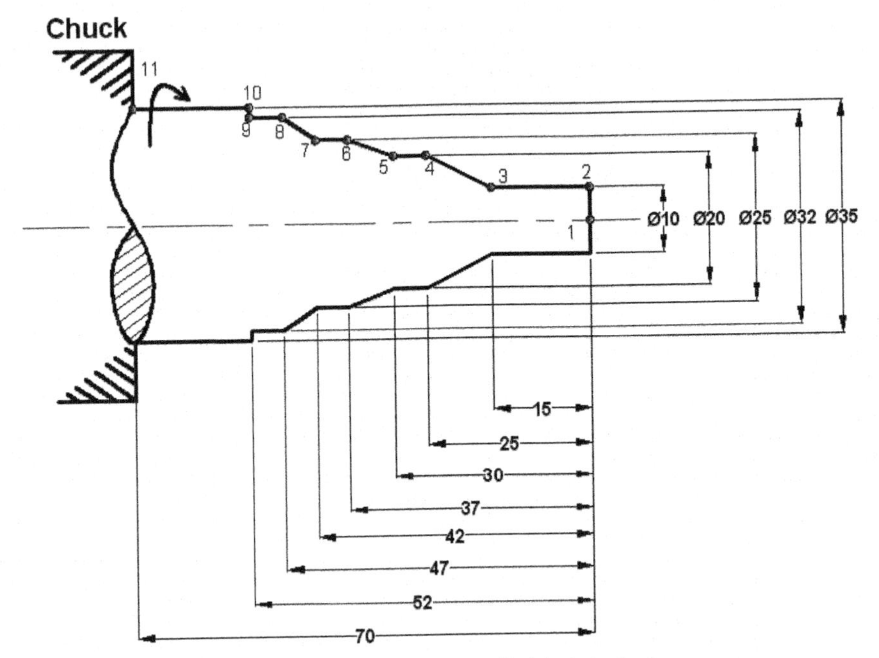

Figure 6.20 2D view of the cylindrical work piece.

6.8 Find the Absolute Coordinates of Given Figure in Radius and Diametrical Method

Following image is showing side view (Figure 6.21).

6.8.1 Radius Method

If you want to use the radius method. You will select/enable the radius parameter setting in the CNC machine. **After enable the setting,** radius method will work in CNC programming.

In Radius method, half diametrical dimension (radius dimension) will take in X-axis of the cylindrical work piece but **Z -axis value will be same**.

(X, Z)	(X, Z)
1. (0, 0)	**7.** (23.5, −24)
2. (9, 0)	**8.** (23.5, −34)
3. (12, −6)	**9.** (25, −37)
4. (12, −10)	**10.** (33, −37)
5. (15, −20)	**11.** (38, −40.25)
6. (21.5, −20)	**12.** (38, −65)

6.8.2 Diametrical Method

In diametrical method, whole diametric dimension (both side of the center) of the cylindrical work piece **will take** but Z -axis value will be same. See Figure 6.21.

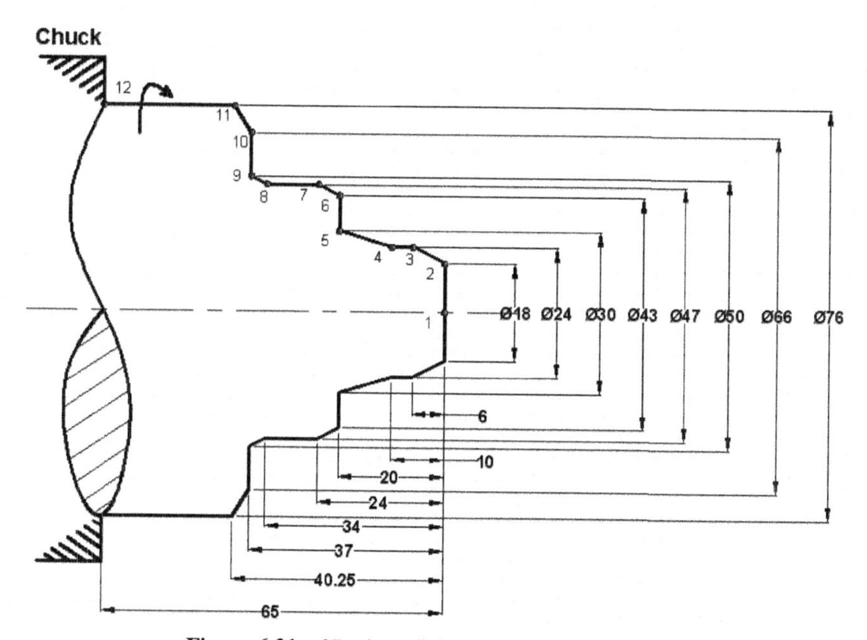

Figure 6.21 2D view of the cylindrical work piece.

(X, Z)	(X, Z)
1. (0, 0)	**7.** (47, −24)
2. (18, 0)	**8.** (47, −34)
3. (24, −6)	**9.** (50, −37)
4. (24, −10)	**10.** (66, −37)
5. (30, −20)	**11.** (76, −40.25)
6. (43, −20)	**12.** (76, −65)

6.9 Find the Absolute Coordinates in the Diametrical Method of Given Figure

6.9.1 During in Final Cut, Cutting Tool Will Follow Figure 6.22 Coordinates Points

(X, Z)	(X, Z)
1. (0, 0)	**5.** (28, −22)
2. (8, 0)	**6.** (34, −22)
3. (20, −6)	**7.** (48, −29)
4. (20, −16)	**8.** (48, −52)

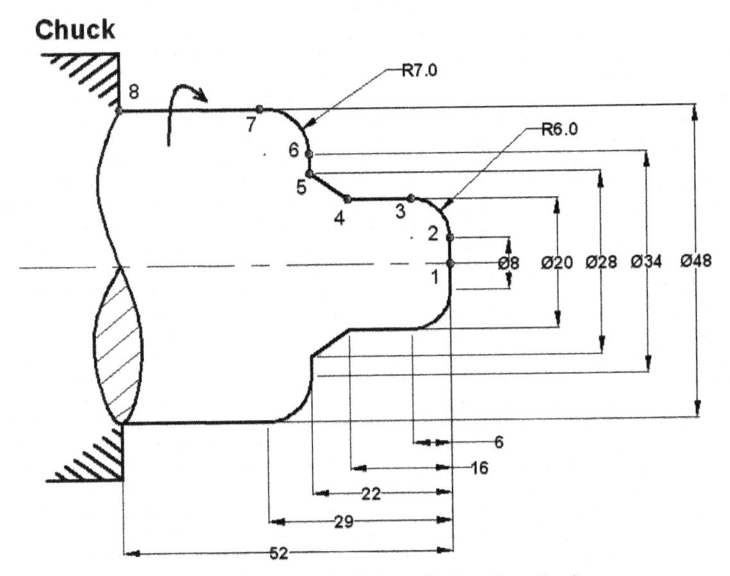

Figure 6.22 2D view of the cylindrical work piece.

6.10 Find the Coordinates in Absolute and Diametrical Method of the Given Figure

Following work piece images are showing side view (Figure 6.23) and also showing 3D view (Figure 6.24) and isometric view with cutting tool path (Figure 6.25).

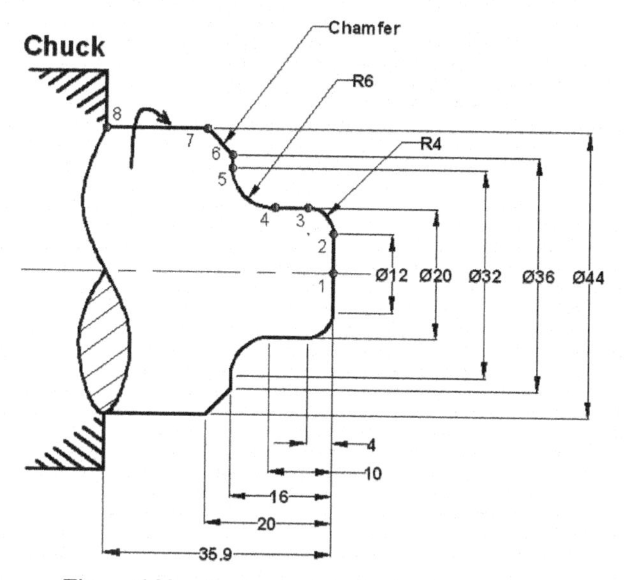

Figure 6.23 2D view of the cylindrical work piece.

Figure 6.24 3D view [4].

Figure 6.25 Cutting tool path with 3D view.

6.10.1 During in Final Machining Operation, Cutting Tool Will Follow Above Coordinates Points

(X, Z)	(X, Z)
1. (0, 0)	**5.** (32, −16)
2. (12, 0)	**6.** (36, −16)
3. (20, −4)	**7.** (44, −20)
4. (20, −10)	**8.** (44, −35.9)

References

[1] CNC programming Manual of MTAB company: (Certificate course on CNC Turning, MTAB Technology Centre, MTAB, Chennai, India).

[2] Fundamentals of Metal Cutting and Machine Tools, B. L. Juneja, G. S. Sekhon and Nitin Seth, Revised Second Edition (2005), New Age International Publisher, India.

[3] https://www.mechaterrain.com/coordinate-system

[4] Drawings and sketches has been generated from Auto CAD 2014.

7

CNC Machine Programming Codes (G-Codes and M-codes)

7.1 G & M Codes are the Main Programming Codes, Which are Used During CNC Programming

7.2 What are G Codes?

Generally G-codes are used in CNC programming for machining of different profile like turning, boring, facing, step turning, threading, chamfer, radius profile, profile turning etc. G-codes are used with numerical numbers like G01, G02, G21, etc. We can say, these codes are used in material removing operations, parameter settings and tool geometry offsets, etc.

7.3 G Codes are of Two Types

7.3.1 Model Command

G codes of this group will be effective in CNC program until replaced by another G code or another G code will not activate. Examples G01, G00, etc.

Programming Example of model commands

G00 X26.5 Z15.0;	The cutting tool is moving in rapid motion, rapid motion means very fast
G00 X26.5 Z1.0;	Here also tool is taking position rapidly with G00 commands
G01 X26.5 Z-38.0 F0.25;	The tool is moving and removing the material in linear motion, linear motion means very slowly

7.3.2 Non Model Command

G codes of this group are effective only once when it is used or when it occurs. It will not effective in the next line, if it is not used again. Examples G02, G03, G04, etc.

7.4 Important and Interesting Facts

1. Maximum spindle speed setting (G50) is valid when the constant surface speed control (G96) is used.
2. The G codes are marked * are set when the power is turned on.
3. The G codes in group 00 are not model. They are effective only in the block in which they are specified.
4. One or more than one G codes are used in one programming line (block) if they are not in the same group. When the number of G codes of the same group is specified, the G code specified last is effective.
5. All the G codes may not apply to each machine.

7.5 Examples of Some G Codes

(similar to Fanuc control system or other control software)

G code		
G code	**Group**	**Function**
G00*	1	Positioning rapid traverse
G01	1	Linear interpolation with Feed
G02	1	Circular interpolation (CW)
G03	1	Circular interpolation (CCW)
G04	0	Dwell
G20	6	Inch system
G21	6	Metric system
G28	9	Machine home position
G32	1	Thread cutting
G40*	7	Tool nose radius compensation cancel
G41	7	Tool nose radius compensation left
G42	7	Tool nose radius compensation rightl
G50	0	Maximum spindle speed
G70	4	Finishing cycle
G71	4	Stock removal cycle (for turning)
G72	0	Stock removal cycle (for facing)
G74	0	Peck drilling in Z axis
G76	0	Thread cutting cycle
G92	1	Thread cutting cycle
G96	2	Constant surface speed On
G97*	2	Constant spindle speed On
G98	11	Feed (mm)/minute
G99	11	Feed (mm)/revolution

Figure 7.1 Table of G codes.

Note: Following codes are similar to **Fanuc Control software** and similar software. **All G codes may not apply to each machine** (Figure 7.1)

7.6 G00 Rapid Positioning/Rapid Traverse/Rapid Motion

G00 command will use for fast tool movement or positioning. When we will apply G00 code, it means it will never use for cut the material only use for tool positioning. We can say, G00 will apply when the tool will come near the job or retract from the job in rapid motion (in this case tool should not be touch from the job/part) [1].

Example

G00 X (coordinate value where the tool takes the position in X-axis) **Z** (coordinate value where the tool takes the position in Z-axis);

OR

G00 X----- Z---- ;

Where,

G00 is used for rapid positioning of the cutting tool.
X and **Z** will be the coordinates in X and Z-axis.

Feed **Simply we can say, the feed is the cutting speed of the tool.**

G00 is used without feed (F) because its feed (F) already defined by the CNC machine manufacturer in meter/minute. That's why we cannot define the feed in the program, if we define the feed with G00 in the CNC program. The machine will give an alarm message.

Note: Generally, the CNC machine manufacturer sets the speed (Feed)/rapid Traverse (G00) of the axis.

Example

CNC machine's rapid motion (speed) value can be in axis (X, Y and Z) like 710 Inch Per Minute (IPM), 945 IPM, 1200 IPM, etc.

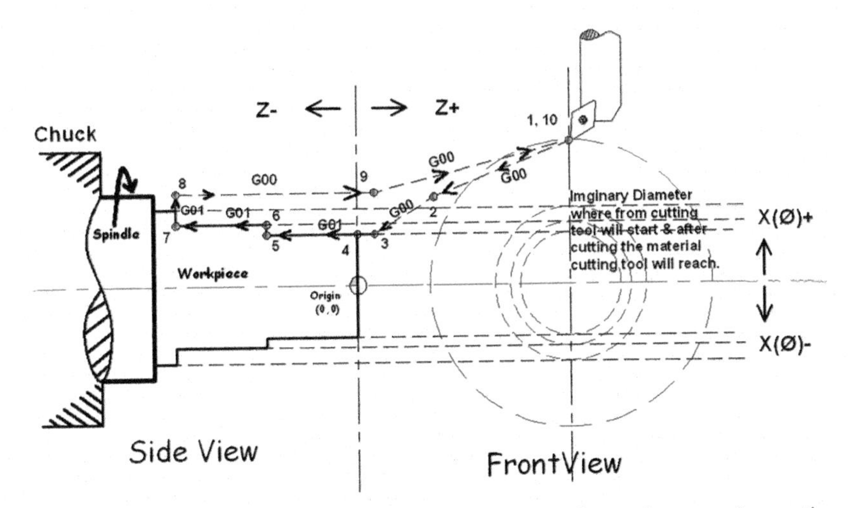

Figure 7.2 Cutting tool path of material cutting during cutting and non-cutting motion of the cutting tool.

Details of Figure 7.2.

1. The continuous line is showing material cutting path using G01 code (Figure 7.2).
2. The dotted line is showing tool path when cutting tool takes the position rapidly near the work piece or cutting tool retracts rapidly from the work piece after material removing for this condition G00 code is used (Figure 7.2).

7.7 G01 Linear Motion with Feed Rate (F)

Command **G01** will apply, when the cutting tool removes or cuts the material in turning operation. Example: straight turning, step turning, facing, chamfering, taper turning operation, etc.

It is very important that when we use the G01 command in CNC programming, we must use **Feed** with G01. Without Feed, G01 command will not work and alarm message will occur in the CNC control screen.

Example

G01 X (coordinate value where the tool takes the position in X-axis) **Z** (coordinate value where the tool takes the position in Z-axis) F;

OR

G01 X----- Z---- F----;

Note:

i. Feed (**F**) must be used with G01 command.
ii. **X and Z** will be the coordinates of the end point of the profile like straight turning, taper turning etc, where the cutting tool will reach.

Example of G00 and G01

Let we assume, we want to cut an apple with a knife. What do we do first? First, knife (cutting tool) takes the position rapidly (**G00**) nearest the apple (job) but knife does not touch the apple and after positioning the knife (tool), knife (tool) cuts the apple very slowly (**G01**) with controlled hand force (Feed) and this *controlled hand force* is called feed (**F**), it means **feed (F)** will use with **G01**. After cutting the apple, now the knife (tool) will retract rapidly (**G00**) from the apple (job) and takes a safe distance far from the apple.

At the same time, when we apply **G00** on the cutting tool, it behaves same like the knife. First, the tool will take the position rapidly (**G00**) nearest the work piece but the tool will not touch from the work piece, after taking the position nearest the job. Now the cutting tool will cut the material very slowly (**G01**). After cutting the material from the job, the tool will retract rapidly (**G00**) from the work piece without touching the work piece. It is an example of the **G00** command.

7.8 G02 Circular Interpolation in Clock Wise (C.W.) Direction

If you want to make a circular/arc shape of any work piece, you will use the **G02/G03** command. But **G02** works only in the clockwise direction (when cutting tool moves in clockwise direction). Whenever we apply **G02**, we use radius (**R**) value together.

Example of G02

G02 X---- Z----- **R-----** F----- ;

Where,

G02 means, the tool will cut the material always in the clockwise direction (cw).

R is the radius value of arc profile.

X and Z are the endpoint coordinates of the arc/radius profile, where the cutting tool will reach after removing of the material.

Note: For making CNC program of radius **or arc profile, always required the coordinates of starting and ending point of the radius profile** (Figure 7.3).

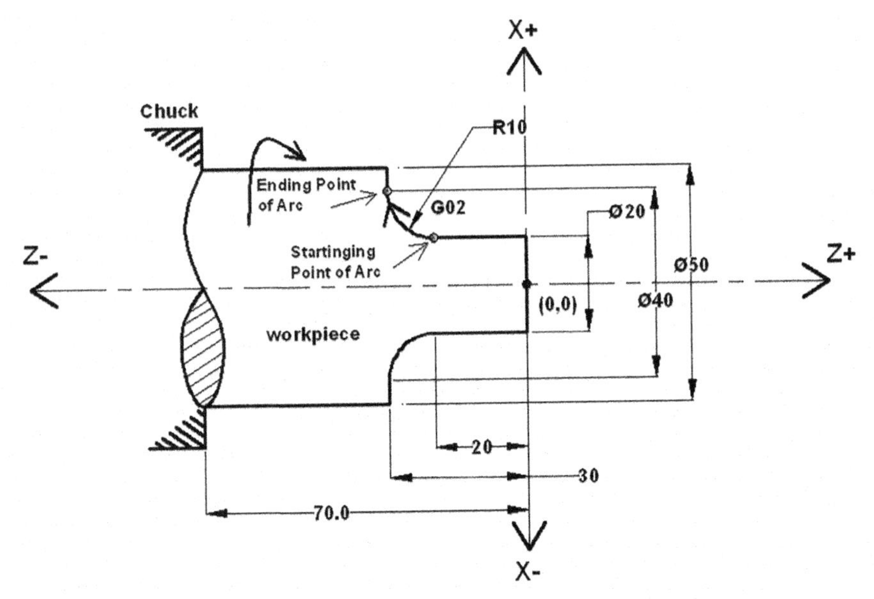

Figure 7.3 Arc profile with clockwise direction – G02.

Note: For making arc profile in the CNC program, CNC programmer required **coordinates of starting point in X-axis** of arc profile and **coordinates of the ending point of Z-axis** of Arc profile with radius (**R**) value.

Example

G01 **X20 Z-20** F0.12;	For making the arc profile, (X20, Z-20) is the starting point coordinates of the arc. Therefore cutting tool will take the position at this coordinate.
G02 X40 Z-30 **R10** F0.1;	Now cutting tool will move and remove the material and reach at the end point (X40, Z-30) of the arc in circular motion G02.

7.9 G03 Circular Interpolation in Counter (Anti) Clock Wise (C.C.W.) Direction

The G03 command works same like to G02 command. Only one difference we will find between **G02 and G03** and it is tool movement direction. When G03 command applies, cutting tool moves and cutting the material in counter (anti) clock wise direction.

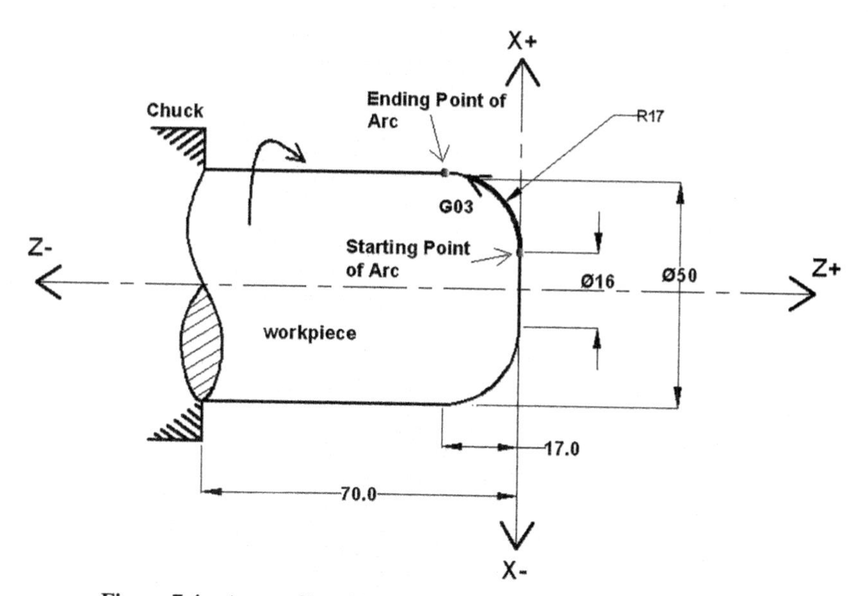

Figure 7.4 Arc profile with counter (anti) clockwise direction (G03).

When we use this command during CNC program, due to this command cutting tool will travel and remove the material in counter clock wise direction.

Example of G03

G03 X---- Z----- **R-----** F----- ;

Where,

G03 means, the tool will cut the material always in a counter clockwise (anti) direction (ccw).
R is the radius value of arc profile.
X and Z are the end point coordinates of the arc/radius profile, where the cutting tool will reach after material removing. See Figure 7.4.

Example

G01 X16 Z0 F0.12;	In this programming line, X and Z are the coordinates of the **starting point** of the radius profile.
G03 X50 Z-17 R17 F0.1;	In this programming line, cutting tool will remove the material from **starting point** to **end point** as an Arc profile (**R**) and reached at coordinates (X50 Z-17).

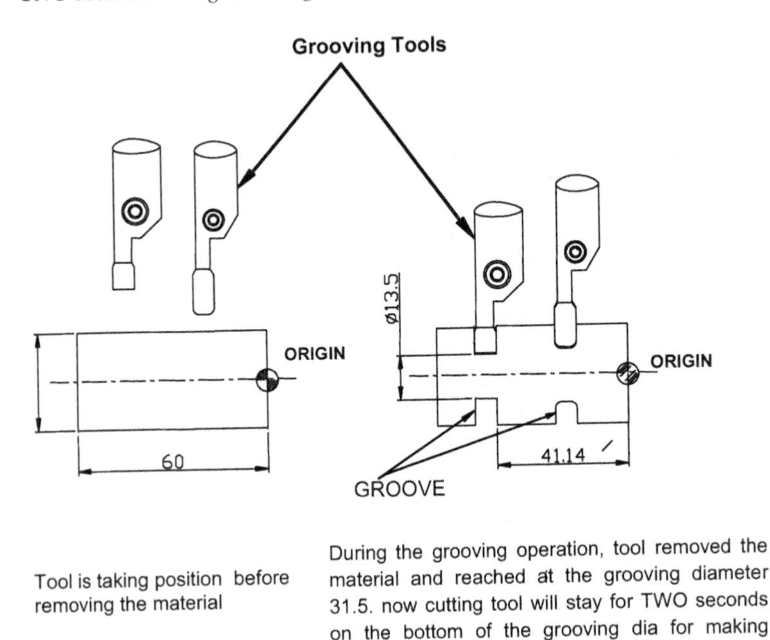

Figure 7.5 Grooving tools are taking position on the work piece surface in the left figure and in the right figure grooving tool is making groove on the work piece surface.

7.10 G04 Dwell Time

Sometimes, especially in **grooving operation**. When we cut the material, the cutting tool **needs to stay** (just for few seconds) after the end of the cutting operation. It is required for achieving accurate surface profile in the bottom of the grooving profile, when the cutting tool is reached there. See Figure 7.5 [2].

G04 command is used with **X/P/U**.

G04 P……;

Where,

G04 is the Dwell
X…/P…/U… is the **Time** in Second

(Example: **X2/P2/U2** it means after cutting the desired profile/shape tool will stay for 2 seconds on that profile for achieving an accurate surface profile.)

OR

X3/P3/U3 means Three Seconds, different-different CNC control systems use different alphabets for dwell time, but working and meaning will be the same.)

Example

G00 X30 Z-41 F0.1;	Cutting tool is taking position in X and Z-axis, before removing the material.
G00 X17 Z-41;	Cutting tool is taking position nearest the work piece in X-axis, before removing the material
G01 X16 F0.25;	Cutting tool is taking final position before moving and removing the material.
G01 X13.5 F0.06;	Tool is removing the material and reaches at diameter 13.5 in linear motion with feed.
G04 X2;	**Dwell Time** (After removing the Material tool will stay at the bottom of the cutting profile for **Two Seconds**)
OR	
G04 P2;	
OR	
G04 U2;	
G01 X16 F0.06;	Now cutting tool is going back (retract in X-axis) to the previous position.

7.11 G17 Selection of XY Circular Plane

G17 code is used, when CNC machine works in **X** and **Y** circular plane (axis) in clockwise or counter clockwise direction. If you want to make circular profile in XY plane. You must activate or you must use the **G17** code in the CNC program. **G17** code is used in **CNC milling** machine.

Example: Milling machine program

G00 X22 Y-10 Z5;	Cutting tool is taking position rapidly (G00)
G01 Z-4 F50;	Tool is taking position slowly (G01) in Z-axis with feed
G01 X22 Y0 F25;	Tool is taking the position very slowly (G01) at starting point of arc with feed
G17 G02 X0 Y22 R22 F15;	Here G17 code will use for making **XY** circular plane
G01 X0 Y160 F25;	Tool is retracting from the work piece and going the to safe distance

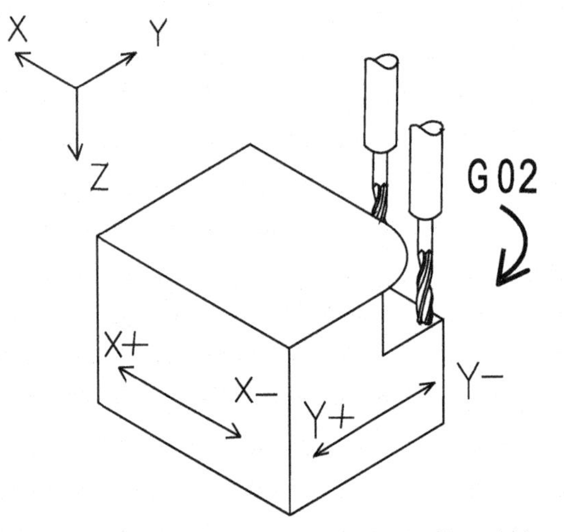

Figure 7.6 Circular motion in XY plane. This command will apply in milling machine.

G17 command is used to select the XY plane during the circular cutting. It can be used in clock wise circular cutting or counter clock wise circular cutting (Figure 7.6).

7.12 G18 Selection of ZX Circular Plane

When the CNC programmer wants to make the circular profile in clockwise or counter clockwise direction in **X and Z**-axis, the G18 code is used in **CNC turning and CNC milling machine**.

If you want to make a circular profile in X and Z-axis plane. You must activate or you must write G18 code in the CNC program.

G18 command is used to select the XZ plane during the circular cutting. It can be used in clock wise circular cutting or counter clock wise circular cutting (Figure 7.7).

Example

G00 X70 Z15;	Cutting tool taking position rapidly (G00)
G00 X16 Z1.0;	Tool is taking position rapidly (G00) near the work piece in X and Z-axis
G01 X16 Z0 F0.10;	Tool is taking position (starting point of the radius) slowly nearest the work piece in Z-axis with feed

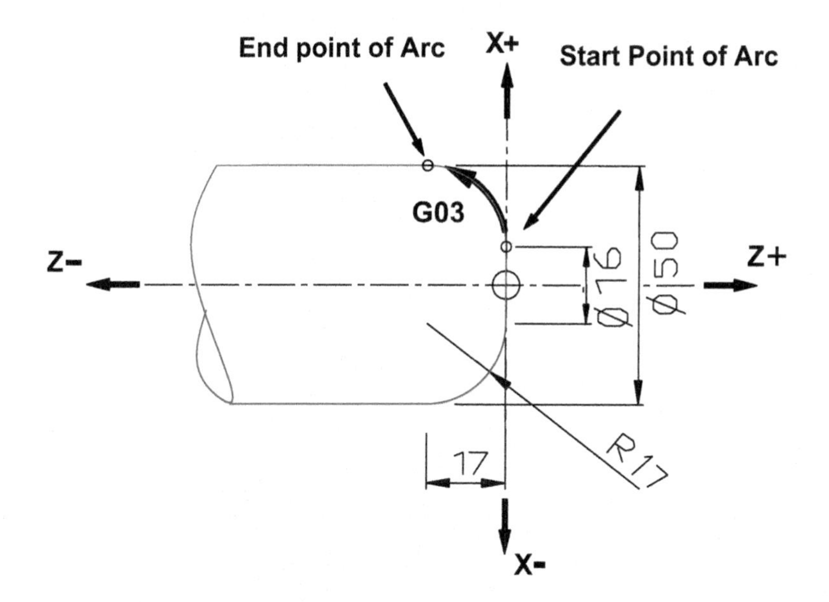

Figure 7.7 Circular motion in the XZ plane. This command will apply in both turning and milling machine.

G18 G03 X50 Z-17 R17 F0.12;	Here G18 code will use for making **ZX** circular plane
G00 X50 Z1.0;	Tool is retracting from the work piece and going to a safe distance in rapid motion

7.13 G19 Selection of YZ Circular Plane

When CNC programmer wants to make a circular profile in Y-axis and Z-axis plane. That time **G19** code is used as a **YZ circular plane**. If you want to make an Arc or circular profile in Y-axis and Z-axis plane. You must activate or you must write G19 code in CNC program [2].

Example

G00 Y70 Z5.0;	Cutting tool taking position rapidly (G00)
G01 Z1.0 F30;	Tool is taking position slowly (G01) in Z-axis with feed
G01 Y70 Z0 F25;	Tool is taking the position very slowly (G01) at starting point of arc with feed

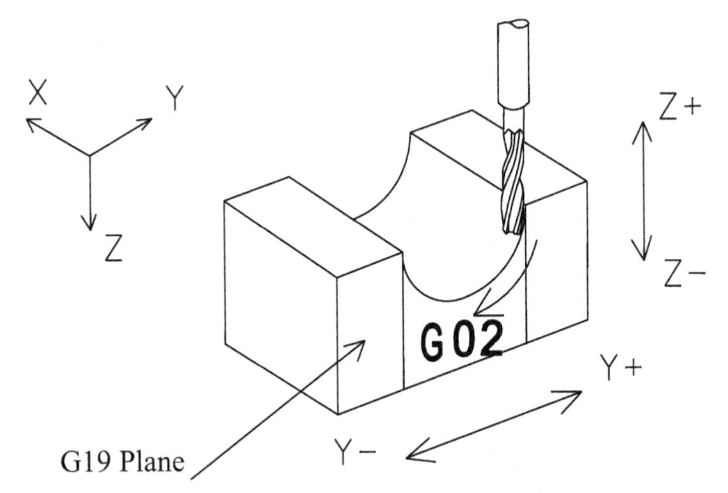

Figure 7.8 Circular motion in the YZ plane. This command will apply in milling machine.

G19 G02 Y30 Z0 R20 F15;	Here G19 code will use for making **YZ** circular plane
G00 X50 Z1.0;	Tool is retracting from the work piece and going to safe distance in rapid motion

G19 command is used to select the YZ plane during the circular cutting. It can be used in clockwise circular cutting or counter clockwise circular cutting. See Figure 7.8.

7.14 G20 Inch Selection System in CNC Programming

G20 code is used for the **Inch system**. If you want to make the CNC program in the Inch method, then you will enable **G20** code in CNC software in parameter setting. **Thou** is the least unit of the Inch system.

$$\text{One Inch} = 1000 \text{ thou}$$

7.15 G21 Metric Selection System in CNC Programming

G21 code is used for **the Metric system**. Metric system means meter/millimeter/micron (μ). When CNC programmer activates or enables the G21 command in CNC software, it means you will write the all CNC program in metric system. In the metric system we write all CNC program in millimeter. Least unit of the metric system is **Micron** (μ) [2].

$$\text{One millimeter} = 1000(\mu)$$
$$\text{One thou} = 0.0254(\mu)$$

7.16 G28 Return to Reference Point (Machine Home Position)

When cutting tool finishes the machining operation, cutting tool goes far from the work piece due to safety reasons or we can say cutting tool goes to the last limit of the positive axis (+). It is called the machine home position (G28). Generally, this command is used for safe loading and unloading of the work piece on the machine.

Example

M09;	Coolant motor will stop after complete machining
M05;	Spindle motor will stop.
G28 U0 W0;	Now cutting tool will go to the home position (G28). This home position is always far from the work piece
M30;	CNC program will stop and Reset

7.17 G32 Thread Cutting Command

The G32 command is used for threading operation, It is a very simple threading method. In this method CNC programmer can control threading depth of cut in each pass.

G32 X..... Z..... F.....;

When the CNC programmer uses the G32 threading command. Threading tool takes fixed (as we need) depth of cut and this procedure repeats again and again with G32 command until threading tool does not reach on the root diameter of the thread profile or does not cut the full thread profile /full depth of cut of the thread.

G00 X36 Z5;	In threading operation, tool is taking position in X-axis and Z-axis.
G32 X36 Z-54.5 F2.5;	Tool is taking first cut and making (removing the material) thread profile.
G00 X46;	After the first cut of the thread, tool retracts in X-axis and takes safe position.
G00 Z5;	Now tool retracts in Z-axis and takes a safe position.

G00 X35.8;	Tool is taking the position again in X-axis (at dia 35.8, where the tool will cut the thread) before cutting the thread.
G32 Z-54.5 F2.5;	Now the threading tool is making the thread in Z-axis. It means the tool is making thread at length 54.5
G00 X46;	After the second cut of the thread, tool retract in X-axis and takes safe position
G00 Z5;	Now tool retracts in Z-axis and takes a safe position.

Note: It will continue until thread profile will not complete. On the other hand we can say, until full thread depth does not achieve.

7.18 G33 Thread Cutting Command

The G33 command works same like to G32 command. Both commands are used in different-different softwares.

7.19 G40 Tool Nose Radius Compensation Cancel

G40 code is used for cancellation of Tool nose radius compensation, when we apply G41 or G42 codes.

7.20 G41 Tool Nose Radius Compensation Left

When we make the CNC program on CNC turning machine. G42 command is used as a tool nose radius compensation left. When we use tool nose radius compensation left. It depends on CNC turning machine bed condition (slant bed/flat bed machine). This code is mostly used during different turning operations like turning operation, taper turning, profile turning, chamfering, arc profile, boring operations etc.

7.21 G42 Tool Nose Radius Compensation Right

It is same like G41 command. The only difference is, both commands are used in different conditions. The G42 command is used as a tool nose radius compensation right. It depends on CNC turning machine bed condition (slant bed/flat bed machine). This code is mostly used during different turning operation like turning operation, taper turning, profile turning, chamfering, arc profile, boring operation etc.

Note: G41 and G42 commands are used for achieving best profile accuracy and complex shapes.

7.21.1 G41 and G42 Will Apply in Following Conditions

Using the G41 and G42 commands depend on CNC turning machine structure (location of the machine bed).

Basically CNC machine has **two types** of bed.

i. **Slant Bed** CNC turning machine (tool post angle will be 45/30 degree)
ii. **Flat Bed** (machine bed will be same like lathe machine) CNC turning machine

7.21.1.1 If CNC turning machine has a slant bed (45°/30°)

Slant bed CNC turning machine means cutting tool will always be situated after machine operator and the work piece. On the other hand, we can say, in this condition cutting tool will be situated between slant bed and the chuck. See Figure 7.9.

i In this case, the **G41** code will use for **Boring (internal)** operation.
ii And **G42** code will use for **External turning**.

Image (Figure 7.10) is a slant bed CNC machine, where turret (cutting tools) mounted on slant bed.

The image (Figure 7.11) is showing cutting view on slant bed CNC machine. We are watching the cutting tool position from the front side of the CNC machine, it means the tool is situated back side of the work piece.

In Figure 7.12, the turret is mounted on the slant bed. It means, when the cutting tool cuts the material that time we cannot see the tool cutting edge because the tool will hide behind the work piece. But in flat bed turning machine, we can see very easily cutting tool edge.

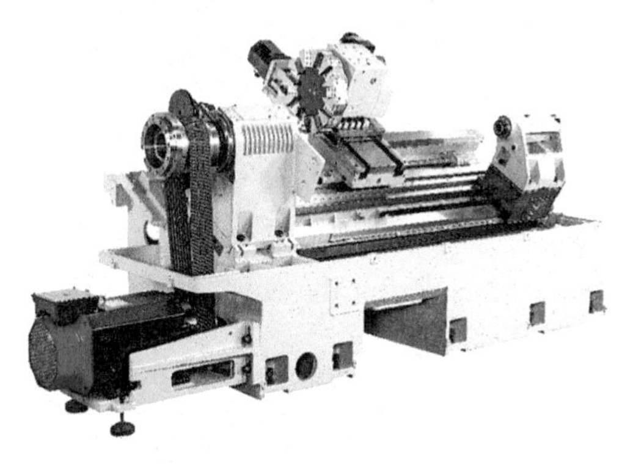

Figure 7.9 Structure of slant bed CNC machine.
(Courtesy: Structure of Slant Bed CNC, Machine Myers Technology Company, USA)

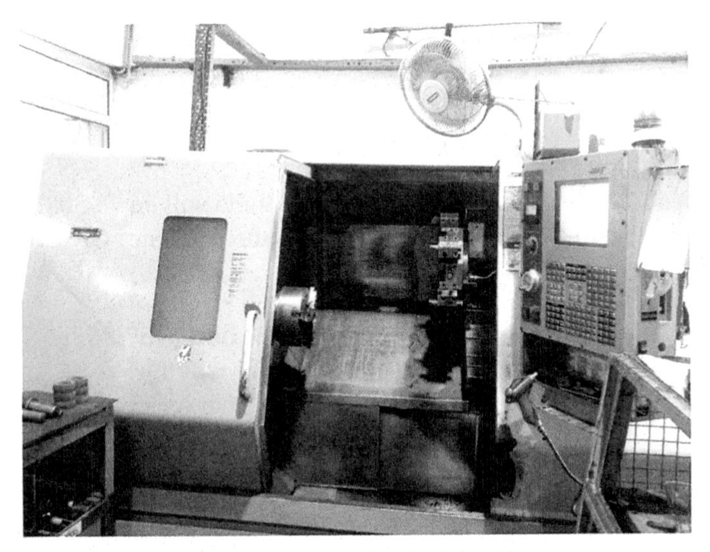

Figure 7.10 Slant bed CNC machine.

(Courtesy: Haas SL 20 CNC Machine, Sara Sae India Pvt Ltd, India)

Figure 7.11 Work piece cutting view on slant bed CNC machine.

(Courtesy: CNC Horizontal Turning machine, JYOTI, Rawat Engg. Tech Pvt Ltd., India)

Figure 7.13 shows the slant bed cutting position. Due to slant bed condition, the cutting tool is positioned after the work piece. That is why we use the G42 command.

Figure 7.14 is showing slant bed cutting condition. Because when we see the boring tool from his back side then we find, the boring tool is situated on the right side of the work piece inside the internal surface. Respectively to the surface (surface wall) boring tool is situated on the left side of the wall surface. That's why we use the G41 command.

Figure 7.12 Closed view of slant bed CNC machine.

(Courtesy: CNC Turning Machine, JYOTI, Rawat Engg. Tech Pvt Ltd., India)

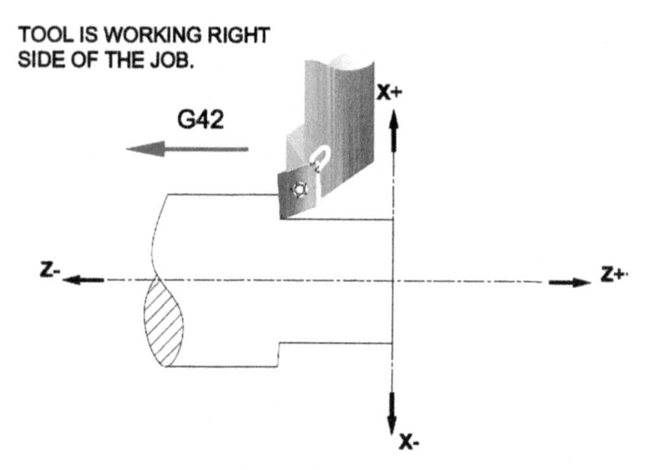

Figure 7.13 Tool is performing external operation in slant bed CNC machine.

In Figure 7.14, the tool is performing the internal operation and figure is showing slant Bed condition for G41

When we see the front face of the work piece from the back side of the tail stock that time we see, where from the cutting tool is cutting the material from the left side of the work piece or right side of the work piece.

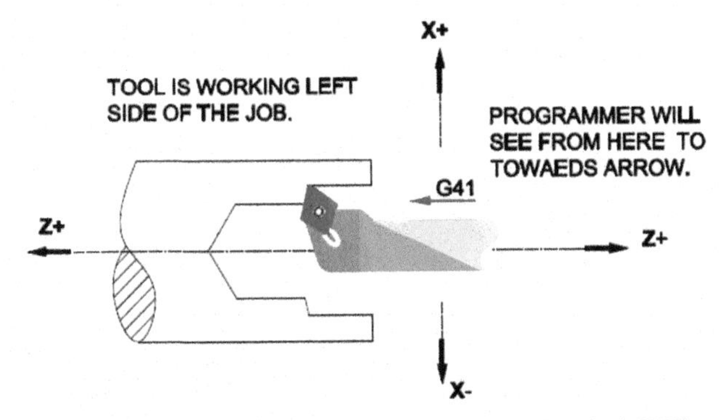

Figure 7.14 Tool is performing internal operation in slant bed CNC machine.

If the tool is cutting the right side material (outer diameter) of the work piece, it means the **G42** code will apply for external turning operation and if the tool is cutting the left side material (inside the hole) of the work piece surface, it means **the G41** code will apply for internal turning operation.

Note: Above condition will applicable for slant bed CNC machine.

7.21.1.2 If turning machine has flat bed same like conventional lathe machine

Flat bed CNC turning machine means cutting tool will always situate between the machine operator and chuck. On the other hand we can say, cutting tool will be situated between the CNC operator and the work piece.

i **G41** code will use for **external turning** operation.
ii And **G42** code will use for **boring (internal)** operation.

Cutting tools mounted on the flat bed. The position of the turret and bed is same like lathe machine. See Figure 7.15.

Figure 7.16 is showing Flat Bed condition same like lathe machine. In this condition we will use the G41 command for external turning operation.

In the Figure 7.17, the cutting tool is turning the material with G41 command and tool is situated our side same like lathe machine.

Figure 7.18 is showing Flat Bed condition same like lathe machine. In this condition we will use the G42 command.

In Figure 7.17, condition in CNC turning machine, when we will see the front face of the work piece from the back side of the tail stock that time we see, where from the cutting tool is cutting the material from the left side of the work piece or right side of the work piece.

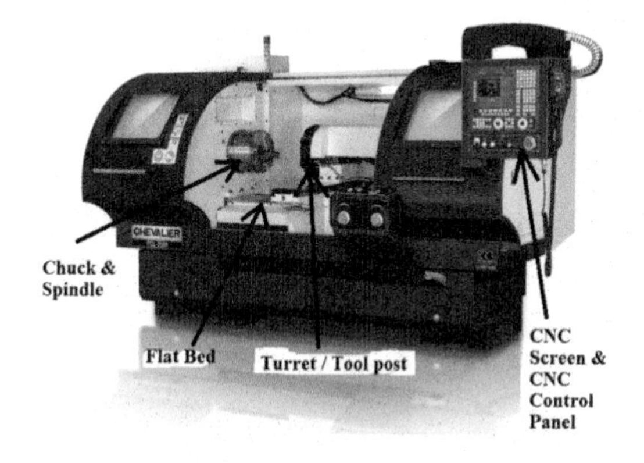

Figure 7.15 Flat bed CNC machine.
(Courtesy: Chevalier FCL-2560, Chevalier_Fagor Automation, Spain)

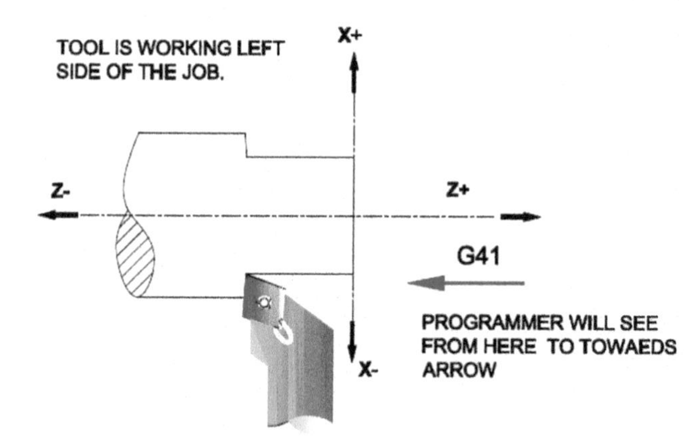

Figure 7.16 Turning tool is machining external operation.

If the tool is cutting the right side material (boring operation) of the work piece, it means the **G42** code will apply for internal turning operation and if the tool is cutting the left side material (external turning operation) of the work piece surface, it means **the G41** code will apply for external turning operation.

Note:

The main key word is where the cutting tool situated, left or right side of the work piece. When you will see this situation (cutting tool and work piece) from the front face of the work piece or back side of the tailstock. Then you can take the right decision about G41/G42 command and you can apply

Figure 7.17 Real machining view of flat bed CNC turning machine.
(Courtesy: Flat Bed CNC machine, HAAS, Rawat Engg. Tech Pvt Ltd., India)

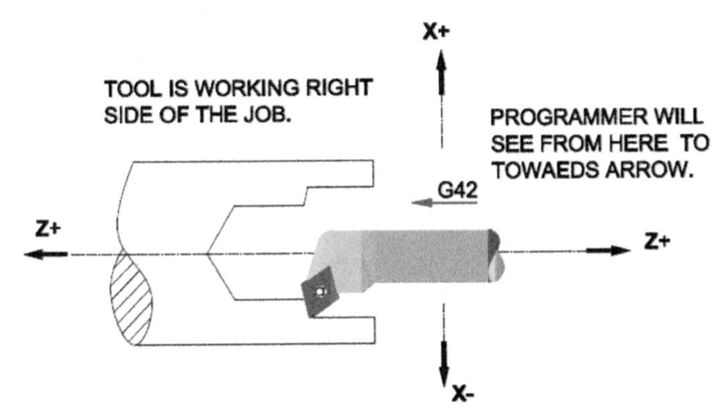

Figure 7.18 Cutting tool is machining internal operation (boring operation).

rightly G41/G42 command in the CNC program. If the tool is left the side of the work piece then G41 will apply, if the tool is the right side of the work piece then G42 will apply.

7.22 Uses of G41 and G42 in CNC Milling Machine

7.22.1 G41 and G42 Codes are Used also in CNC Milling Machine for Cutter Compensation

Figure 7.19 shows the vertical milling machine.
Figure 7.20 shows milling machine operation.

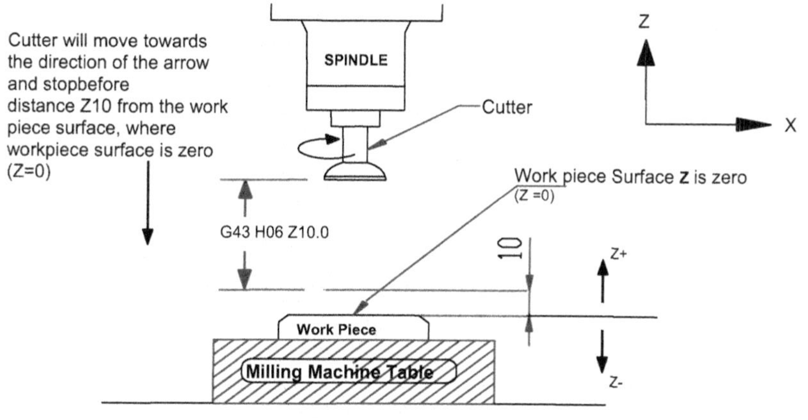

Figure 7.19 Vertical milling machine.

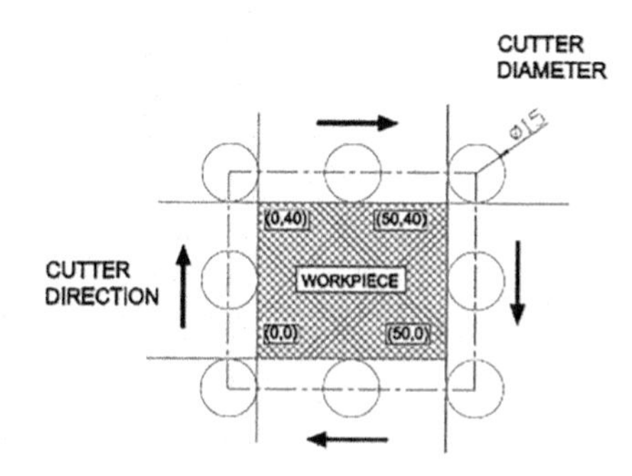

Figure 7.20 The cutter is removing the material as a side milling operation in the milling machine condition where if cutter takes the cut from the left side of the work piece then the G41 command will apply. When we use cutter compensation, it means cutter diameter will be compensated (adjusted) automatically by the CNC milling machine. Cutter compensation is used for accurate dimensions and complex profile. During the machining process, different types (different diameters) of cutters are used for milling profile. So cutter compensation is used for adjusting the cutter diameter during the machining process, without choose right cutter compensation (G41/G42), the cutter cannot cut the correct profile as per drawing. When we take the correct cutter compensation (G41/G42), cutter follows the correct profile/coordinates as per the following drawing.

7.22.1.1 Condition No. 1

Cutter should be positioned at the left side of the work piece. When the cutter moves on the **left side of the work piece during cutting operation if that time we follow the cutter from the back side of the cutter towards to cutter direction, during this period cutter will go far from us. In this condition, the G41** command will use as a left cutter compensation. See Figure 7.21.

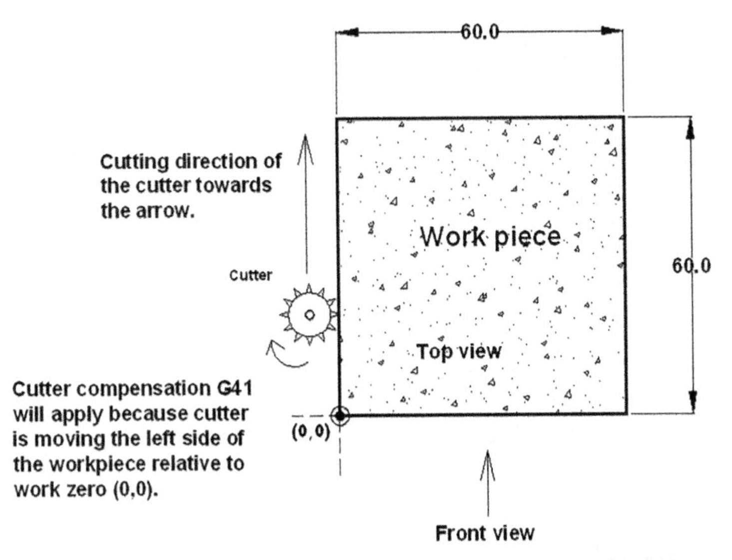

Figure 7.21 Cutter is taking cut from the left side of the work piece with G41 command.

7.22.1.2 Condition No. 2

Cutter should be positioned at the right side of the work piece. When cutter starts the movement and moves to the **right side of the work piece during the cutting operation and this time if we follow the cutter from the back side of the cutter towards the cutter direction. In this case, the cutter will go far from us. In this condition, we use the G42** command as right cutter compensation. See Figure 7.22.

7.23 G43 Tool Length Compensation Positive

7.24 G44 Tool Length Compensation Negative

7.25 G49 Tool Length Compensation Cancel

7.26 G50 Maximum Spindle Speed (for Control Maximum Spindle Speed)

G50 command is used for controlling the maximum spindle speed of CNC machine. When we use G50 (example: **G50 S1600**) command in CNC program, it means spindle RPM (revolution per minute) will not exceed more than 1600 rpm.

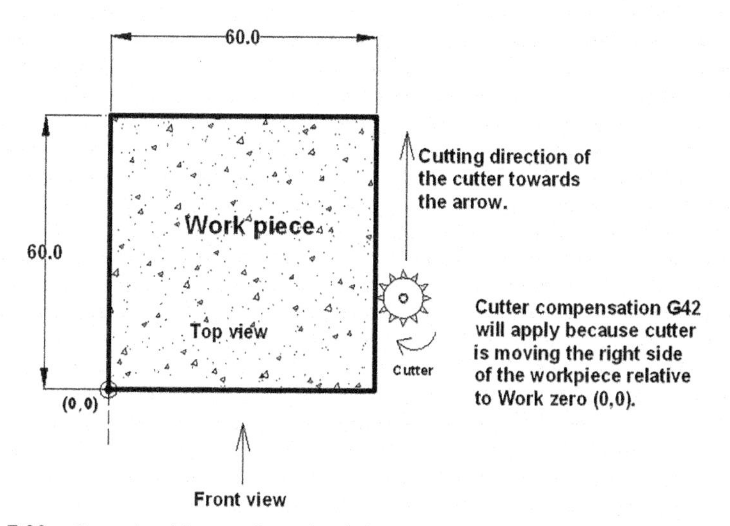

Figure 7.22 Cutter is taking cut from the right side of the work piece with G42 command.

When we use G50 command in CNC program, in the next line (block), we must apply G92/G96 command with cutting speed (rpm).

Example

G91 G28 U0 W0;	
G50 S1600;	This line is telling spindle will not rotate more than a maximum of 1600 rpm.
T0202;	
G96 M03 S210;	Spindle will rotate in clockwise direction (M03) on 210 rpm, it means 210 is minimum spindle speed.
G00 X15 Z50;	Cutting tool is taking position rapidly in X and Z-axis.

G50 command is used when turning tool is turning on different diameters, during this process, we get the fine surface finish on each turning diameter.

On the other hand, we can say when the turning tool reaches different diameters; every diameter of the work piece shows different RPM at that particular time. It happened only due to the use of **G50 and G92/G96** commands. That's why spindle speed varies on different diameters.

If CNC programmer wants only one spindle speed (RPM) on the work piece surface. Then programmer must use the **G97 command. A** constant surface speed is used during the drilling, threading, grooving operation, etc. and ***if the work piece is very heavy or work piece diameter is very big***.

7.27 What are the Benefits of the Constant Surface Speed (G96)?

When we apply the constant surface speed on CNC turning machine.

We get the following benefits:

i. Easier programming: we do not calculate rpm for each diameter.
ii. We get the constant surface finish.
iii. Optimum cutting tool life.
iv. Optimum machine cycle time: faster spindle runs the faster machining (the tool machine).

$$\textbf{Work piece Diameter } \alpha \ \frac{\mathbf{1}}{\textbf{Spindle Speed (RPM)}}$$

7.28 Why We Use G50 and G96 Command in Turning Operation?

We use G50 command to control maximum spindle speed. In CNC turning machine, work piece diameter is inversely proportional to rpm due to this reason when the cutting tool comes towards to center spindle rpm exceeds. This exceeded rpm may be the cause of the accident. To control to this exceeding rpm we use G50 command in CNC program. We can say G50 does not exceed the rpm over the given rpm in the CNC program.

When diameters vary on CNC turning machine during machining G96 command gives the facility to change the rpm as per diameter but this rpm will be between G50 and G96. Using of G50 and G96 command work piece gets the better surface finish and work piece rpm will be under control.

7.29 G54 Work Coordinate System (Work Zero Offset/ Work Zero)

G54 is called work zero (0, 0) position of the work piece. In the figure (Figure 7.23) X = 0 and Z = 0 will meet on the center of the front face of the work piece. It means the value of this center will be (0, 0), where X = 0 and Z = 0.

Generally, we take the front face of the work piece as a Z = 0 and X = 0,

Z = 0 (front face of the work piece)

X = 0 (center of the work piece)

Figure 7.24 is showing different position of the cutting tool in X and Z-axis with sign. If the tool travels above the horizontal axis

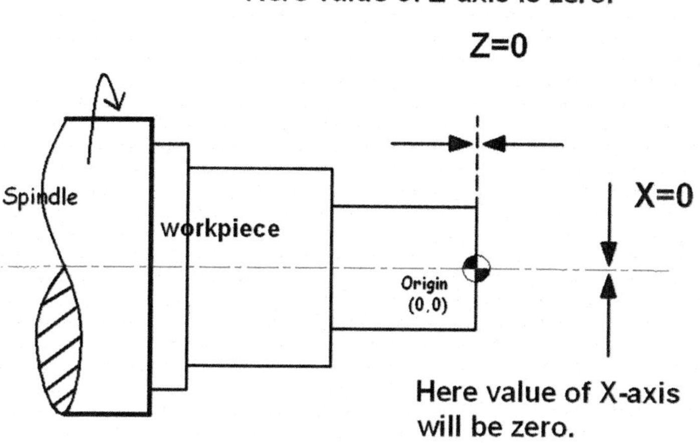

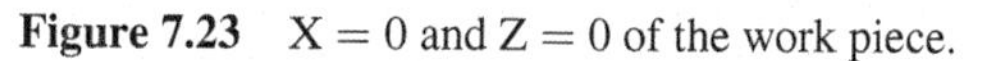

Figure 7.23 X = 0 and Z = 0 of the work piece.

Figure 7.24 Cutting tool is showing X and Z-axis directions with sign.

(central axis of the cylindrical work piece), it means the tool will show positive value of **X**-axis (**X**+). If the tool will travel below the central axis of the work piece, it means the tool will show the negative value of **X**-axis (**X**−). If the tool will stay/touch on the center line of the work piece, it means **X**-axis value will be zero (**X** = **0**).

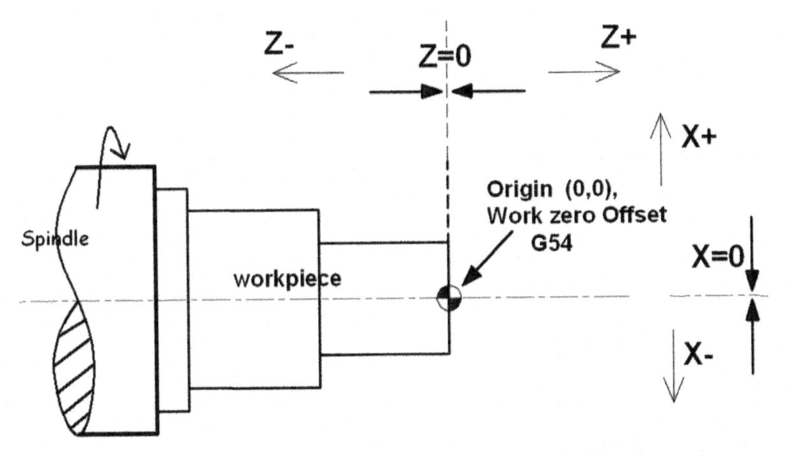

Figure 7.25 This image is showing X and Z-axis directions.

Figure 7.24 is also showing Z-axis position with sign. If the tool travels towards to spindle direction from the work piece face it means the value of **Z** will be negative and if the cutting tool will travel towards to tail stock direction it means the value of **Z** will be positive. If the tool will stay/touch on the front face of the work piece, it means **Z**-axis value will be zero (**Z** = **0**).

So according to the Figure 7.25, where **X-axis and Z-axis meet or intersect each other** at one point and that point (the intersecting point of the X and Z-axis) is called origin/work zero (0, 0).

7.30 G55 Work Coordinate System (Work Zero Offset)

7.31 G56 Work Coordinate System (Work Zero Offset)

7.32 G57 Work Coordinate System (Work Zero Offset)

7.33 G58 Work Coordinate System (Work Zero Offset)

7.34 G59 Work Coordinate System (Work Zero Offset)

Commands **G55** to **G59** is used during mass (bulk) production because that time we need more work zero offset so that we can take the origin of the maximum work piece. We can take 6 work zero offset for machining. These commands are used mostly in CNC milling machine. Some time in special condition, we use more than one work zero offset in CNC turning machine.

Example

G54, G55, G56, G57, G58, G59 etc.

7.35 G70 Finishing Cycle (Finishing Cycle for G71 and G72 Command)

In the machining process, first we do the rough machining operation after the rough machining, CNC programmer takes finishing cut (final cut) of the work piece. For this final cut programmer apply the G70 code. G70 code is called the finishing cycle. When we use this cycle as a finishing cycle, cutting tool cuts very less material and work piece gets an accurate profile with high accuracy [3].

G70 P---- Q---- F----;

Where,

G70 = Finishing cycle command
P = First programming line number of G71 cycle
Q = Last programming line number of G71 cycle
F = feed (millimeter/revolution or millimeter/minute), we can say the cutting speed of the tool

7.36 G71 Stock Removal Cycle (for External Diameter/Internal Diameter)

This command is used to remove extra material from the external or internal surface of the work piece during turning operation.

The G71 command is very popular between FANUC control users.

Using of this command programmer can make error free CNC program, can save machining time and gets mass production. See Figure 7.26.

G71 U -- R -- ;
G71 P -- Q -- U -- W -- F -- ;
G71 U -- R -- ;

Where,

U = Depth of cut (one side) in X-axis
R = Tool retraction (tool clearance position in X-axis after every rough cut)

G71 P -- Q -- U -- W -- F -- ;

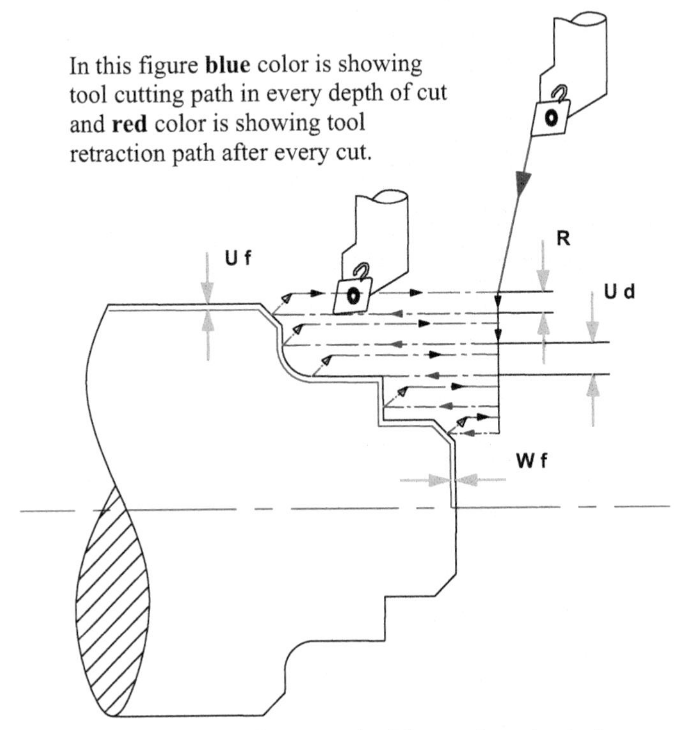

Figure 7.26 Cutting too path of the rough turning tool.

Where,

- **P** = First programming line number of G71 cycle
- **Q** = Last programming line number of G71 cycle
- $\mathbf{U}_f$ = finishing allowance in X-axis (it Leaves the material in X-axis for final cut)
- $\mathbf{W}_f$ = finishing allowance in Z-axis (it Leaves the material in Z-axis for final cut)
- **F** = feed (millimeter/revolution or millimeter/minute), we can say the cutting speed of the tool

Example

GOO X30 Z2;
G71 U1.5 R1.0;
G71 P10 Q11 U0.20 W0.10 F0.25;
N10 G00 X9.0;
G01 Z0;

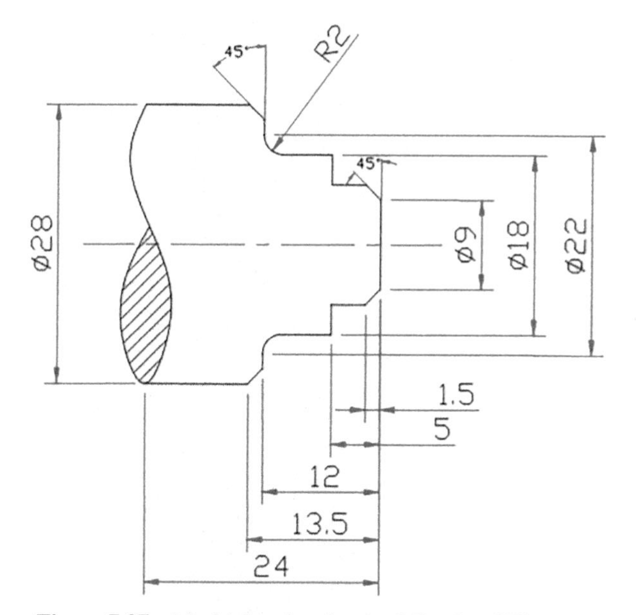

Figure 7.27 Machining drawing for following **G71** cycle.

G01 X12 Z-1.5;
G01 Z-5;
G01 X18;
G01 Z-10;
G02 X22 Z-12 R2.0;
G01 X25;
G01 X28 Z-13.5;
G01 Z-24;
N11 G01 X30;
G00 Z2;

In the **G71** cycle, the programmer makes only finishing program (final cutting program) of the work piece. According to drawing. See Figure 7.27.

7.37 G72 Face Stock Removal Cycle or Facing Cycle

This command is used for maintaining the work piece length as per drawing and this operation is called facing operation. By using this facing cycle, we can maintain the work piece length as per given drawing. See Figures 7.28 and 7.29.

We have extra material in length as per drawing. So we will take G54 (origin) inside (1.0 mm) the material from the front face. Now this extra material will be in Z+ axis. when the tool will cut the extra material. This extra material cutting is called Facing operation.

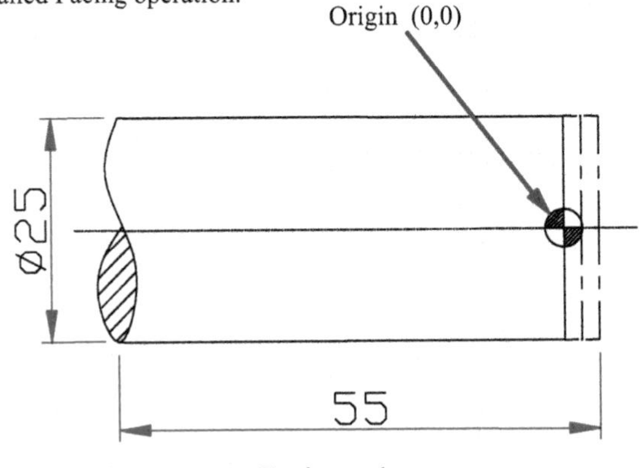

Facing = 1mm.

Final Length = 54 mm.

Figure 7.28 Extra material on length will remove by using G72 facing cycle.

G72 W -- R -- ;

G72 P -- Q -- W -- F -- ;

G72 W -- R -- ;

Where,

$\mathbf{W}_d$ = Depth of cut in **Z**-axis (removing the extra material for maintaining the length as per drawing)

R = Tool retraction (tool clearance position in **Z**-axis after every cut on the face)

G72 P -- Q -- W -- F -- ;

Where,

P = First, program line number in the G72 cycle

Q = Last, program line number in the G72 cycle

$\mathbf{W}_f$ = Finishing allowance in Z-axis (during rough cutting tool will leave the material in Z-axis for final cut)

F = Cutting Feed in (millimeter/revolution or millimeter/minute), we can say the cutting speed of the tool

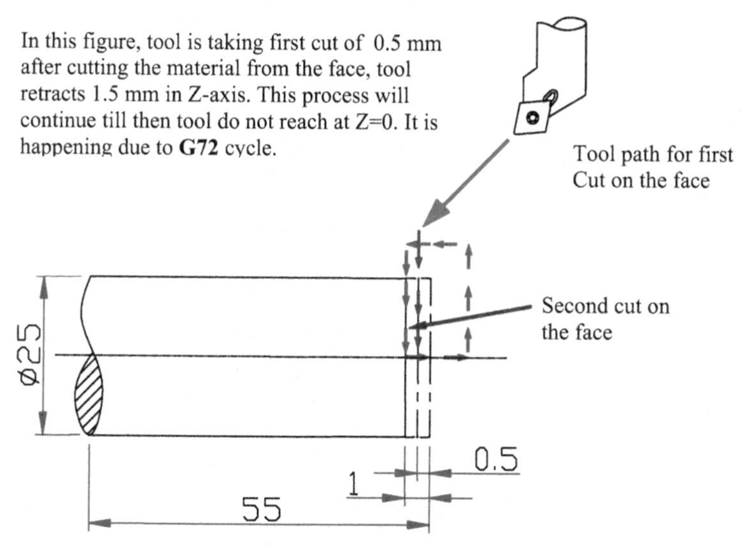

Facing length = 1mm.

After facing final length = 54 mm.

Figure 7.29 Cutting tool path of the G72 facing cycle.

Example:

G00 X28 Z2;
G72 W0.5 R1.5;
G72 P21 Q22 W0 F0.15;
N21 G00 Z0;
G01 X0.0;
N22 Z00 Z2;
G00 X28 Z2;

7.38 G74 Peck Drilling Cycle

G74 is the Peck drilling cycle for the deep hole drilling operation. In this cycle tool will start drill in small segments until the total drill length not drilled. This cycle is used during mass (bulk) production in industry [3].

G74 R.......;

G74 Z......Q....... F.......;

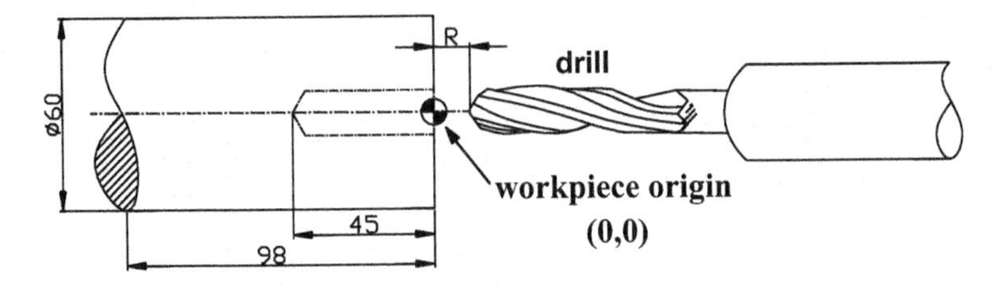

Drill tool will start drilling operation from the reference point (R) and drill continue after drilling tool will back to reference point.

Figure 7.30 Drilling operation using by drilling cycle G74.

Where,

Z = Total depth of drill hole in Z-axis
R = Retraction amount of the drill after each depth of cut due to this facility drill and the cutting material does not heat up because coolant flows properly in to the hole also scrap come out from the hole when tool retracts.
Q = Depth of cut
F = Cutting Feed in (millimeter/revolution or millimeter/minute)

Example:

$$\begin{aligned} Q\,500 &= Q\,0.5(\text{mm}) \\ Q\,5000 &= Q\,5.0(\text{mm}) \\ 1000\text{ microns} &= 1.0\text{mm} \end{aligned}$$

Note: In lathe machine or CNC turning machine, drilling operation will be performed always in the center of the work piece. See Figure 7.30.

Example

O12342; (Drill diameter 15 mm)
G54;
G21 G90 G98;
G91 G28 U0 W0;
T0505;
G97 M03 S400;

G90 G00 X100 Z100;	
G00 X0 Z50;	Drill is taking position in the center of the work piece face in **X = 0** axis.
G00 Z5 M08;	Drill is taking position nearest the work piece face in **Z**-axis before drilling operation
G74 R1.0;	Tool will start the drilling operation but after every 6.0 mm drilling (in depth) in **Z**-axis Drill will retract (go back) 1.0 mm.
G74 Z-45.0 Q 6.0 F0.07;	Now drill tool will start the drilling operation, after drilling **drill hole length** will be 45mm.
G00 Z5 M09;	Now drill will retract from the hole and will go back at **Z5 safe distance in Z-axis**.
G91 G28 U0 W0;	
M05;	
M30;	

In the first block of the G74 command, R is showing retraction of the threading tool. It means the tool will **retract 1.0 mm (R 1.0)** after every depth cut (depth of cut 6.0 mm) for proper coolant flow into the hole.

In the second block, Z is showing total drill hole length 45.0 mm and Q is showing the depth of cut of the drill. Q6.0 means, the drill will take maximum 6.0 mm cut in every pass until drill does not reach at the end point of the drill hole.

7.39 G76 Thread Cutting Multiple Repetitive Cycles

This cycle is used for threading operation. Except of the G76 cycle, we can use G32, G33, and G92 thread command for making thread profile.

But we can say the G76 cycle is an advanced automatic canned cycle. When we use G76 threading cycle, the size of the threading program will be very short, it means it will take very less space.

On the other hand, we can say G76 threading cycle occupies very less space respectively G32, G33, and G92 threading command. By using of G76 command we can save time, errors less program and can do work very fastly. See Figure 7.31.

Example

G76 $P_{(a)(b)(c)}$Q.....R.....;

G76 X.....Z.....P....Q......R....F.....;

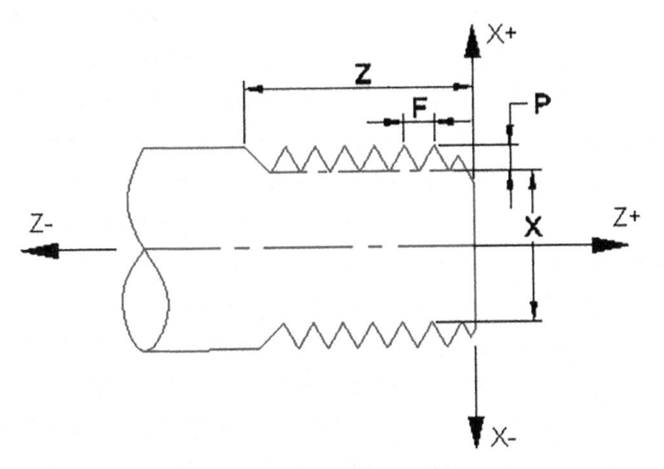

Figure 7.31 Threading drawing and its parameters.

Where in the first block

G76 P……Q….. R …..;
P(a)(b)(c)

P: Parameter P defines the thread behavior
a: First two digits represent the number of finish passes (1 to 99).
b: Second two digits represent Pull out the angle, it means when the threading tool comes out after each threading pass or retraction angle of the thread tool after each threading pass.
c: Third two digits represent the angle of threading insert (0°, 29°, 60°, 90° etc), we can say it is represents thread profile angle.
Q: Minimum depth of cut Example Q0015 (microns) = 0.015 (mm)

Note: Above value will be entered without a decimal point.

Q = 200 (Q = 200 micron)
R: Finishing allowance Example- 0.010 mm

Where in the second block

G76 X….. Z….. P…. Q…… R…. F…..;

X: Minor diameter of the thread (core diameter)
Z: Value of the threading length in Z-axis
P: Thread height (one side) Example- **P** 1280 microns = 1.280 mm
Q: First threading cut Example- Q 60 (0060) microns = 0.060 mm

R: Thread taper value (if thread is in tapered form, then external thread taper value will be in R minus (R– 0.0) and internal thread value will R plus (R 0.0).

F: Threading Feed = Thread pitch

7.39.1 Thread Formulas

$$\text{Thread height} = 0.613 \times \text{Pitch in mm (For metric thread)}$$
$$\text{Minor diameter} = \text{Major diameter} - 2 \times \text{Thread height}$$

7.39.2 Comparison Between Threading Cycle G76 and Thread Command G33/G32

Example

See Figure 7.32 for threading operation.

```
O14007;
G54;
G21 G90 G98;
G91 G28 U0 W0;
N1 T0606;
G90 G00 X50 Z100;
G97 M03 S650;
```

When we use the **G97** code, work piece will rotate in constant RPM.

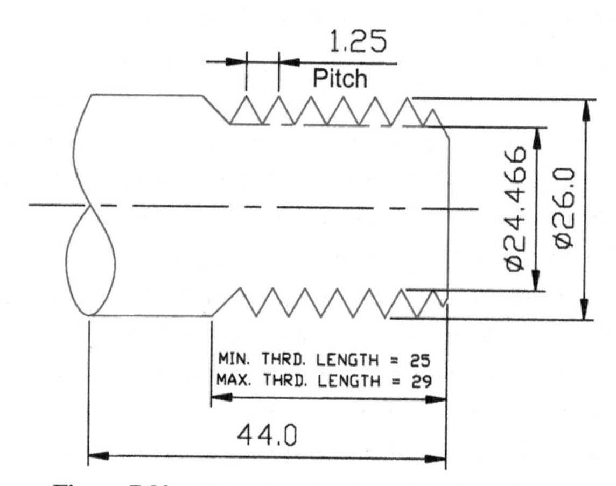

Figure 7.32 Threading drawing with dimensions.

G00 X36 Z10 M08;	Threading tool is taking position rapidly in **X** and **Z** axis before cutting the thread.
G76 P011060 Q100 R0.02;	Now tool will take cuts automatically according to threading parameters which are defined in the canned cycle (G76).
G76 X24.466 Z-27 P767 Q200 R0.0 F1.25;	
G00 X36 Z10 M09;	Tool is retracting and going back after threading.
G91 G28 U0 W0;	
M05;	
M30;	

7.39.3 Where We Use G32/G33 Thread Command and What are the Benefits of G32/G33 Commands?

G32/G33 thread commands are simple and free hand commands. It means thease are using as a simple turning program. If we use these commands, we have the liberty to take thread depth as well as we can take tool movements according to thread cutting condition. also we can take even single thread cut by using the G32/G33 command.

We need four programming lines for one threading cut. In **first line** tool takes position in X and Z-axis, in **second line** tool cuts the thread profile, In **third line** tool retract in X-axis and last **fourth line** tool will retract in the Z axis. This procedure will continue until tool will not reach at minor diameter.

We can perform following thread cutting operations by using G32/G33 thread commands.

i. Longitudinal threading
ii. Transverse threading
iii. Taper threading
iv. Variable threading

Note: For threading operation, must use constant surface speed (G97)

Threading with the help of G33 thread command. See Figure 7.31 for threading.

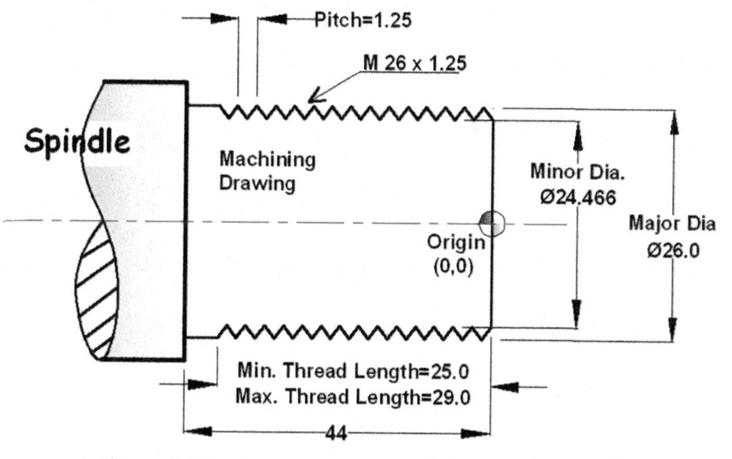

Figure 7.33 Threading drawing with dimensions [4].

Example

See Figure 7.33 for threading operation.

O14007;
G54;
G21 G90 G98;
G91 G28 U0 W0;
N1 T0606;
G90 G00 X50 Z100;

G97 M03 S650;	When we use the **G97** command, work piece will rotate in constant RPM.
G00 X36 Z5 M08;	Threading tool is taking position rapidly in **X** and **Z** axis for thread cutting.
G00 X25.8 Z5;	Now Threading tool took the position in X-axis before cutting the thread.
G33 Z-27 F1.25;	Now the tool is making thread on Length Z-27.0
G00 X36;	Tool is retracting in X axis after threading or tool is releasing from the thread Surface after first threading pass
G00 Z5;	Now the tool is retracting in Z axis after retracting in X-axis.
G00 X25.6;	**Again** threading tool took the position before cutting the thread.
G33 Z-27 F1.25;	Now again same procedure will be repeated again back to back. During this procedure thread profile will be improved.

G00 X36;
G00 Z5;
G00 X25.4;
G33 Z-27 F1.25;
G00 X36;
G00 Z5;
G00 X25.2;
G33 Z-27 F1.25;
G00 X36;
G00 Z5;
G00 X25.0;
G33 Z-33 F1.25;
G00 X36;
G00 Z5;
G00 X24.8;
G33 Z-27 F1.25;
G00 X36;
G00 Z5;
G00 X24.6;
G33 Z-27 F1.25;
G00 X36;
G00 Z5;
G00 X24.466;
G33 Z-27 F1.25;
G00 X36;
G00 Z5 M09;
G91 G28 U0 W0;
M05;
M30;

Note: G32 threading command will be use same like G33 command.

7.40 G80 Canned Cycle Cancel

G80 command is used to disabling the effect of canned cycle like G73, G81, G82, G83, G84 etc.

7.41 What is Canned Cycle?

Canned is multiple repetitive material removing cycle, its mean tool will repeat same machining operation in several time, until tool will not reach his final dimensions (profile) as per drawing.

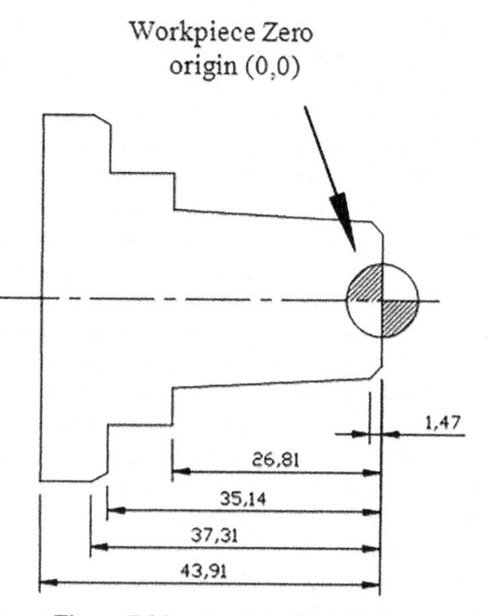

Figure 7.34 Absolute dimensions.

7.42 G90 Absolute Command

When we use the absolute command (G90), all dimensions are measured from the one point and that point is called origin point (reference point). When we apply G90 command, absolute command activates in the CNC program. See Figure 7.34.

7.43 G91 Incremental Command

In CNC programming, when we will apply G91 command, the reference point (origin) will shift with every dimension in CNC programming. On the other hand we can say, when we take dimensions it comes in segments (parts). You can compare both figures (G90 and G91). See Figure 7.35.

When we make CNC program by using the G91 command, then that case, reference point will shift for every dimension, for an example in Figure 7.35, work piece face will take as a reference point for the first dimension (13.73). End of the first dimension will be the reference point for the second dimension (9.33). End of the second dimension will be the reference point for the third dimension (7.5) and End of the third dimension will be the reference point for the fourth dimension (8.58).

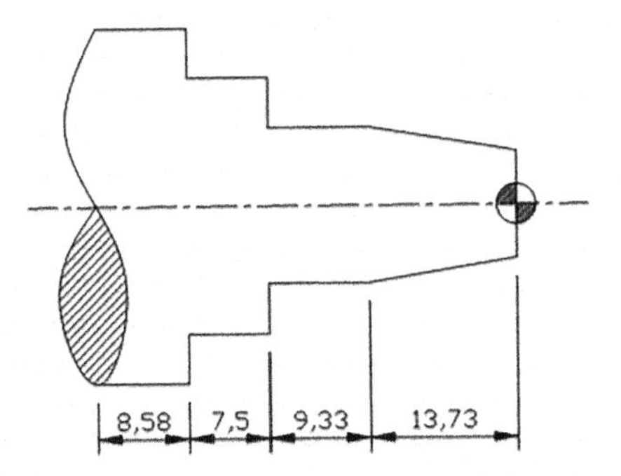

Figure 7.35 Incremental dimensions.

7.43.1 Very Important Note

Do not confuse when you are applying (using) G90 and G91 command in CNC program. Otherwise it can be the cause of machine accident or human injury.

7.44 G92 Thread Cutting Cycle

This is a very simple threading command. The G92 command gives the freedom to the programmer to control the depth of cut in each pass during programming. If you are facing some problem due to depth of cut during the threading process, you can use G92 thread command.

G92 X-- Z-- F-- ;

Where,

X = Major diameter of the threading work piece
Z = End position of the thread length in Z-axis (thread length)
F = Threading feed (pitch)

7.44.1 Note

During the threading operation thread pitch will be equal to thread feed. Pitch = Feed.

Example

See Figure 7.36 for threading operation.

O12078;
G54;

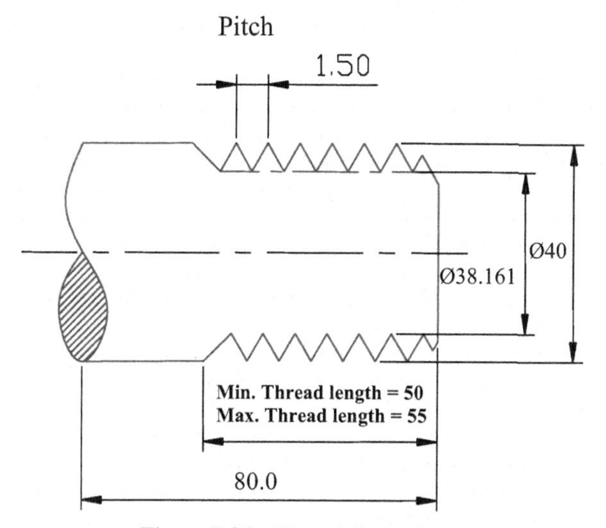

Figure 7.36 Thread dimensions.

G21 G90 G98;
G91 G28 U0 W0;
T0202;
G90 G00 X100 Z100;
G00 X60 Z10;
G97 M03 S650;

G00 **X40** Z5.0M08;	Tool is taking position rapidly before threading operation in X and Z-axis.
G92 Z-55.0 F1.5;	Now the threading tool is taking first cut in Z-axis (thread length 55) on the diameter 40.0 with feed 1.5
X39.8;	Threading tool is taking the second cut in X-axis and going to in Z-axis (length 55).
X39.6;	Threading tool is taking the third cut in X-axis and going to in Z-axis (length 55).
X39.4;	Threading tool is taking the fourth cut in X-axis and going to in Z-axis (length 55).
X39.2;	Threading tool is taking the fifth cut in X-axis and going to in Z-axis (length 55).
X39.0;	This process will continue, until the threading tool will not reach at the minor diameter(X37.0).

X38.8;
X38.6;
X38.4; Does not need to give thread length Z-55 (Z-axis value) in each block because CNC machine software will take automatically from G92 command.
X38.2; This threading process will continue until tool does not reach on minor diameter.

X38.161;

G00 X40 Z10 M09;
G00 X100 Z100;
G91 G28 U0 W0;
M30;

7.45 G94 Feed Per Minute

G94 command (some control system use G94 command and some control system used the G98 command for feed rate) is used for feed rate (cutting speed of the tool). Generally in CNC machine programs millimeter per minute is used as a feed unit or It depends on measuring unit. In some CNC control systems are used the **G98** command as a **feed per minute**.

Feed: During in turning operation, if we measure, how much distance travel by the cutting tool on the cylindrical surface in one minute if we measure that travelling distance on the cylindrical surface. It is called feed rate. On the other hand we can say, if we measure the cutting length of the material, which is cut by the cutting tool in one minute during the machining process, is called feed per minute.

7.46 G95 Feed Per Revolution

G95command is used for feed (mm per revolution) rate in CNC program. "During machining of the work piece, how much distance travel by the cutting tool in one revolution of the circumference of the work piece is called feed." In some CNC control systems, the **G99** command is used as a **feed per revolution**.

Generally we use the G95/G99 command in turning machine but also we can use the G94 command in turning machine.

Feed rate For milling machine

In milling machine, generally we use the G94 command (millimeter per minute).

7.47 G96 Constant Surface Speed On

Constant surface speed means constant spindle rpm on constant diameter, when the tool cuts the material from the diameter (constant) that time spindle rpm will be constant, it means it will not exceed on same diameter.

On the other hand we can say, when the cutting tool removes the material in different diameters then that time, spindle rpm will vary. We are cutting different diameters from the work piece surface so we have different rpm. It means rpm will vary in different diameters/we have different-different rpm on different-different diameters.

Note:

i. If the diameter is constant spindle rpm will be constant.
ii. If the diameter is not constant (if diameter will be changed) during material cutting, then spindle rpm will not be constant. It means whenever diameter will increase or decrease during material cutting, spindle rpm will vary.

7.47.1 The Relation Between Diameter and Spindle RPM for CNC Turning the Machine

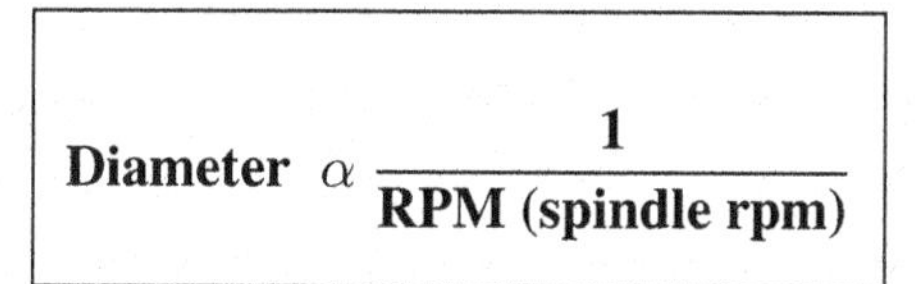

$$\textbf{Diameter}\ \alpha\ \frac{\textbf{1}}{\textbf{RPM (spindle rpm)}}$$

Constant surface speed (spindle speed) is used for the better surface finish.

Before applying G96 command, CNC programmer must use the G50 command for maximum spindle speed control. When the programmer applies G50 command, spindle speed (RPM) does not exceed the given spindle rpm.

Example

G50 S1700; Due to G50 command, When work piece will rotate, it will not exceed more than 1700 rpm and will not reduce less than 230 rpm.

G96 M03 **S230**; Using G96 command spindle speed (rpm) will vary between 1700 rpm to 230 rpm according to diameter.

Now spindle will rotate in a clockwise direction (M03), due to using of G96 Command, the spindle will rotate in constant surface speed at constant diameter. If cutting tool changes the cutting diameter then spindle rpm will change and will be constant until, diameter does not change again. It happened only when we use G50 and G96 command in the program. Most important thing is, these commands (G50 and G96) are used for the best surface finish.

The conclusion is, whenever cutting tool will change the cutting diameter of the work piece, spindle rpm will change automatically, if we use both G50 and G96 command in CNC program [3].

7.48 G97 Constant Spindle Speed

This command is used as constant spindle speed, it means work piece will rotate on constant (same) spindle rpm on different diameters. We can say RPM will not change at any time until G97activate. This command is used generally in the drilling operation, parting operation, grooving operation, threading operation and big, long and heavy diameters.

Example

G97 M03 **S400;** In this block spindle will rotate in clockwise direction at constant rpm 400.

7.49 M-Codes

(Miscellaneous Code/Machine Operating Code)

7.50 What are M Codes?

M codes are called machine operating code or machine code. For example- when programmer/engineer wants to change cutting tool, open/close the CNC machine door, forward/reverse of tailstock, coolant on/off etc., that time M codes will be used. We can say, **M** codes are used for spindle stop (M05), tool change (M06), spindle rotation with direction (M03/M04), coolant motor on/off (M08/M09), chuck clamp/unclamp (M10/M11), spindle gear change (M41,M42, M43, M44), etc. in CNC program. M codes are used with numerical numbers. Example: M00, M01, M30, etc.

Generally, M codes are not model and non-model but some of these codes behave like model command. Some M codes are effective in CNC program until replaced by another M code or another M code will not activate. M codes are vary machine to machine. It depends on machine model.

Examples M03, M04, M05, M08, M98, etc.

7.50.1 Important and Interesting Facts

Only one M code can apply/come in one CNC programming line. It means more than one M code can not execute in one programming line (block) of the CNC program.

7.51 M00 Program Stop

When we use M00 command, CNC program will be stop but CNC program will not stop permanently. We can say this command does not use for permanent CNC program stop. Generally some time due to some problem, It is used between two cutting tools in the CNC program.

7.52 M01 Optional Program Stop

The M01 option always available on CNC control panel as a M01 switch/key. When we **ON** M01 switch on the control panel than M01 will work/activate in the program but M01 should be written in the program where we need.

7.53 M02 End of Program

As its name. M02 command is used for permanently stop the CNC Program. Once M02 applied than all machining function and machining operation will stop permanently. This command is used in few special conditions.

7.54 M03 Spindle Rotation in Clockwise C.W. Direction

When we used **M03** command in the CNC program, the spindle rotates in clockwise direction. It means work piece will also rotate in clockwise direction and in milling machine cutter will rotate in clockwise direction. For working of this command spindle speed value must give in the same block [1].

Example

M03 S700; in this line spindle will rotate in c.w direction with 700 rpm.

7.55 M04 Spindle Rotation in Counter Clockwise (C.C.W.) Direction/Rotation in Anti Clockwise Direction

M04 command is used for spindle rotation in **counter clockwise (C.C.W.) direction**. Using of this command CNC turning machine spindle will rotate in counter clock wise direction and in milling machine cutter will rotate in counter clockwise direction. For working of this command spindle speed value must give in the same block.

Example

M04 S1300; in this line spindle will rotate in C.C.W direction with 1300 rpm.

7.56 M05 Spindle Stop

M05 command is used for spindle stop.

7.57 M06 Tool Change

M06 is an automatic tool change command. Generally CNC programmer does not apply M06 command in CNC turning machine. When the programmer gives T0404 command, the machine automatically changes the cutting tool.

Example:

O52107;
G21 G90 G40 G80;
G91 G28 Z0;
G28 X0 Y0;
T02 M06; When this line will execute, cutting tool will change automatically and tool number 02 will come at the cutting position.
G90 G00 X0 Y0 Z35;

7.57.1 Note

Generally **M06** command is used in **CNC milling machine**.

7.58 M07 Coolant Motor ON with Mist

When the CNC programmer uses this command, coolant comes thru spindle.

7.59 M08 Coolant Motor ON

This command is used for a coolant motor **ON**, so that coolant can flow properly on the tool and cutting surface of the work piece for reducing heat and achieve better surface finish during the machining operation [3].

7.60 M09 Coolant Motor OFF

After completion of machining operation we use M09 command for coolant motor stop.

7.61 M10 Chuck Open/Vice Open

This command is used for chuck open/vice open, when the operator wants to open the chuck he will use the M10 command. Generally foot pedal is used in CNC machine for chuck open.

7.62 M11 Chuck Clamp/Vice Close

The M11 command is used. When we need to clamps/holds the work piece between jaws/vice. Generally foot pedal is used in CNC machine for chuck clamp.

7.63 M19 Spindle Orientation/C-Axis Engage

M19 is the spindle orientation command. It means spindle rotates in angularly (angular motion) in circular direction. We can say if we want to rotate the work piece in circular direction in any angle for example 20 degree, 64 degree, 89 degree or any other angle. We will apply **M19** command. This circular orientation called **C-Axis**.

Figure 7.37 Face drilling operation with spindle orientation command M19.

(Courtesy: CNC Horizontal Turning machine, JYOTI, Rawat Engg. Tech Pvt Ltd., India)

Example

M19 C73;

In the some CNC control systems are used A/C/S in the program for rotation angle.

Example

If you want to perform drilling operation on different angles on the work piece face on CNC turning machine. In that case you need another axis and that another axis is called **C-Axis**. Without **C-Axis**. you cannot perform drilling operation on the face of the work piece on turning machine. These different angles can be like 7 degree, 26 degree, 42 degree, 65 degree, 77 degree, 86 degree etc.

Figure 7.37 shows angular drilling operation in CNC machine and it is showing face drilling drawing.

Figure 7.38 will machining on Turn-mill CNC machine using by M19.

Example

```
O14005;
G54;
G21 G90 G98;
G91 G28 U0 W0;
```

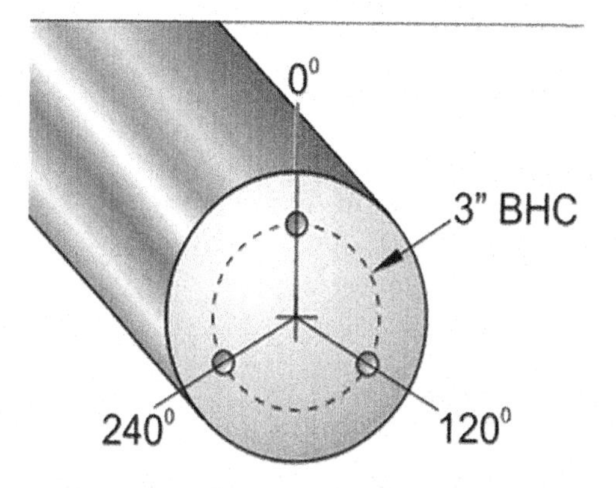

Figure 7.38 Face drilling drawing.

T0404;	Drilling tool diameter, Ø10 mm
G90 G00 X100 Z100;	
M154;	C-axis motor ON (engage)
G00 X50 Z50;	
M133;	Live tool rotation forward
G00 Z2 M08;	
M19 C120;	In C120, **120** is the orientation angle of the spindle where drilling operation will perform.
G01 Z-11.5 F0.06;	
G00 Z2;	
M19 C 240;	In C240, **240** is the orientation angle of the spindle where drilling operation will perform.
G01 Z-11.5;	
G00 Z2;	
M19 C 0;	In C0, **0** is the drilling angle where the machine will drill
G01 Z-11.5;	
G00 Z2 M09;	
M135;	Live tool rotation stop
M155;	C-axis motor off (disengage)
M05;	
G00 X150 Z250;	
M30;	

Where,

Code description

M154 is used for C-axis motor ON (engage)
M155 is used for C-axis motor off (disengage)
M133 is used for live tool rotation forward
M135 is used for live tool rotation stop

7.64 M21 Tail Stock Forward

M21 code is used for automatic movement of tailstock in the forward direction for engage the work piece.

This code is used for hold the heavy or lengthy work piece from the center of the front face of the work piece.

7.65 M22 Tail Stock Reverse

M22 code is used for automatic movement of tailstock in reverse direction for disengage the hold work piece.

7.66 M23 Chamfer Out Thread ON

If you want to make the thread and you want the **chamfer also at the end of the thread length** then you will use **M23** command before threading operation. This command works in **G76**, G92 threading operation.

Figure 7.39 shows chamfering or threading tool retraction mark after every threading cut. Due to this facility (chamfering), like nut easily pass on one end to another end or nut easily comes out from the last end of the thread.

Figure 7.39 shows thread screw image, where chamfer is showing at the end point of the thread, towards the screw's head. Generally we take 45° chamfer angle in M23 command during parameter setting. Thread tool makes the 45° angle at the end point of the thread with the help of M23 command. You can see this chamfer angle on thread image Figure 7.40.

Example

Let, we take one shaft and perform threading operation on CNC turning machine but before threading operation we use the **M23** command in CNC programming. After completion of threading operation when we use

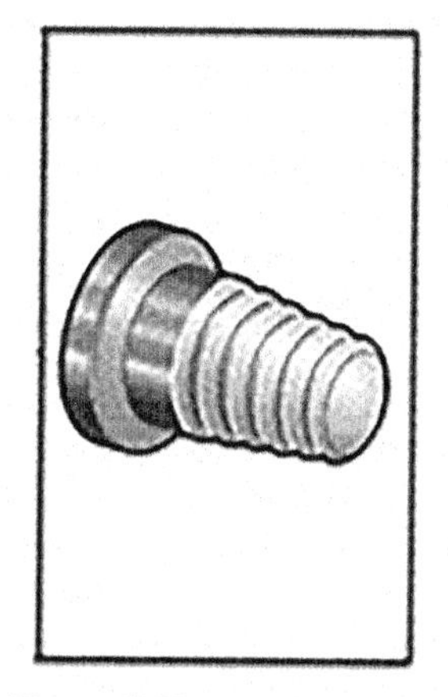

Figure 7.39 Thread screw.

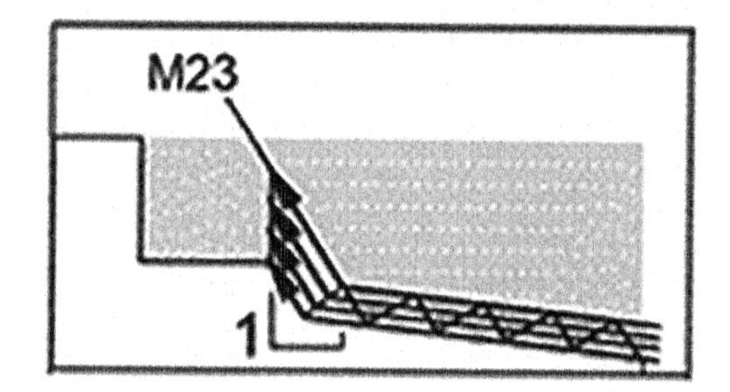

Figure 7.40 45° thread chamfer at the end of thread profile by using the M23 command.

compatible nut on this thread shaft, the nut will move (twist) smoothly on the thread profile from starting end to opposite end and nut can come out from the opposite end of the thread shaft (if, after opposite end, shaft diameter less than thread minor diameter). It is possible only due to using of **M23** command because chamfer does not performed at the end (opposite) of the thread during threading operation by threading tool.

7.67 M24 Chamfer Out Thread Off

If you want to make the thread but you don't want the **chamfer at the opposite end of the thread length** then you will use **M24** command before threading operation (Figure 7.41).

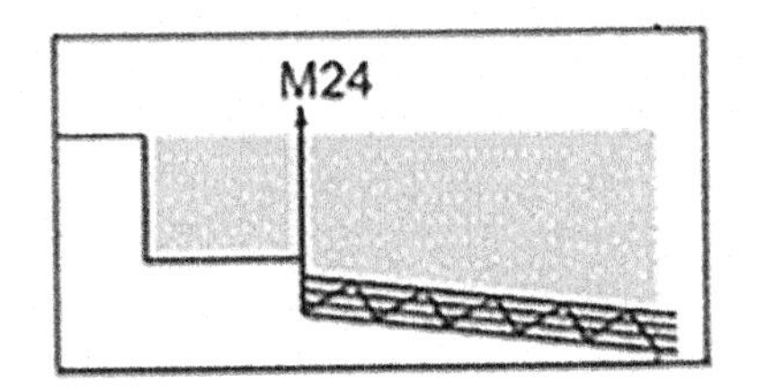

Figure 7.41 Thread profile without chamfer by using the M24 command.

Example

Again let, we take one shaft and perform threading operation on CNC turning machine but before threading operation we use the **M24** command in CNC programming. After completion of threading operation when we use compatible nut on this thread shaft, nut will move (twist) smoothly on the thread profile from starting end to opposite end but nut will not come out from the opposite end of the thread shaft due to chamfer less thread. It is possible only due to using of **M24** command because chamfer does not perform at the opposite end of the thread during threading operation by threading tool.

7.68 M30 Program End and Return to Program Top

M30 command is used for CNC program end and rewind. When all machining operation is completed, this command is used at the end of the CNC program. After using M30 command blinking cursor will go at first programming line. On the other hand we can say, **M30** command is used for program end and reset. It means blinking cursor, which highlights the executing programming line during machining (operation) through specific color (like light yellow/light orange, etc). After the end of the program, highlighting cursor comes back again at the starting line of the program. So that operator can start the CNC machine again for production.

7.69 M38 Door Open

M38 command is used for door open in CNC machine.

7.70 M39 Door Close

M39 command is used for door close in CNC machine.

7.71 M41 Selection of Spindle Gear 1

M41 command is used as a neutral gear. It means spindle will be free without any tightness.

7.72 M42 Selection of Spindle Gear 2

This command is used for low gear change. It means spindle will rotate at low rpm.

7.73 M43 Selection of Spindle Gear 3

This command is used for medium gear change. It means spindle will rotate at medium rpm.

7.74 M44 Selection of Spindle Gear 4

This command is used for high gear change. It means spindle will rotate at high rpm.

7.75 M98 Subprogram Call in the Main Program

In CNC programming, M98 command is used for subprogram calling. Subprogram is the part of the main CNC program. It works with in main CNC program but it writes separately under different program number. It means subprogram saves in the CNC memory with his own program number.

Let us assume, we have drawing and work piece and we want to machining on that work piece but on that work piece, *same operation* (profile) is repeating again and again at one location or different-different location. In this case we will make two different programs with different program numbers. One program is called main program and another program is called subprogram. In the main program we write main programming code with cutting tool number, spindle rpm, coolant motor ON, etc. with coordinates positions to be machined and at last we write program end code (M30).

In the subprogram, programmer will write only machining program, which machining operation will repeat one time or more than one time as per drawing requirements at one location/different locations and at the end of subprogram we write subprogram end code (M99).

M98 P0000000;

Where,

M98 = Subprogram call
P000 = First three digits are number of repetitions
P0000 = Last four digits are subprogram number

7.75.1 Main Program

When we talk about the main program and subprogram, then we can say, program is divided in two parts. One is called the main program and another is called sub program.

Always we give M98 (sub program command) command in the main program for sub program calling but it is very important that sub program writes separately under new program number in CNC machine memory. When the M98 command executes in the main program, M98 calls the sub program that is already kept/written under new program number in CNC machine memory by the programmer.

7.75.2 Sub Program

When profile/pattern repeats one or more than one time as per drawing, then we write and call the subprogram from the main program for machining.

O0008; **(main program)**
.... ;
.... ;
M98 P0000000; (sub program)
..........;
..........;
M30; (main program end and reset)

Where,

M98 = Subprogram call
P000 = First three digits are number of repetitions
P0000 = Last four digits are subprogram number
M99 = Subprogram exit

Following example is showing programming format of M98 and M99.

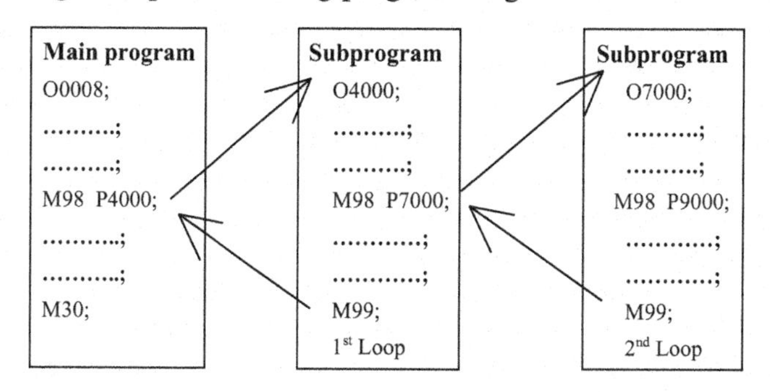

If we see the above example, then we find here 'P' is denoting the subprogram number and number of repetition (It means how many times same profile will machine with different depth.). The left first three digits (example: 007) in 'P' are showing number of repetition and last four digits (1000) in 'P' are showing the sub program number [2].

Example

M98 P0071000;

Here in the above example, Subprogram number **1000** will be executed **07** times, if required.

Example

O10255;	(Main program number)
G54;	
G21 G90 G98;	
G91 G28 U0 W0;	
T0101;	
G90 G00 X150 Z150;	
G00 X60 Z10;	
M08;	
G00 Z2;	
G00 Z-31;	
M98 P0020100;	(Call Sub Program 0100 two times in Main Program)
G00 X60;	
G00 Z-45;	
M98 P0040100;	(Call Sub Program 0100 four times in Main Program)

```
G00 X60;
G00 Z5 M09;
G91 G28 U0 W0;
M05;
M30;
```

```
O0100;                           (Sub Program for Main Program)
G90 G01 U-4.0 F0.25;             (G90 is Absolute Command)
G01 U-2.0 W-1.0 F0.12;           (Chamfering Before Radius)
G02 U2.0 W-4.0 R2.0 F0.12;       (Radius Profile)
G01 U2.0 W-1.0 F0.12;            (Chamfering After Radius)
G00 W6.0;
M99;                             (End of Sub Program)
```

7.76 M99 End of the Subprogram/Return to the Main Program

The M99 command is used at the end of the subprogram to return in the main program. You can see above sub-programming example where subprogram completes his task (it can be one time or repeat more than one time), and the **M99** command is used at end of the subprogram. We can say after using the M99 command, the machine returns to in the main program.

7.77 What are the Advantages of Using of Subprogram?

When we use subprogram **as per drawing requirements,** CNC program makes very easy and short, machining of complex profile makes very easy and error free.

References

[1] Production Technology (Manufacturing Process, Technology and Automation), 17th Edition, 2009, R.K.Jain, Khanna Publishers, India).

[2] CNC Technology & Programming, Tilak Taj, 2016, Dhanapat Rai Publishing Company, India.

[3] CNC programming Manual of MTAB company: (Certificate course on CNC Turning, MTAB Technology Centre, MTAB, Chennai, India).

[4] Drawings and sketches has been generated from Auto CAD 2014.

8

CNC Programming Method

8.1 For the Manufacturing of Any Product in CNC Turning Machine, We should have Following Facilities

1. Production drawing of the work piece
2. Raw material as per drawing requirements
3. Compatible CNC turning machine and control system. (here you will learn about **FANUC control system** or similar control systems)
4. CNC programmer/Engineer/CNC operator
5. Cutting tools
6. **CNC program as per drawing**
7. Measuring instruments

8.2 Mostly G00 & G01 Codes are Used in Very Large Scale During CNC Programming and CNC Operations in Maximum CNC Control Software [1].

8.2.1 G00

When cutting tool takes the position for cutting the material or retract from the work piece that time G00 code will use in CNC programming.

On the other hand we can say, **G00** command is used for rapid tool positioning. It is used for non-cutting movement of the cutting tool.

G00 X---Z---;

Detailed programming line

G00 X (coordinate of X-axis, where cutting tool will go) **Z** (coordinate of Z-axis, where cutting tool will go);

8.2.2 G01

G01 code is used for material cutting/removing with feed rate.
Here, we will make CNC programming line for material cutting.

G01 X---Z--- F---;

Detailed programming line
G01 X (value of X-axis coordinate) **Z** (value of Z-axis coordinate) **F** (Feed value)**;**

8.3 Programming of Straight Turning Operation by Using G01 Code

Here, we will learn how to cuts the material through cutting tool in CNC programming.

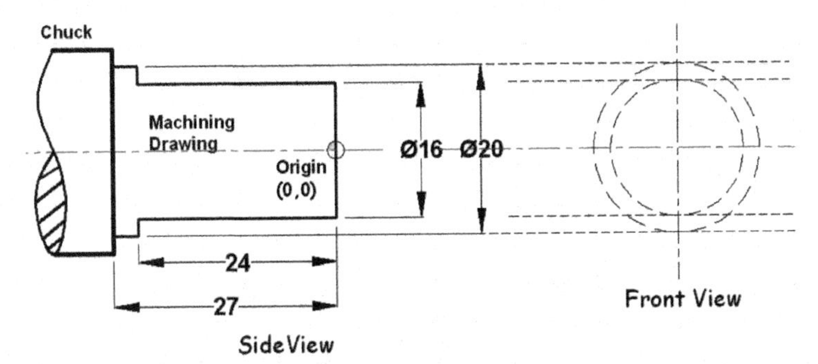

Figure 8.1 Cylindrical drawing is showing dimensions with side view and front view.

Figure 8.2 Actual machining operation in turning machine.

(Courtesy: CNC Horizontal Turning machine, JYOTI, Rawat Engg. Tech Pvt Ltd., India)

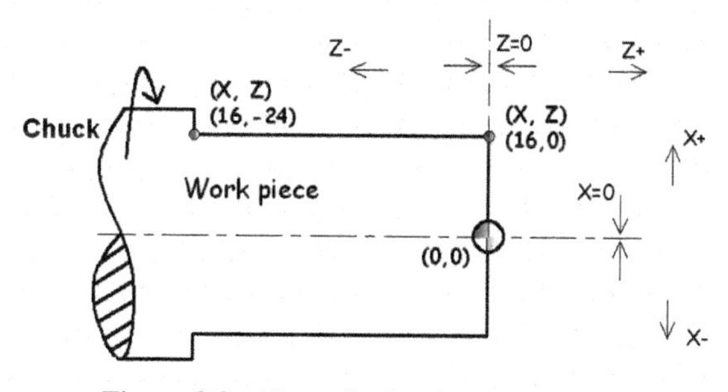

Figure 8.3 Figure is showing coordinates.

In Figure 8.3, X-axis shows the diameter (round shape) of the work piece and length of the work piece shown in the Z-axis.

Figure 8.3 Coordinates of the work piece in the term of X & Z from where from the material can be removed in turning operation.

(X, Z)

1. **(16, 0)** starting point of straight turning operation on the face of the work piece where cutting tool will take position before cutting.
2. **(16, −24)** Ending point of straight turning operation, where the cutting tool will go and turn the material.

In Figure 8.4, the cutting tool will follow the above coordinates in linear motion and removes the material

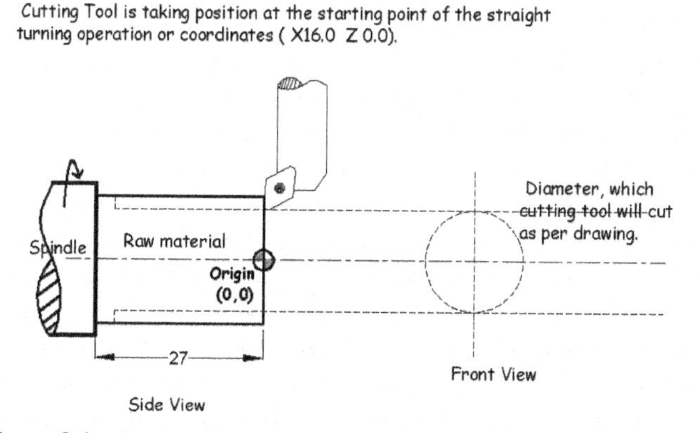

Figure 8.4 Cutting tool is taking position during work piece rotation.

In Figure 8.4, the cutting tool is taking a position in a linear motion (G01) at the coordinates (16, 0) before cutting the material with slow feed

G01 X16 Z0 F0.12; starting point of straight turning operation

In Figure 8.5, the cutting tool is cutting the material and **going toward at the coordinates X16 Z-24** in linear motion with very slow feed (F).

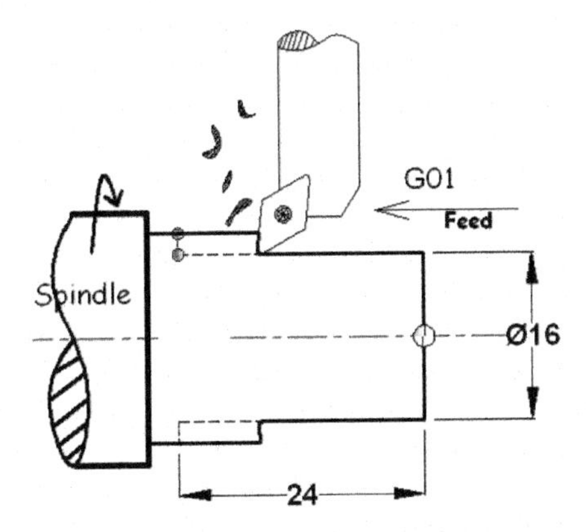

Figure 8.5 Tool is removing the material.

In Figure 8.6, cutting tool is producing the 16.0 mm diameter till 24 mm length in Z-axis negative direction.

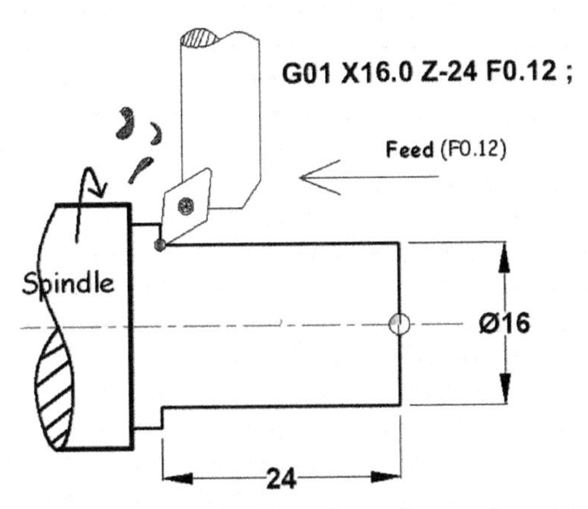

Figure 8.6 Cutting tool reached at the end coordinates after material removing.

G01 X16 Z-24 F0.12; ending point of straight turning operation

Cutting tool removed the material **at diameter 16 mm** and **reached at the coordinates X16 Z-24 in linear motion (G01) with feed (F).**

8.4 Programming of Arc (Radius) Profile in Clockwise Direction (G02)

In Figure 8.7, following dots are showing coordinates of the tool path, from where the cutting tool will go and cut the material.

Coordinates of the above work piece in the term of X & Z

(X, Z)

1. **(9, 0)**
2. **(9, –5)** **starting point of radius**
3. **(13, –7)** **Ending point of radius**

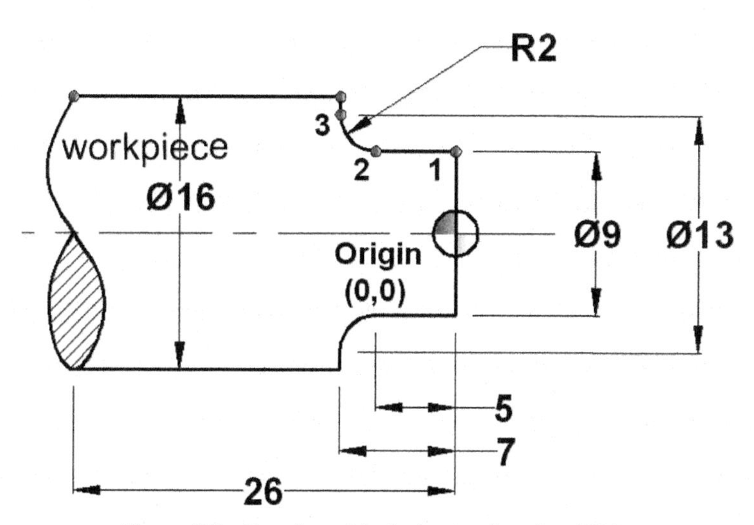

Figure 8.7 Drawing with clock wise direction (G02).

Coordinates are shown in Figure 8.8 where cutting tool will go and cut the material during machining.

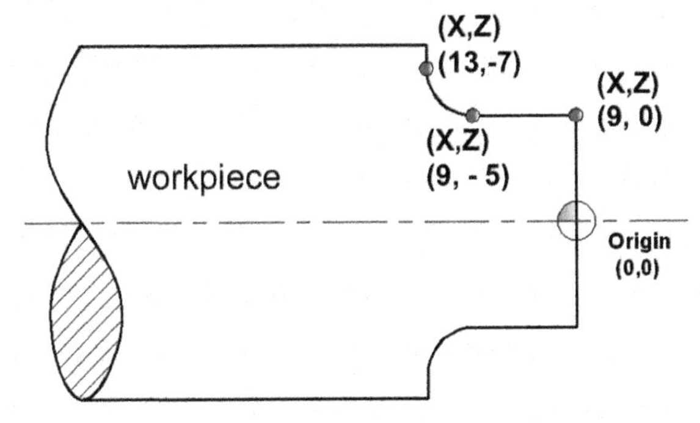

Figure 8.8 Figure is showing coordinates.

In Figure 8.9, the cutting tool is taking a position at the starting point coordinates (9, 0) for straight turning operation.

G01 X9 Z0 F0.12;

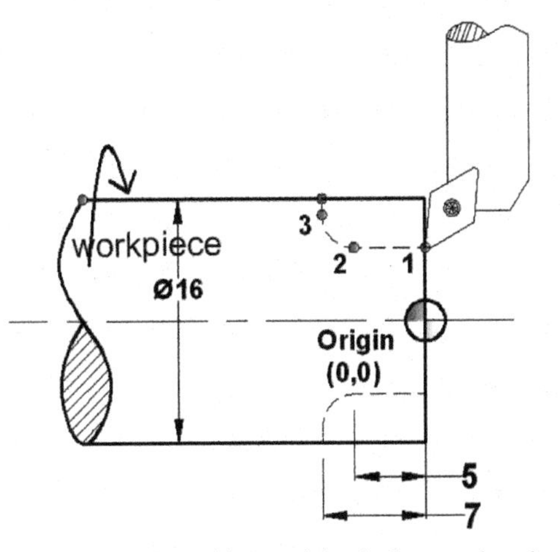

Figure 8.9 Cutting tool is taking position before cutting the material.

In Figure 8.10, cutting tool removed the material **at diameter 9.0 mm** and **reached at the coordinates X9 Z-5 in linear motion (G01) with slow feed (F) but this coordinates also known as the starting point of the radius.**

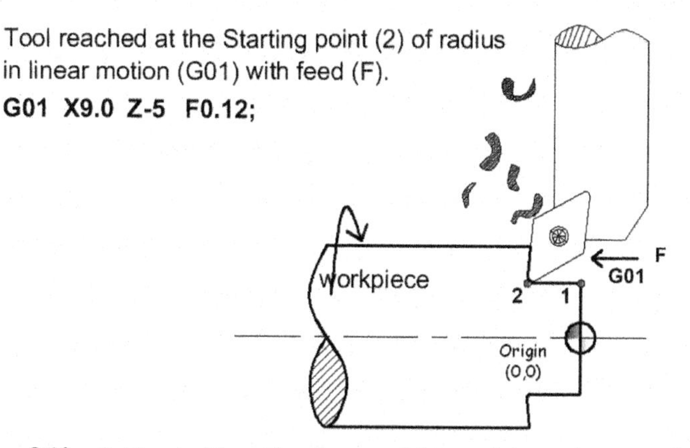

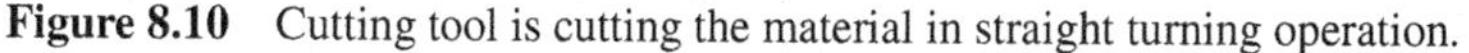
Figure 8.10 Cutting tool is cutting the material in straight turning operation.

G01 X9 Z-5 F0.12; this is the **starting point of the radius,** it is also known as an ending point of straight turning operation.

In Figure 8.11, the cutting tool **reached the coordinates X13 Z-7 in circular motion (G02) with slow feed (F) during this machining cutting tool** removed the material and made the radius profile in clock wise direction.

G02 X13 Z-7 R2.0 F0.12; this is an ending **point of the radius.**

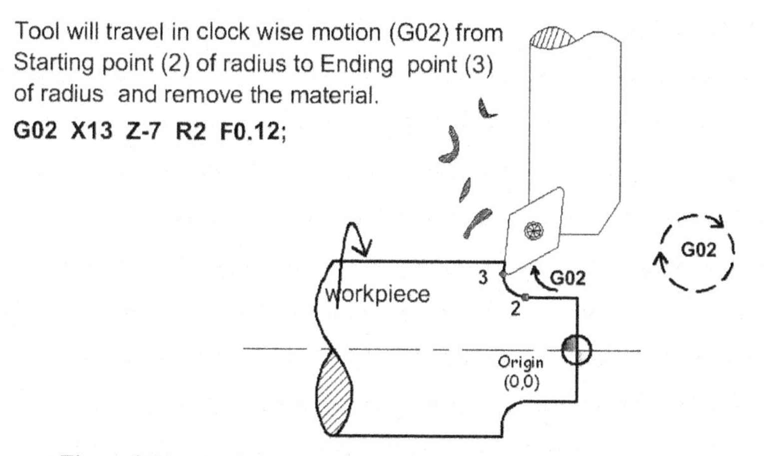

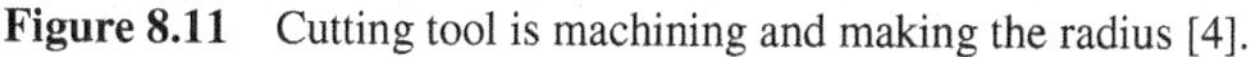
Figure 8.11 Cutting tool is machining and making the radius [4].

8.5 Programming of Arc (Radius) Profile in Counter Clock Wise Direction (G03)

Figure 8.12 shows machining drawing.

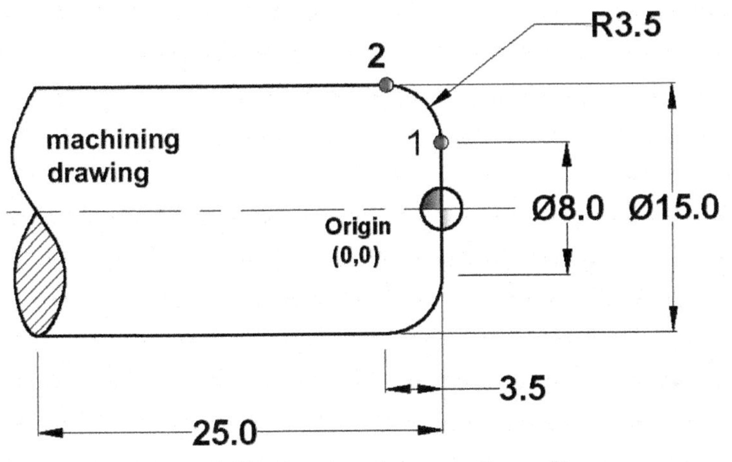

Figure 8.12 Drawing of corner radius profile.

Coordinates are shown in Figure 8.13 where the cutting tool will go and cut the material during machining.

Figure 8.13 shows the coordinates, where from the material will be removed in turning operation.

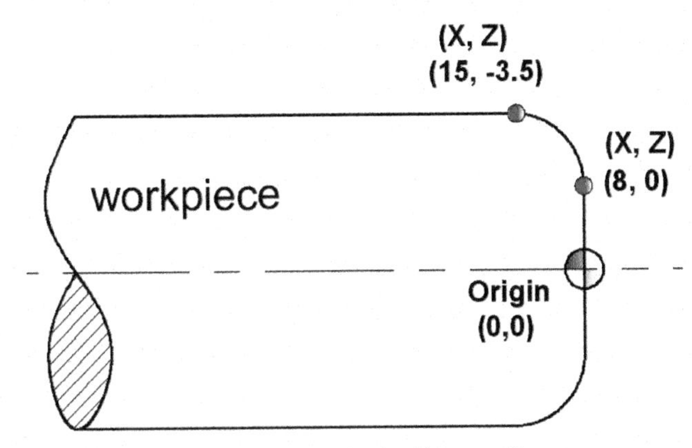

Figure 8.13 Figure is showing coordinates.

Coordinates of the above work piece in the term of X & Z

(X, Z)

1. **(8, 0)** **Starting point of the radius**
2. **(15, −3.5)** **Ending point of the radius**

In Figure 8.14, the cutting tool is taking position at the starting point of radius for making radius profile, coordinates of the starting point are (8, 0).

G01 X8 Z0 F0.15; Starting point of radius [cutting tool will touch at the coordinates (8, 0)].

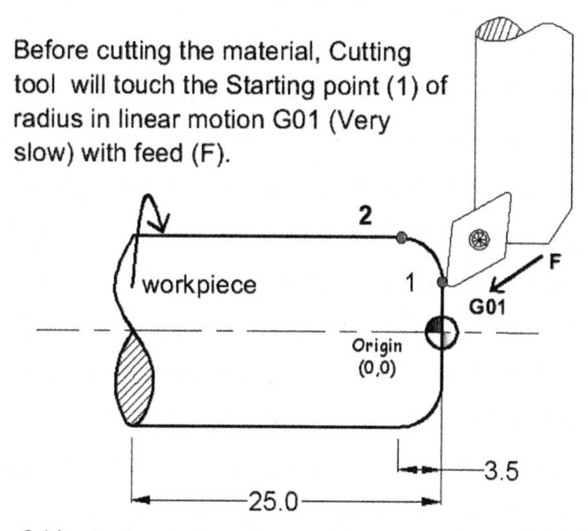

Figure 8.14 Cutting tool is **taking position at coordinates X8.0 Z0.0.**

In Figure 8.14, the cutting tool is **taking position at coordinates X8.0 Z0.0** near the starting point of radius before removing the material, during this, the spindle is rotating.

In Figure 8.15, the cutting tool is cutting the material and **going towards the end point of radius coordinates X15 Z-3.5** in CCW circular motion with slow feed (F).

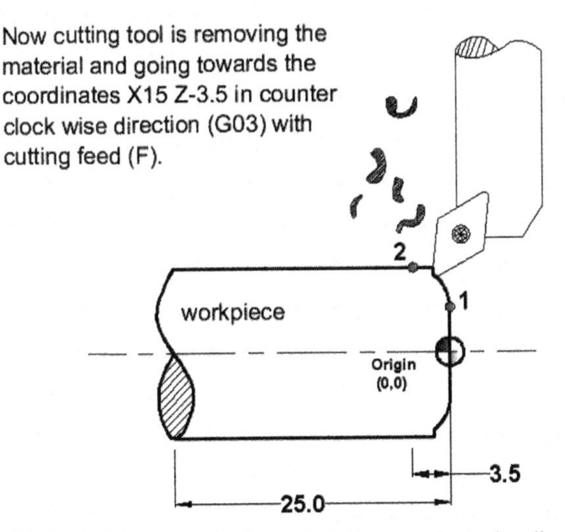

Figure 8.15 Tool is removing the material in anti clockwise direction.

In Figure 8.16, cutting tool **reached at the coordinates X15 Z-3.5 in Anti clock wise circular motion (G03) with slow feed (F) during this machining, cutting tool** removed the material and made the radius profile in anticlock wise direction.

G03 X15 Z-3.5 R3.5 F0.12; Cutting tool has removed the material and reached at the end point of radius (X15, Z-3.5).

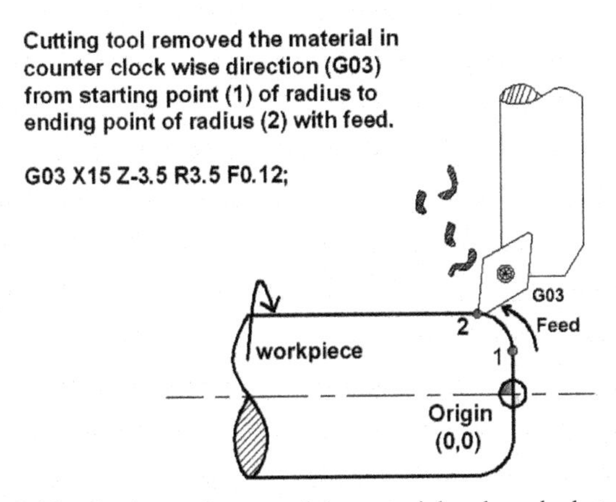

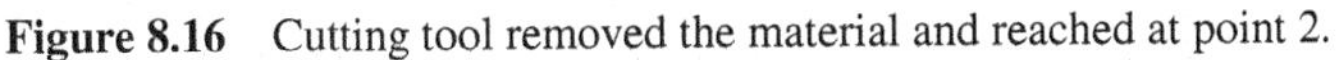
Figure 8.16 Cutting tool removed the material and reached at point 2.

8.6 Programming of Taper Turning Operation [2]

Figure 8.17 shows the raw material drawing.

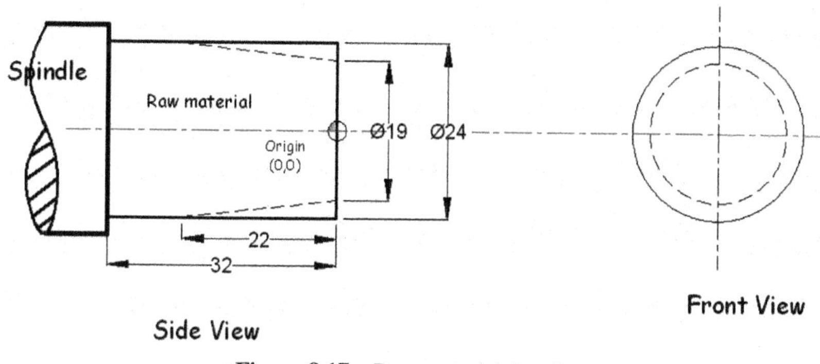

Figure 8.17 Raw material drawing.

Figure 8.18 shows the machining drawing.

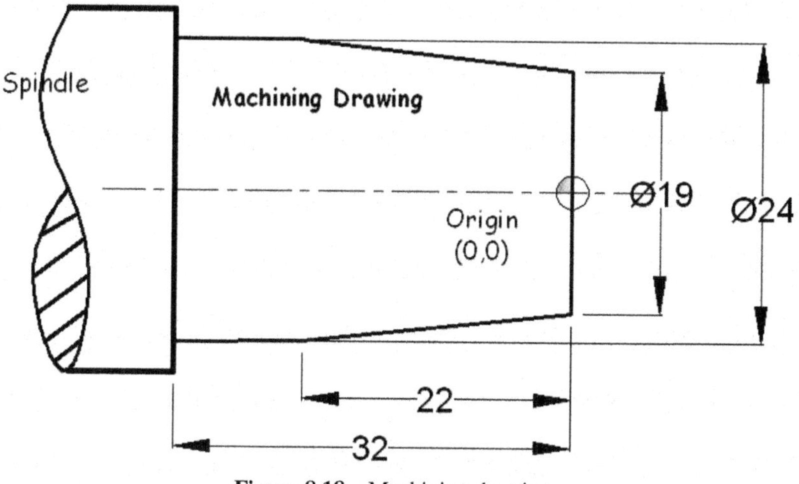

Figure 8.18 Machining drawing.

Coordinates are shown in Figure 8.19 where the cutting tool will go and cut the material during machining.

Coordinates are shown in Figure 8.19 where from the cutting tool will remove the material.

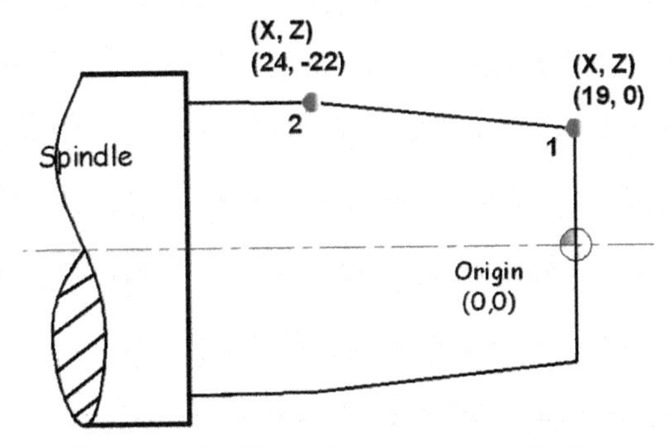

Figure 8.19 Figure is showing coordinates.

Coordinates of the Figure 8.18

(X, Z)

1. **(19, 0)** Starting Point of taper turning operation
2. **(24, −22)** Ending Point of taper turning operation

In Figure 8.20, the cutting tool is taking position at the starting point of radius for making radius profile, coordinates of the starting point are (19, 0).

In Figure 8.20, the cutting tool is taking position on the starting point of Taper turning.

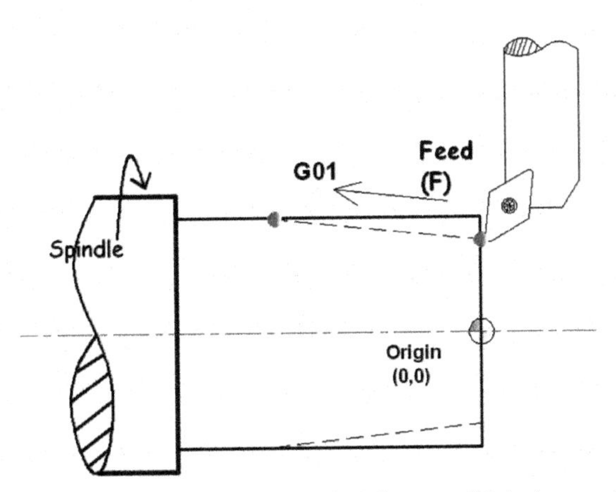

Figure 8.20 Tool is taking position before cutting the material during work piece rotation.

In Figure 8.21, the cutting tool is cutting the material and **going towards at the coordinates** X24 Z-22 in taper turning operation with slow feed (F).

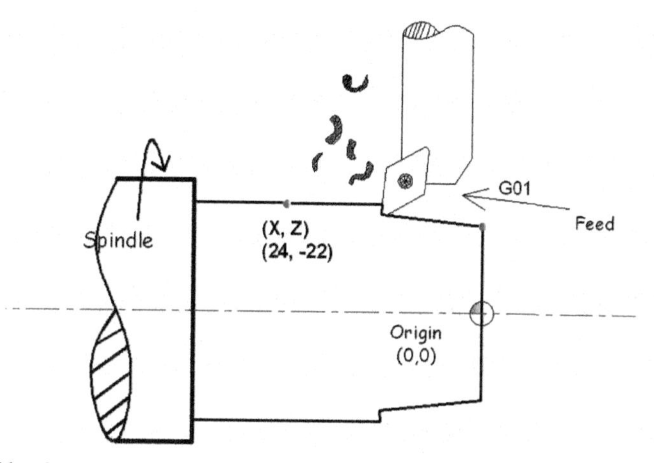

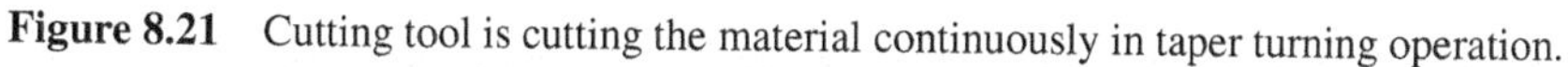

Figure 8.21 Cutting tool is cutting the material continuously in taper turning operation.

In Figure 8.22, the cutting tool removed the material and reached the end point of the coordinates.

G01 X24 Z-22 F0.15; Cutting tool reached at the end coordinates (24, −22) of the taper turning.

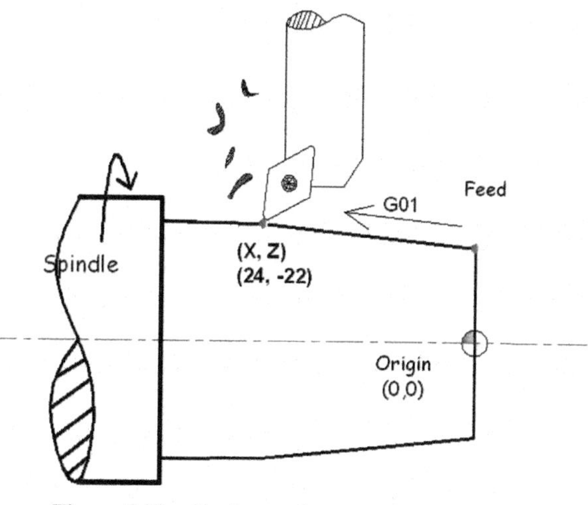

Figure 8.22 Cutting tool removed the material.

Now in Figure 8.22, the cutting tool removed the material and reached at the endpoint of Taper turning (X24, Z-22) in linear motion (G01) with feed.

8.7 CNC Programming Procedure Step by Step with Straight Turning Operation

Here, we will see different figures, where the cutting tool comes near the work piece and cuts the extra material for making the desired Turning profile as per drawing.

8.7.1 First, We Need Product Drawing Which We Want for Machining (Figure 8.23)

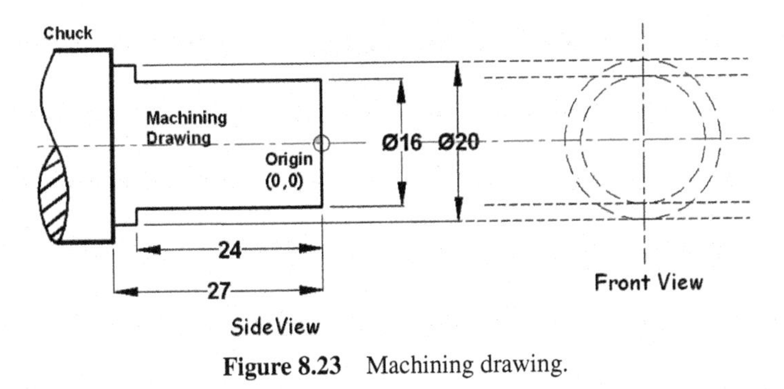

Figure 8.23 Machining drawing.

8.7.2 Second, We Need Raw Material with Required Dimensions (Figure 8.24)

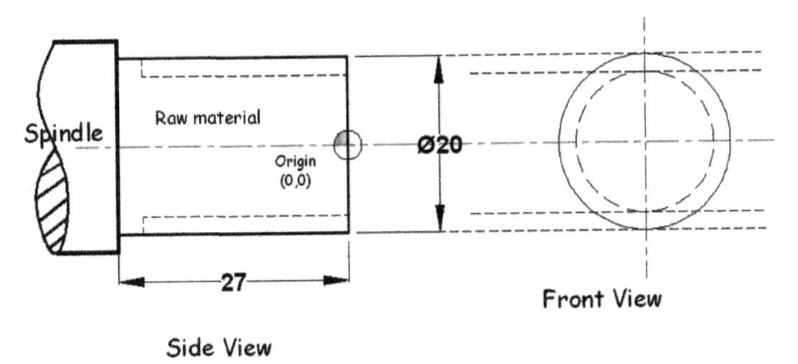

Figure 8.24 Drawing of raw material.

8.7.3 Third, We Need Coordinates of the Drawing for Programming (Figure 8.25)

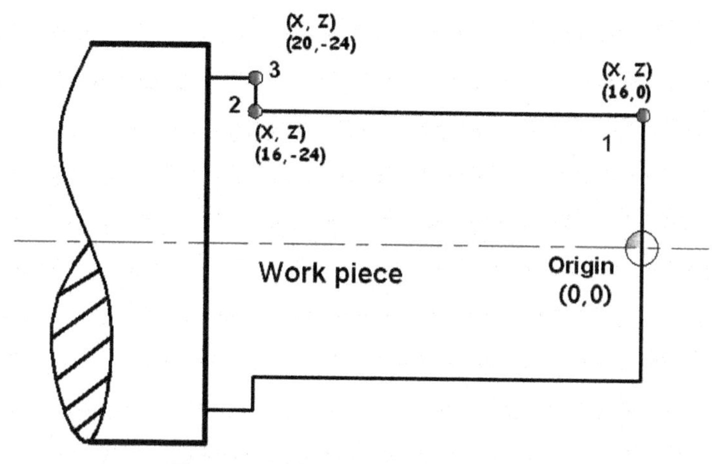

Figure 8.25 Drawing with coordinates.

Before CNC programming, we find out the coordinates of the Figure 8.25 in the respect of work zero (0, 0). This origin will be also reference point for all cutting tools in this program. So we can say, coordinates value in X & Z-axis of the cutting tools will be zero in the respect of the center (0, 0) of the work piece face.

We can understand the turning machine axis position through following comments.

- Value of the Z in Z-axis at the front face of the work piece, **Z = 0**
- Towards the spindle from the front face of the work piece, Z will be **Z−**
- Towards the tailstock from the front face of the work piece, Z will be **Z+**
- Value of the X in X-axis above the center line, where the cutting tool will cut the material, X+
- Value of the X in X-axis below the center line, where cutting tool generally does not go, X−

Now, the coordinates are in the terms of X & Z in the Diametrical and Absolute method.

Coordinates are taken from the Figure 8.23.

(X, Z)

1. (16, 0)
2. (16, −24)
3. (20, −24)

In Figure 8.26, for material cutting, the cutting tool will come near the work piece in rapid motion (G00) and takes the imaginary position (X30 Z15) but does not touch with the work piece.

G00 X30 Z15;

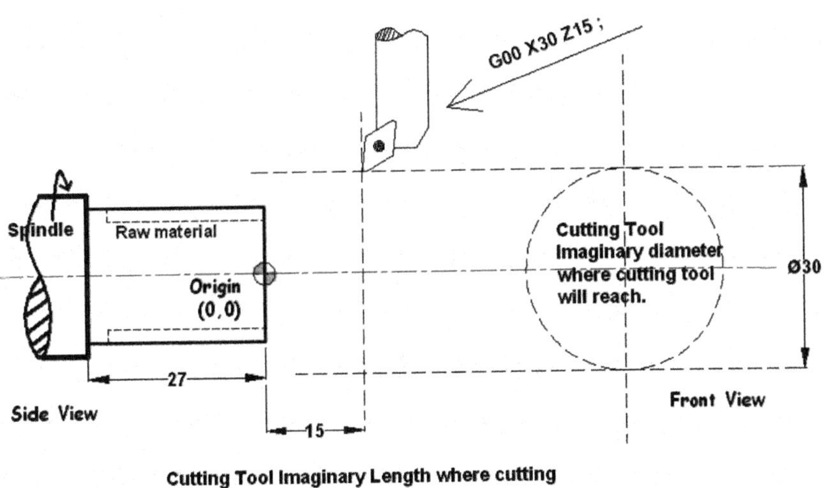

Figure 8.26 Turning tool is coming near the work piece in rapid motion.

In Figure 8.27, again the cutting tool came nearest the job and took the final position (X16 Z1.0) before cutting the material.

G00 X16 Z1.0;

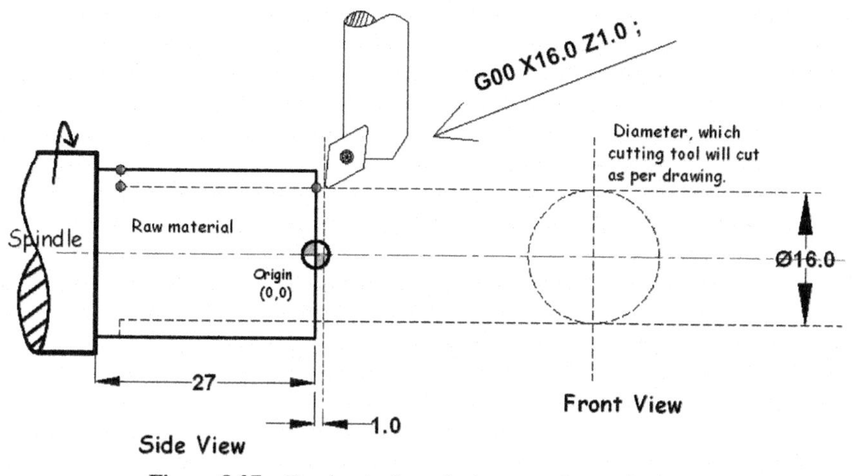

Figure 8.27 Turning tool reached nearest the work piece.

In Figure 8.28, now cutting tool is removing the material in linear motion (G01) with feed in straight turning operation and continuously going towards the desired coordinates (X16 Z-24) as per drawing.

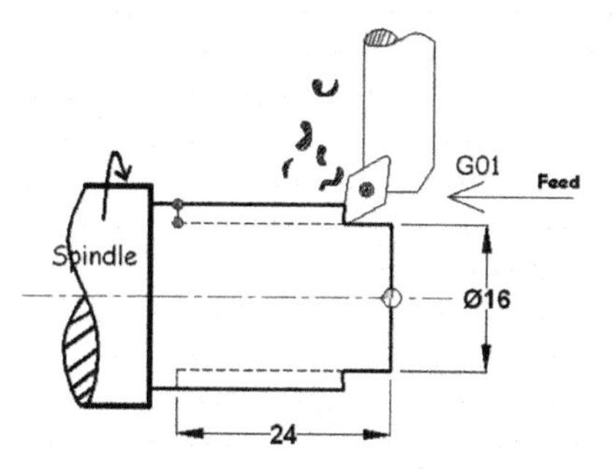

Figure 8.28 Turning tool is removing the material.

In Figure 8.29, till now cutting tool is removing the material in linear motion (G01) as per drawing and continuously going toward the coordinates X16 Z-24.

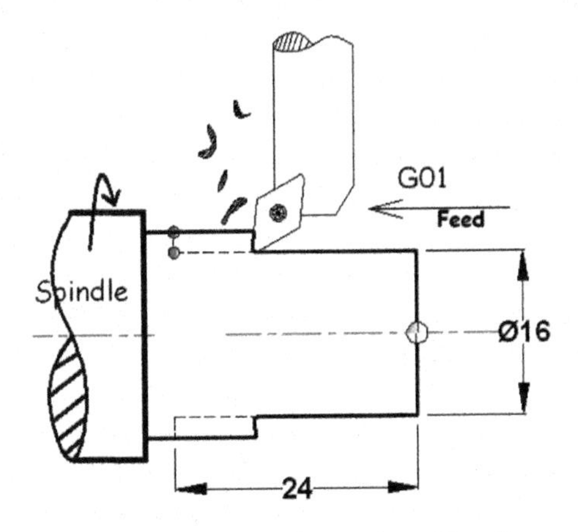

Figure 8.29 Cutting tool is removing material continuously in linear motion (G01) with feed.

In Figure 8.30, the cutting tool removed the material as per drawing as a straight turning operation and reached at the desired coordinates X16 Z-24.

G01 X16 Z-24 F.12;

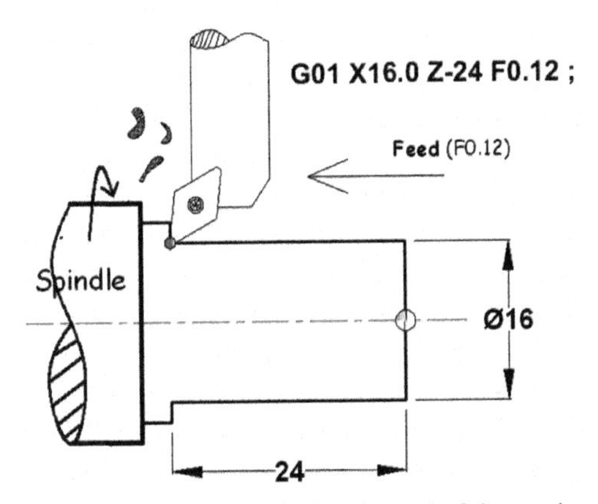

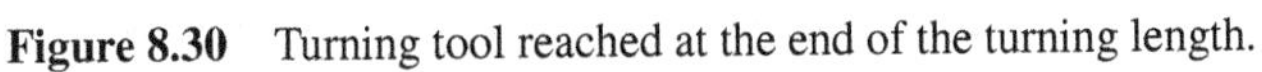

Figure 8.30 Turning tool reached at the end of the turning length.

In Figure 8.31, after cutting the material as per drawing tool will make the step. During step turning process cutting tool will go towards the bigger diameter X20 and this time value of Z will be same like the previous value (Z-24).

After making the step in the same time, cutting tool will not stay at the diameter X20. The tool will take safe clearance in X-axis after step turning operation. Safe clearance means cutting tool will release (exit) slowly with G01from the material surface.

G01 X21 Z-24 F0.12;

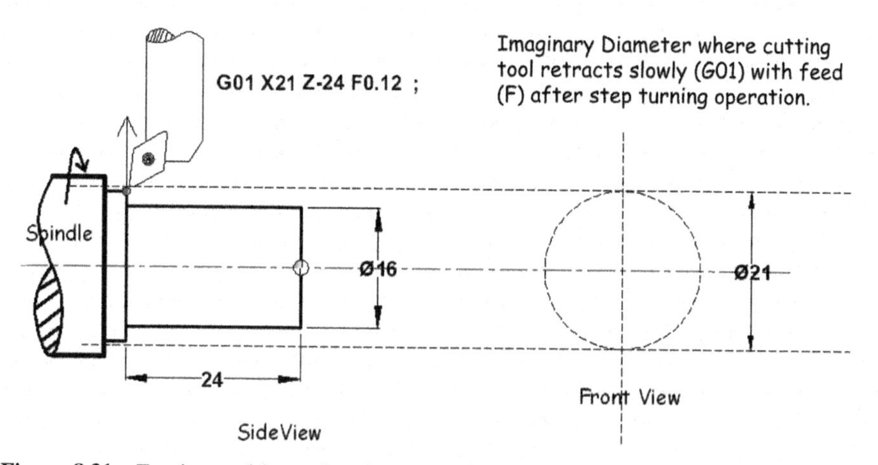

Figure 8.31 Turning tool is turning the step and maintaining some clearance between the material surface and cutting tool for safe retraction.

In Figure 8.32, now cutting tool will retract (go away/take safe clearance from the job) in Z-axis at same previous diameter ø21. **It means cutting tool will travel and reach at Z1.0 in Z-axis and that time X will be same like the previous diameter.**

G00 X21 Z1.0;

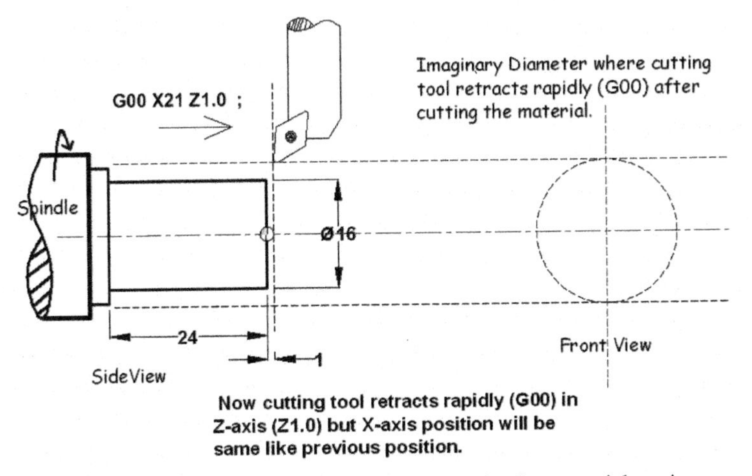

Figure 8.32 Cutting tool is retracting in Z-axis after material cutting.

In Figure 8.33, finally cutting tool retracts rapidly from the job and reached on imaginary coordinates (safe distance) in X & Z-axis.

G00 X50 Z60;

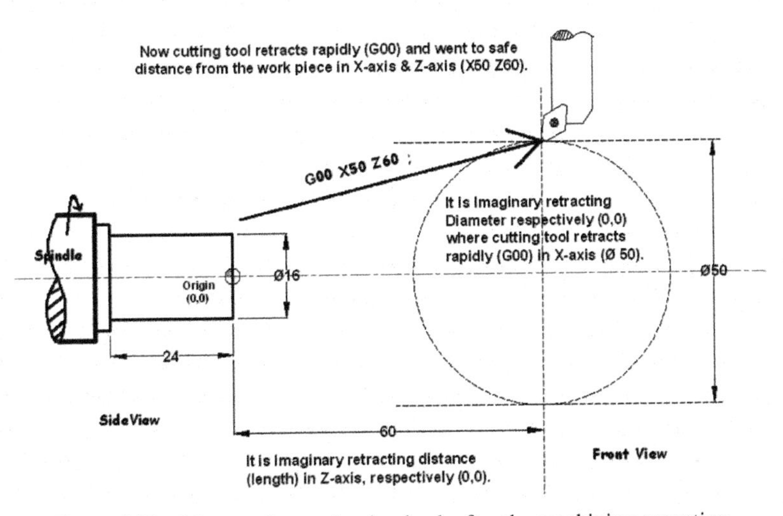

Figure 8.33 Now cutting tool going back after the machining operation.

Note: Following codes are the CNC programming codes which are used in CNC machine.

We can write above machining process of straight turning operation as a CNC program, the following codes are showing above machining process as a CNC program.

```
G00 X30 Z15;
G00 X16.0 Z1.0;
G01 X16.0 Z-24.0F0.12;
G01 X21.0 Z-24 F0.12;
G00 X21.0 Z1.0;
G00 X50 Z60;
```

Note: Above CNC program is not a complete CNC program. It is just a material removing part. In the next chapter, we will learn and make complete CNC program.

8.7.4 Some Few Machining Operations Step by Step with Figures Like Simulation

8.7.4.1 Step turning operation: only material removing program

Figure 8.34 shows raw material drawing with machining drawing.

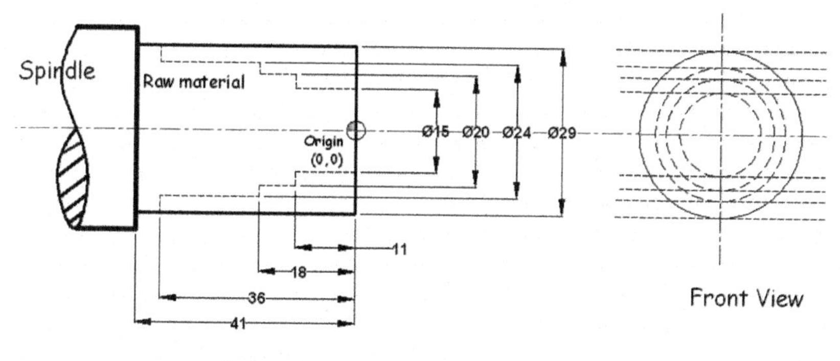

Figure 8.34 Raw material drawing with machining drawing.

Figure 8.35 shows the points where the cutting tool will travel and remove the material as per drawing.

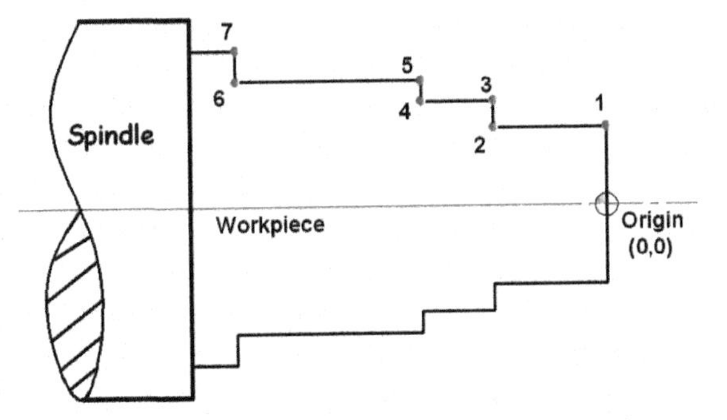

Figure 8.35 Cutting tool path where the cutting tool will travel and remove the material as per drawing.

Figure 8.36 shows points and coordinates where the cutting tool will travel and remove the material as per drawing.

Figure 8.36 shows coordinates points and the cutting tool will follow these coordinates and remove the extra material from the round surface as per given drawing, during the machining process.

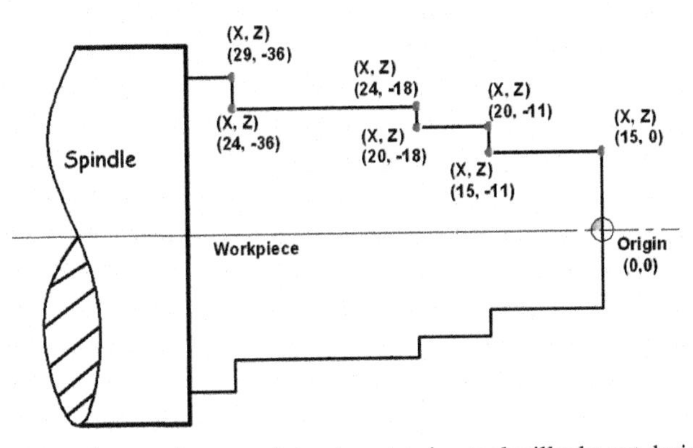

Figure 8.36 Shows the coordinates points where turning tool will take cut during machining.

Coordinates are in the term of X & Z-axis as per drawing.

(X, Z)	(X, Z)
1. (15, 0)	5. (24, −18)
2. (15, −11)	6. (24, −36)
3. (20, −11)	7. (29, −36)
4. (20, −18)	

Spindle is rotating continuously, when the cutting tool comes near and removes the material.

G00 X24 Z1.0;

In Figure 8.37, before cutting the material, the cutting tool is taking position nearest the work piece in rapid motion (G00).

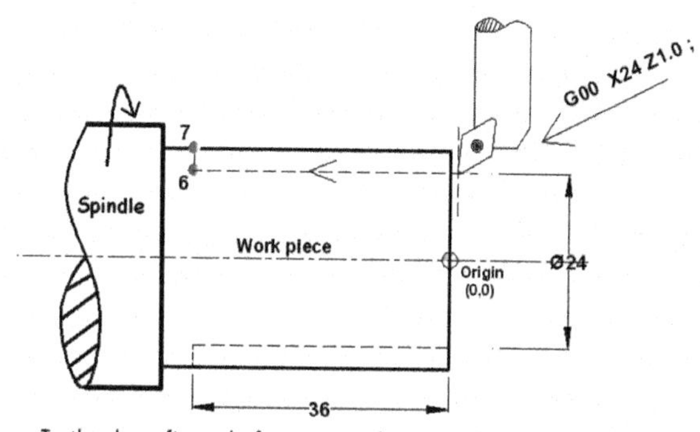

Figure 8.37 Turning tool is coming near the work piece before material removing, this time work piece is rotating.

In Figure 8.38, the tool is cutting the material and going continuously towards the coordinates **X24 Z-36** in linear motion (G01) with cutting feed.

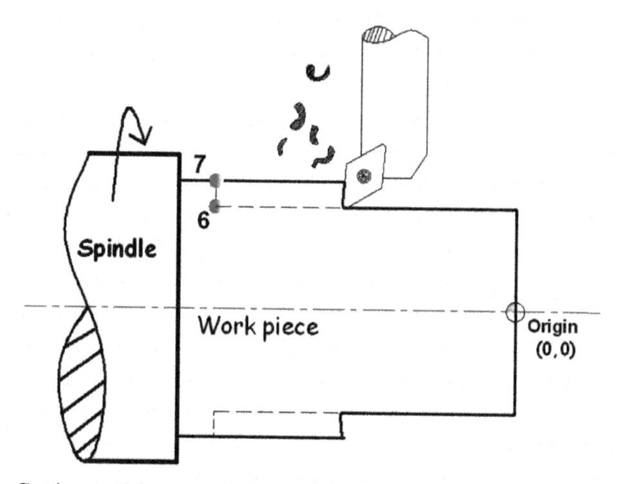

Figure 8.38 Cutting tool is removing material continuously with linear motion G01 and feed [3].

G01 X24 Z-36 F0.12;

In Figure 8.39, the cutting tool cut the material and reached at the coordinates **X24 Z-36** in linear motion (G01) with cutting feed.

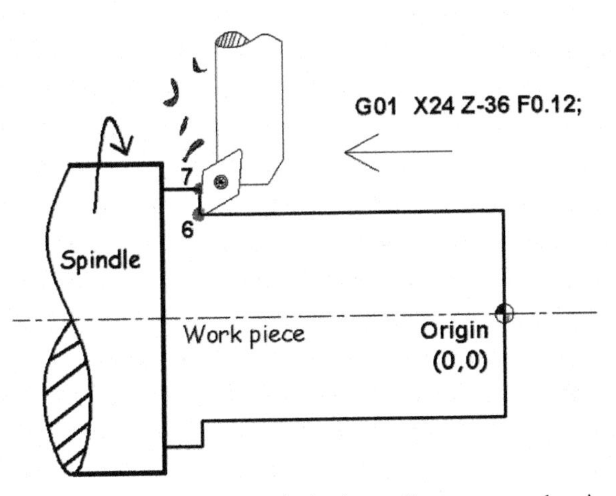

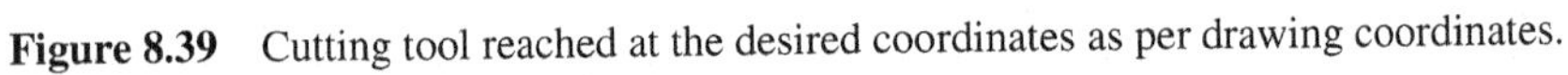

Figure 8.39 Cutting tool reached at the desired coordinates as per drawing coordinates.

G01 X30 Z-36 F0.12;

In Figure 8.40, the cutting tool is removing the material and making the step towards the diameter 29.0 (X29), after making the step cutting tool will

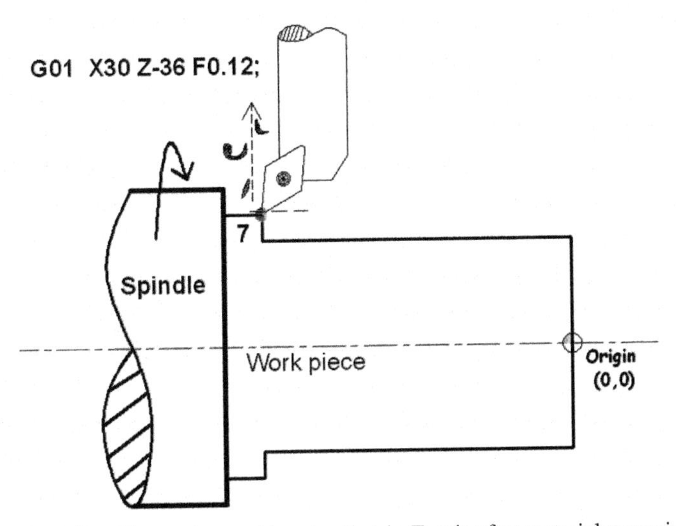

Figure 8.40 The cutting tool is retracting in Z-axis after material removing.

retracting in linear motion (G01) in X-axis direction at diameter 30 mm after step turning and take safe clearance.

G00 X30 Z1.0;

In Figure 8.41, after retracting in X-axis (X30.0) cutting **tool will retract in Z+1.0 mm in Z-axis positive direction in rapid motion (G00)** for making the next step.

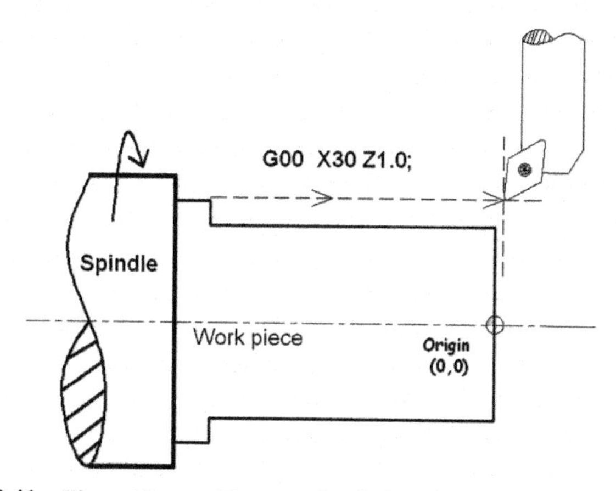

Figure 8.41 The cutting tool is retracting in Z-axis after material removing.

G00 X20 Z1.0;

In Figure 8.42, for **making the next step** cutting tool will **take the position in rapid motion (G00) at diameter 20 mm** and Z-axis position will be the same as the previous position (Z1.0 mm).

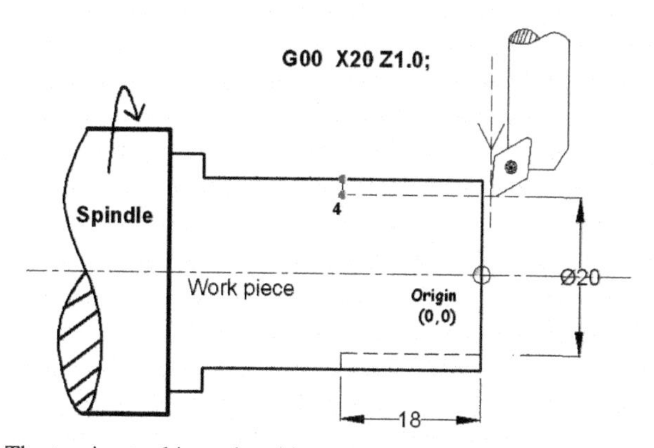

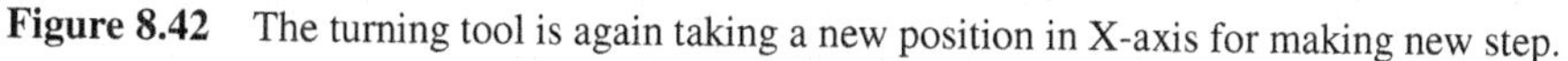

Figure 8.42 The turning tool is again taking a new position in X-axis for making new step.

In Figure 8.43, the cutting tool is cutting the material **continuously** and **going towards the coordinates X20 Z-18** in linear motion (G01) with feed (F).

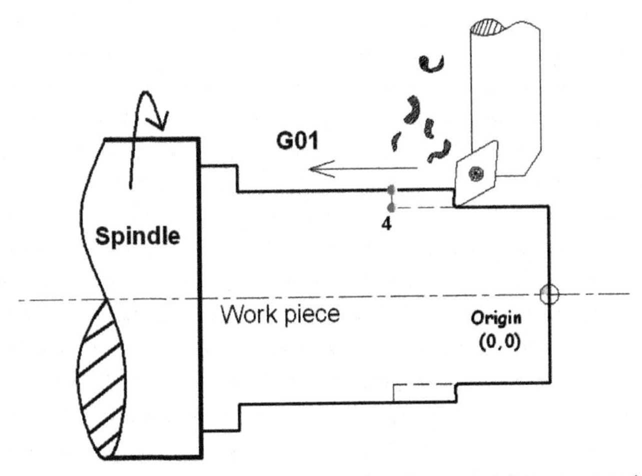

Figure 8.43 Turning tool is continuously removing the material in step turning operation and going towards the desired coordinates as per drawing.

G01 X20 Z-18 F0.12;

In Figure 8.44, the cutting tool **cut the material and reached at the coordinates X20 Z-18** in linear motion (G01) with feed (F).

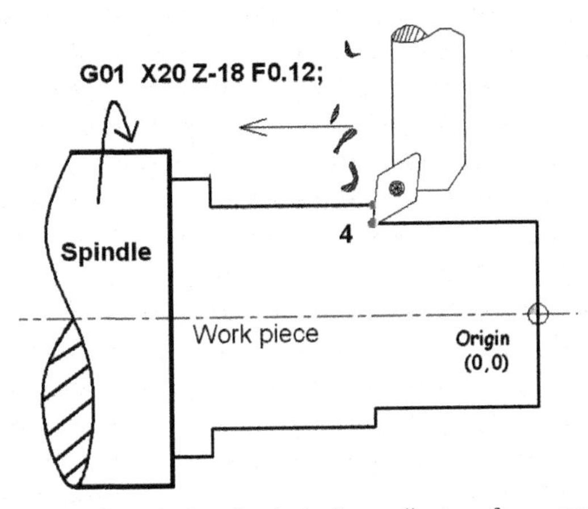

Figure 8.44 Cutting tool reached at the desired coordinates after removing the material.

G01 X25 Z-18 F0.12;

In Figure 8.45, the cutting tool is removing the material in step turning operation **and retracts in X-axis direction** at the diameter 25 mm **in linear motion (G01)** with feed. At the same time, Z-axis value will be same as previous (Z-18).

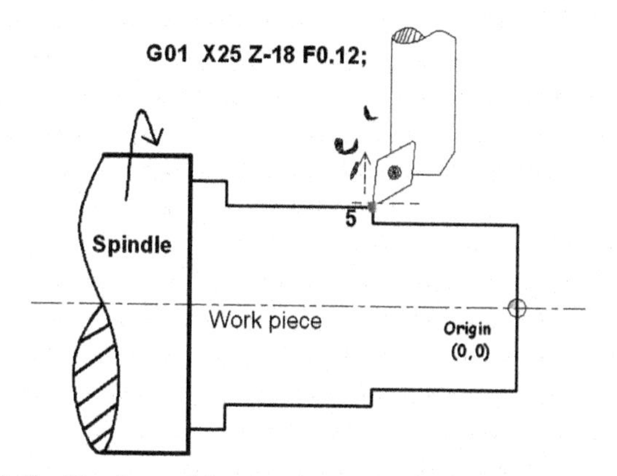

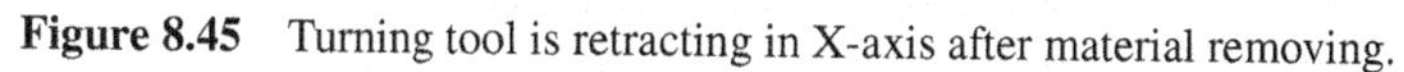

Figure 8.45 Turning tool is retracting in X-axis after material removing.

G00 X25 Z1.0;

In Figure 8.46, after retracting in X-axis now tool will **retract in Z + 1.0 mm in Z-axis positive direction respect of** the work piece face (Z = 0) **in rapid motion (G00) for making the next step.**

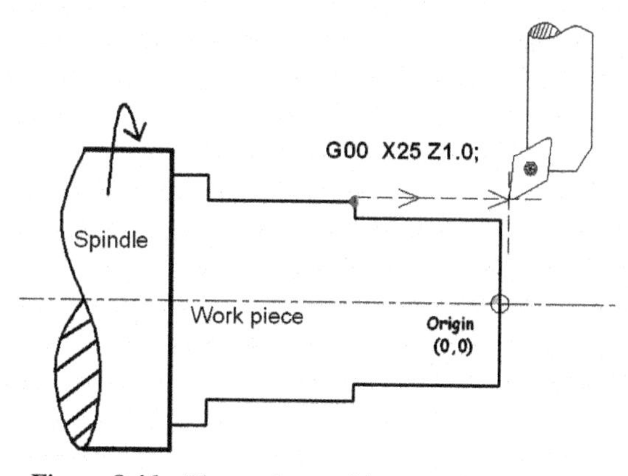

Figure 8.46 The cutting tool is retracting in Z-axis.

G00 X15 Z1.0;

In Figure 8.47, the cutting tool is **taking position in X-axis at diameter 15 mm in rapid motion (G00)**, during this Z-axis position will be same like previous position (Z1.0).

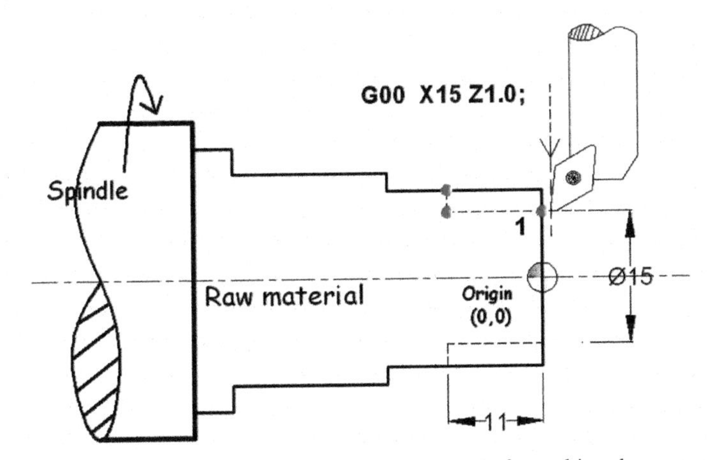

Figure 8.47 Turning tool is taking position in X-axis for making the next step.

In Figure 8.48, the cutting tool is cutting the material continuously and **going towards the coordinates X15 Z-11** in linear motion (G01) with feed (F).

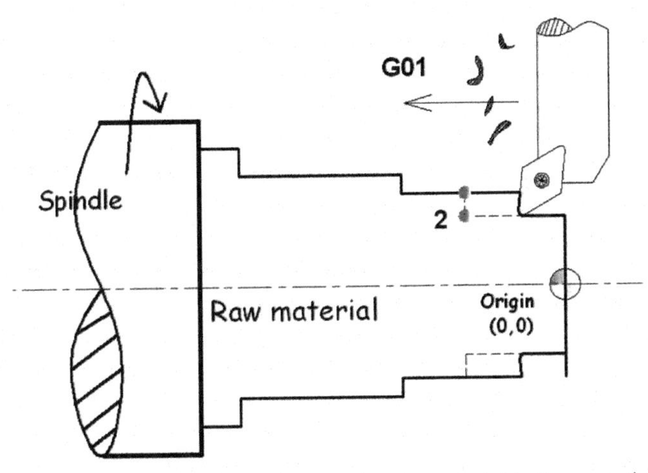

Figure 8.48 Turning tool is continuously removing the material in linear motion with feed.

G01 X15 Z-11 F0.12;

In Figure 8.49, the cutting tool has **removed the material and reached at the coordinates X15 Z-11** in linear motion (G01) with feed (F).

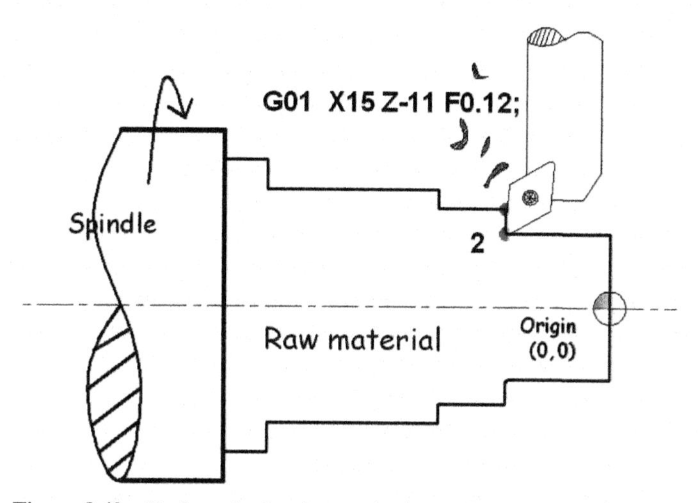

Figure 8.49 Tool reached at the required coordinates after machining.

G01 X21 Z-11 F0.12;

In Figure 8.50, the cutting tool is cutting the material in step turning operation and during this tool is retracting in X-axis direction at diameter 21 mm. During this Z-axis position will be the same like the previous position.

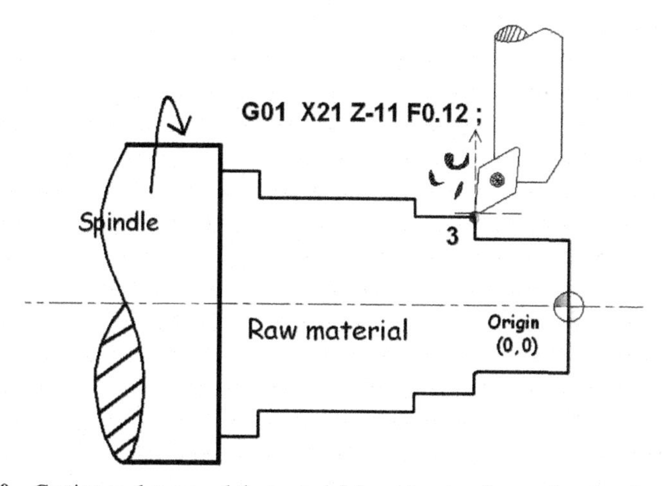

Figure 8.50 Cutting tool removed the material from the step face and took safe retraction in X-axis after machining.

G00 X21 Z1.0;

In Figure 8.51, the cutting tool is retracting Z + 1.0 mm in Z-axis positive direction respect of the front face (Z = 0) of the work piece in rapid motion (G00).

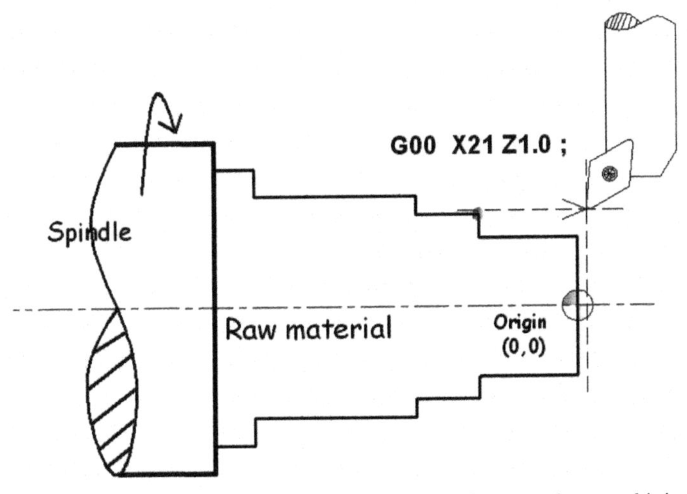

Figure 8.51 Turning tool is retracting in Z-axis after complete machining.

In Figure 8.52, the cutting tool is **retracting and taking safe distance in X & Z-axis** from the front face of the work piece in rapid motion (G00).

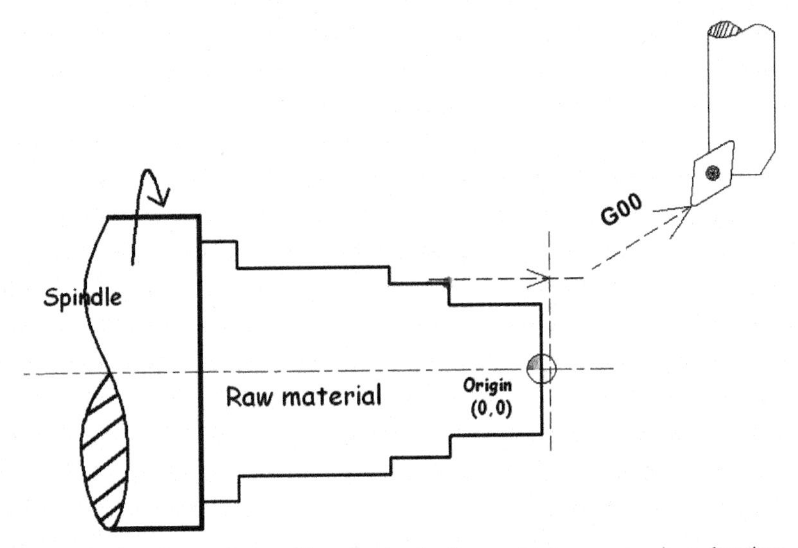

Figure 8.52 The turning tool is going back after machining as per given drawing.

Note: Following codes are the CNC programming codes which are used in CNC machine.

We can write above machining process of Step Turning operation as a CNC program, the following codes are showing above machining process as a CNC program.

```
G00 X24 Z1.0;
G01 X24 Z-36 F0.12;
G01 X30 Z-36 F0.12;
G00 X30 Z1.0;
G00 X20 Z1.0;
G01 X20 Z-18 F0.12;
G01 X25 Z-18 F0.12;
G00 X25 Z1.0;
G00 X15 Z1.0;
G01 X15 Z-11 F0.12;
G01 X21 Z-11 F0.12;
G00 X21 Z1.0;
```

Note: Above CNC program is not a complete CNC program. In the next chapter, we will learn and make complete CNC program.

8.7.5 CNC Programming of Corner Radius Operation Step by Step with Rough Cut by Using G03 Code

8.7.5.1 Important note

We cannot cut the material in one cut when the amount of rough material is large. In this case we will take small rough machining cuts before final machining cut.

Following drawing is for the machining operation. See Figure 8.53.

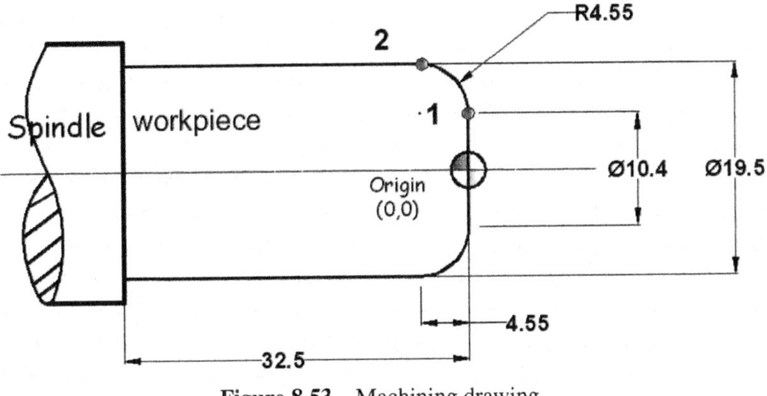

Figure 8.53 Machining drawing.

Figure 8.54 shows the coordinates points. The cutting tool will follow these coordinate points and remove the extra material from the round work piece as per given drawing, during the machining process.

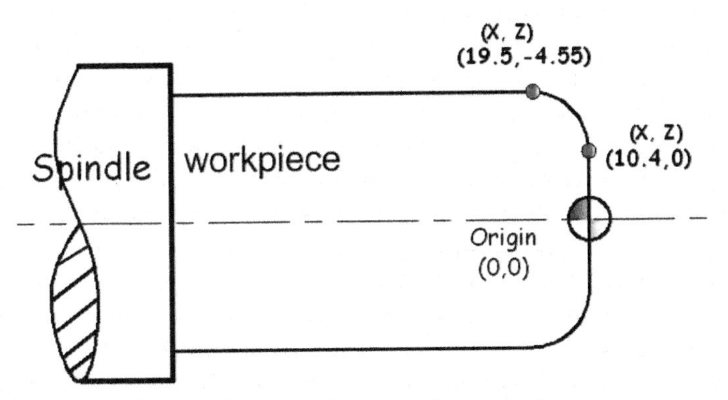

Figure 8.54 Coordinates position in the figure.

Coordinates of the above work piece as per drawing. In the term of.... X & Z

(X, Z)

1. (10.4, 0) Starting point of the corner radius
2. (19.5,-4.55) Ending point of the corner radius

G00 X50 Z60;

In Figure 8.55, the turning tool is taking position at coordinates X50 Z60 in rapid motion (G00), respect of work zero (0, 0).

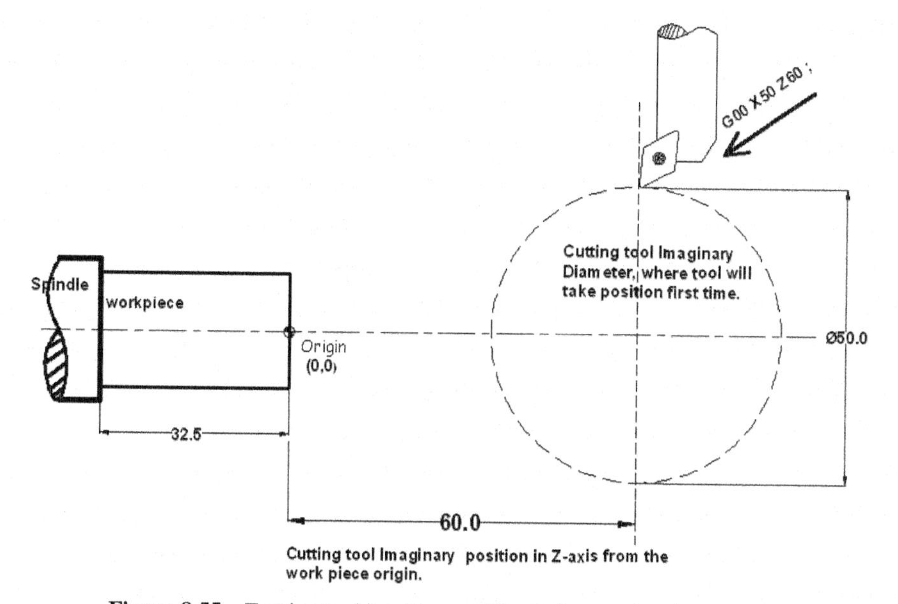

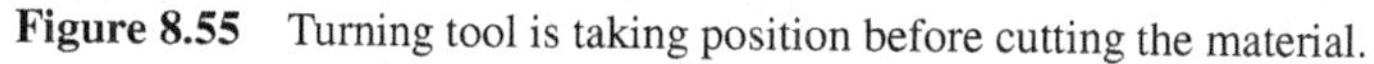

Figure 8.55 Turning tool is taking position before cutting the material.

G00 X25 Z10;

In Figure 8.56, the cutting tool is moving rapidly (G00) and taking position near the work piece at X25 Z10 coordinates in X & Z-axis.

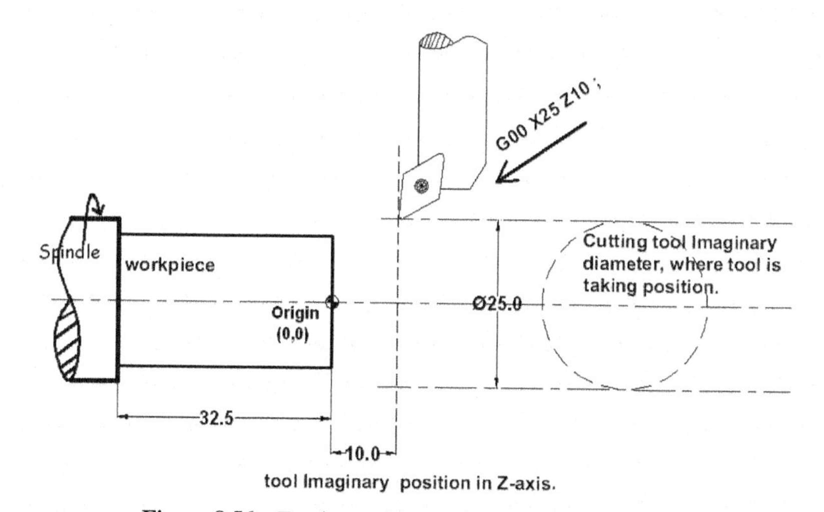

Figure 8.56 Turning tool is coming near the work piece.

G00 X15 Z1;

In Figure 8.57, before removing the material as a radius profile cutting tool will come rapidly (G00) nearest the work piece at coordinates X15 Z1.0, Where X15 is the starting diameter of the **roughing radius**.

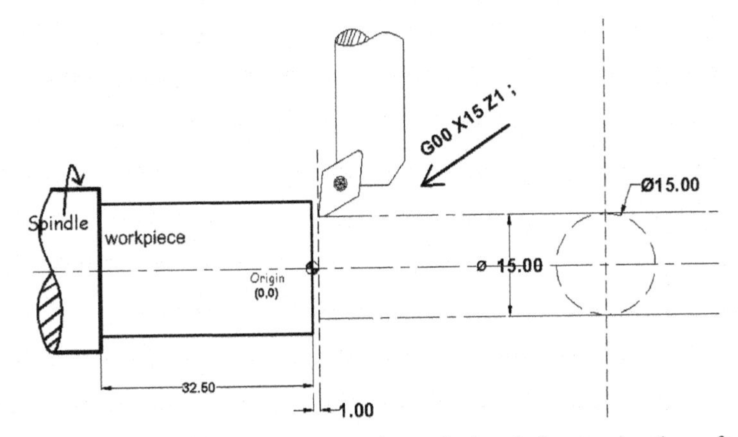

Figure 8.57 Turning tool came nearest the work piece before turning the surface.

G01 X15 Z0 F0.12;

In Figure 8.58, the cutting tool is touching the material coordinates (X15 Z0) of the work piece face before cutting the rough radius profile at diameter 15 mm in linear motion (G01) with slow feed (F). It is a starting point of the rough radius profile.

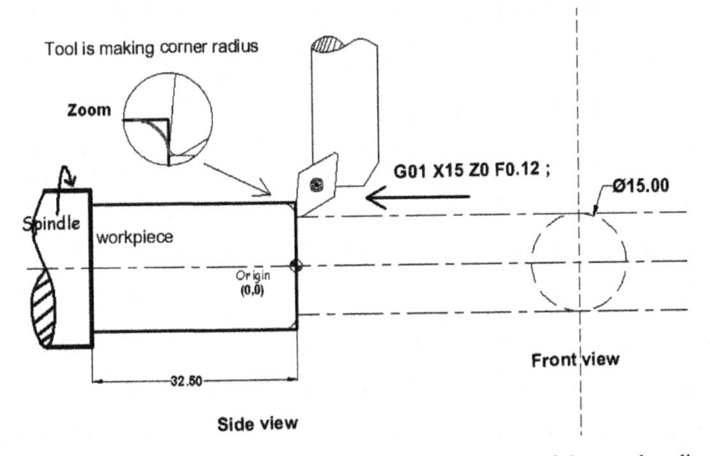

Figure 8.58 Turning tool is taking position at the starting point of the rough radius profile.

In Figure 8.59, the tool **is continuously removing the material in counter (anti) clockwise direction (G03)** and going towards the coordinates X19.5 Z-2.25 at diameter 19.5 mm.

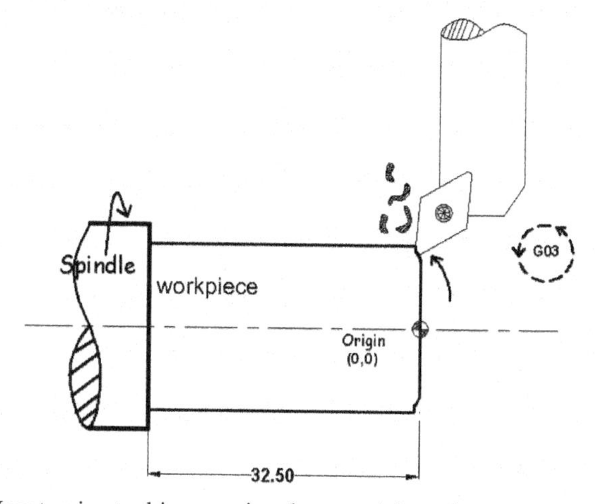

Figure 8.59 Now turning tool is removing the material continuously in counter clock wise direction and making rough radius profile.

G03 X19.5 Z-2.25 R2.25 F0.12;

In Figure 8.60, the cutting tool removed the rough material in C.C.W circular motion and reached the coordinates X19.5 and Z-2.25 in X and Z.

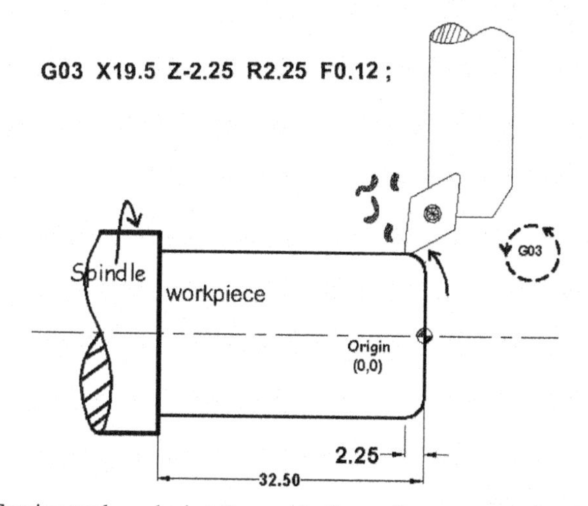

Figure 8.60 Turning tool reached at the required coordinates as per drawing after making rough radius profile.

G00 X19.5 Z1.0;

In Figure 8.61, after removing the rough radius profile material, the tool is retracting in Z+ positive axis for final radius profile cutting

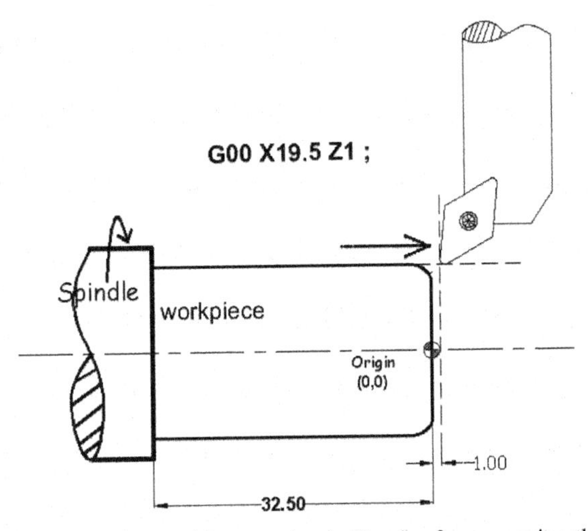

Figure 8.61 The cutting tool is retracting in Z-axis after removing the material.

G00 X10.4 Z1;

In Figure 8.62, before **final cutting**, the tool is taking the position at coordinates **X10.4 Z1.0** in X & Z-axis.

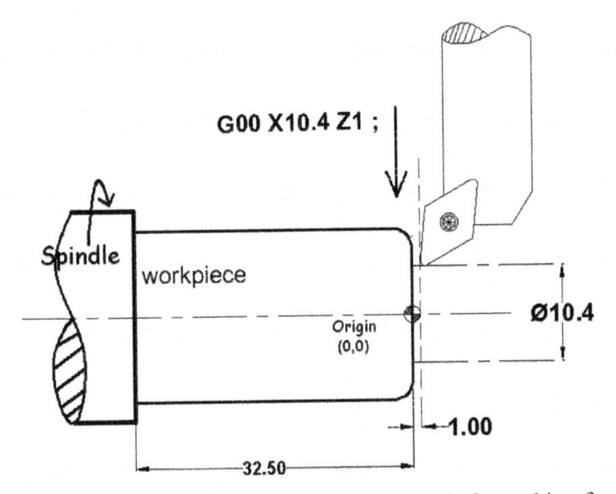

Figure 8.62 The turning tool is taking new position in X-axis for making final radius profile.

G01 X10.4 Z0 F0.12; (starting point of radius)

In Figure 8.63, now finally cutting tool **reached and touched at the coordinates X10.4 Z0.0 of the starting point of the radius** in linear motion (G01) with feed (F).

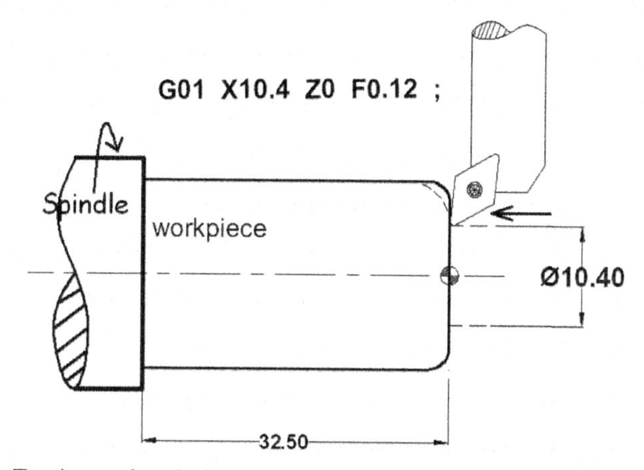

Figure 8.63 Turning tool took the position at the starting point of the final radius profile before cutting the material.

In Figure 8.64, the cutting tool is continuously cutting the material and **making final radius profile** towards the **coordinates X19.5 Z-4.55 in the anticlockwise direction in circular motion** with feed (F).

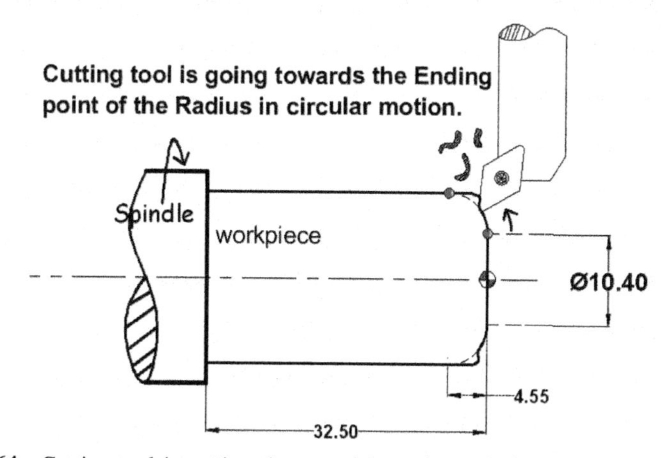

Figure 8.64 Cutting tool is cutting the material continuously in counter (anti) clockwise direction and making final radius profile.

G03 X19.5 Z-4.55 R4.55 F0.12; (Ending point of radius)

In Figure 8.65, cutting tool removed the material of radius 4.55 mm in C.C.W circular motion and reached at the coordinates X19.5 Z-4.55 in counter (anti) clockwise direction.

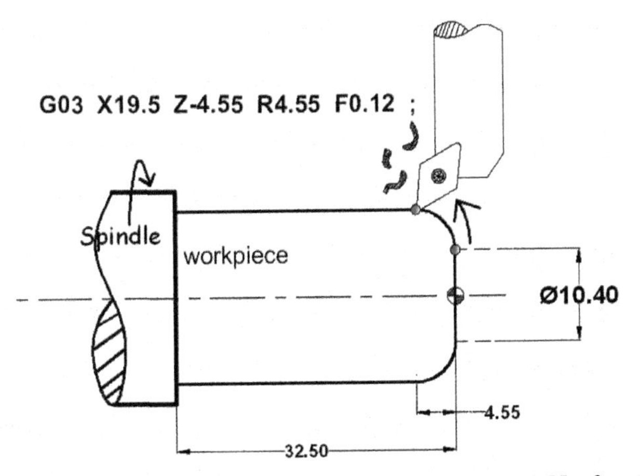

Figure 8.65 Turning tool reached at the coordinates X19.5 Z-4.55 after removing the material as per drawing.

G00 X19.5 Z1;

In Figure 8.66, the tool is retracting in **Z+ positive axis at distance 1.0 mm** from the front face of the work piece but **X-axis value will be same like previous** value.

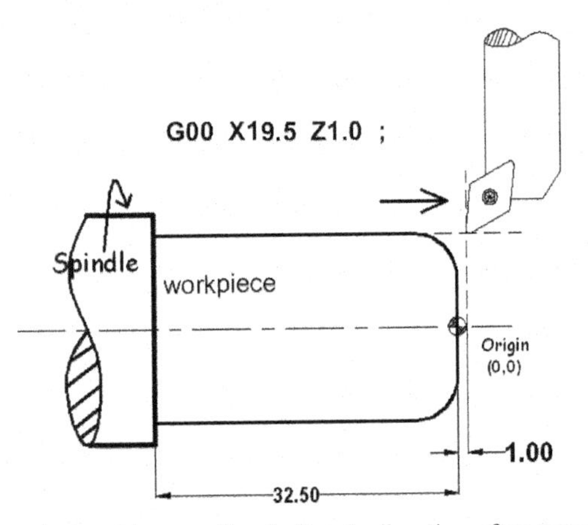

Figure 8.66 The turning tool is retracting in Z-axis direction after removal of the material.

In Figure 8.67, finally cutting tool is going at a safe distance in X & Z-axis coordinates after complete machining process in rapid motion (G00).

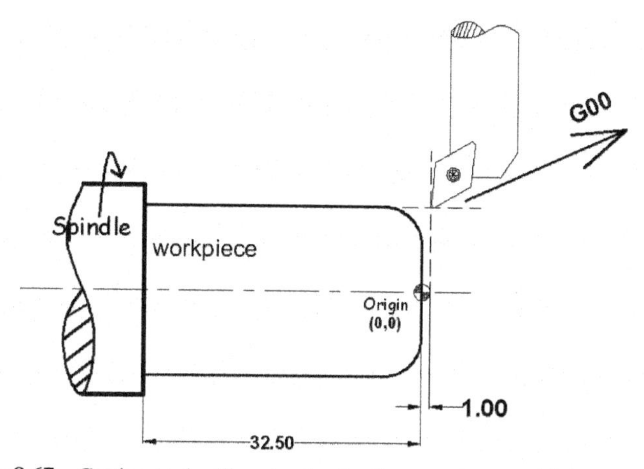

Figure 8.67 Cutting tool will go back after the complete machining operation.

8.7.5.2 Following codes are showing above machining process as a CNC program [4]

G00 X50 Z60;	Cutting tool is taking position near the job at coordinates X50 Z60 in rapid motion (G00).
G00 X25 Z10;	Tool is coming nearest the job in rapid motion.
G00 X15 Z1.0;	Finally, the cutting tool comes near the job in rapid motion.
G01 X15 Z0 F0.12;	Tool touched (tool touched the starting point of radius) the work piece in linear motion (G01) with feed.
G03 X19.5 Z-2.25 R2.25 F0.12;	Tool is making **rough** radius in counter clock wise motion with feed.
G00 X19.5 Z1.0;	Tool is retracting in **Z-axis** from the work piece in rapid motion.

G00 X10.4 Z1.0;	Tool is retracting in **X-axis** from the work piece in rapid motion.
G01 X10.4 Z0 F0.12;	Tool touched (tool touched the starting point of radius) the work piece face for final radius cutting in linear motion (G01) with feed.
G03 X19.5 Z-4.55 R4.55 F0.12;	Now cutting tool is taking final cut from the material in circular motion (counterclockwise) with feed.
G00 X19.5 Z1.0;	Tool is retracting in **Z-axis** from the work piece in rapid motion.

We will make.... CNC program with the complete industrial format in the next chapter.

8.8 What is the Safe Position (Distance) of the Cutting Tool in the Respect of Work Piece During Machining or After Machining?

The place where, cutting tool stays after machining or before machining is called cutting tool's safe position. Generally, after material cutting, the cutting tool takes position far from the work piece. This safe position depends on work piece size and loading/unloading procedure of the work piece.

References

[1] CNC Technology & Programming, Tilak Taj, 2016, Dhanapat Rai Publishing Company, India
[2] CNC programming Manual of MTAB company: (Certificate course on CNC Turning, MTAB Technology Centre, MTAB, Chennai, India)
[3] Drawings and sketches has been generated from Auto CAD 2014.
[4] Fanuc control system or similar control system

9

Complete CNC Programming: Industrial Format

In the previous chapter, we tried to explain to our beginners a very simple CNC programming language. In this chapter we are increasing the level of their understanding of CNC programming. We will make CNC program like an industrial CNC program or similar to industrial CNC programming.

We will add few important programming lines about **before removing the material** and **after removing the material. These added programming lines will be common in the CNC program.**

1. In the first line of the CNC program, we write CNC program number (example: O58061), where O is the program number and digits show the drawing number of the drawing's title box. There are so many program numbers (program names) feed in the CNC memory (CNC control panel), that's why we need one particular CNC program name or number.
   Example **O-----;**
2. Next programming line writes as a work zero
   Example **G54;**
3. In this programming line we will take maximum spindle speed control command.
   Example **G50 M03 S------;**
4. Now this line we take the cutting tool with tool offset number.
   Example **T-------;**
5. Now in this line tool takes the position rapidly (G00) in X and Z-axis co-ordinates.
   Example **G00 X----- Z-----;**
6. In this programming line spindle rotates with constant surface speed (**G96**) in clock/anti clock wise direction (M03/M04) with required rpm.
   Example **G96 M03 S-----;**

7. Now in this programming line cutting tool will come rapidly nearer and takes the position in X and Z-axis.
   Example **G00 X----- Z-----;**
8. In this programming line coolant motor will start.
   Example **M08;**
9. In this programming line just before cutting the material, cutting tool will come nearest the work piece in rapid motion.
   Example **G00 X----- Z-----;**
10. **From here, we will write machining procedure (material cutting procedure) with cutting and non-cutting movements of the tool as per drawing.**
    Example **G01 X----- Z---- F-----;**
11. After machining process cutting tool will retract from the work piece surface and takes safe position in X and Z-axis direction.
    Example **G01 X----- Z----- F-----;**
    G00 X----- Z-----;
    G01 X----- F-----;
12. After the machining process in this line coolant motor will stop.
    Example **M08;**
13. Now in this line spindle motor will stop.
    Example **M05;**
14. After coolant motor and spindle motor stop, cutting tool will retract and take safe distance in X and Z-axis (in some cases, we use the G28 command for machine home position).
    Example **G00 X---- Z-----;**
15. And this line CNC program will stop and reset.
    Example **M30;**

9.1 CNC Programming Procedure With Industrial Format

9.1.1 Straight Turning Operation With CNC Programming

Figure 9.1 shows the raw material drawing.
Figure 9.2 shows the machining drawing.

In Figure 9.3, above points are coordinates points. The cutting tool will follow these coordinates and remove the extra material from the round surface as per the given drawing, during the machining process.

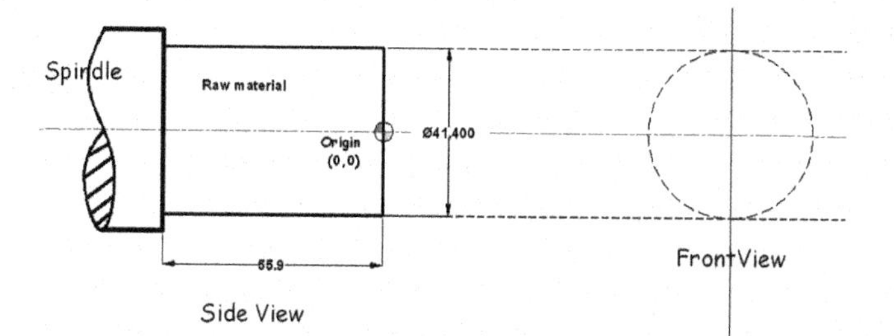

Figure 9.1 Drawing of raw material.

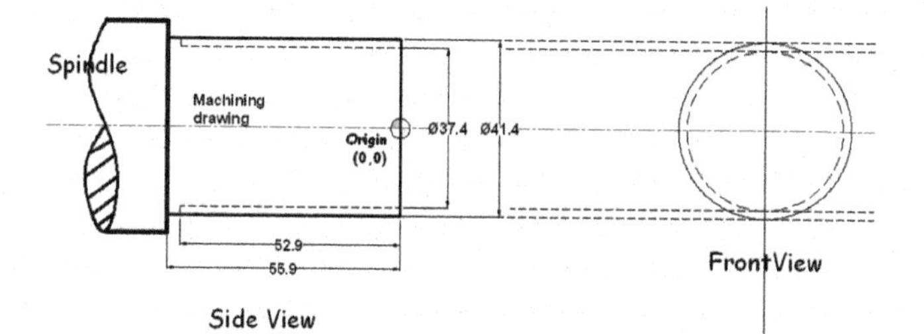

Figure 9.2 Machining drawing.

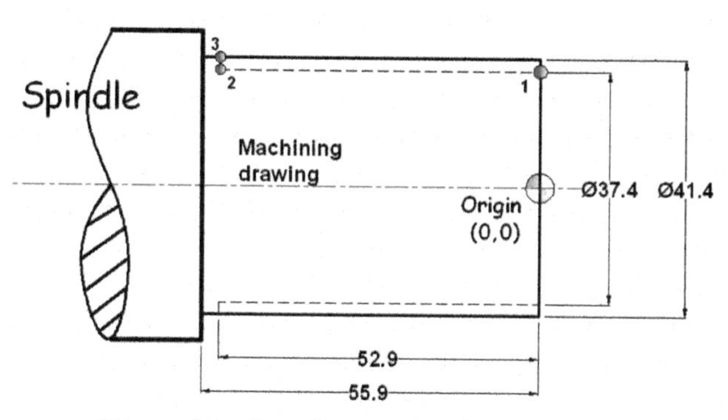

Figure 9.3 Coordinates points for machining.

Coordinates
(X, Z)

- (37.4, 0)
- (37.4, −52.9)
- (41.4, −52.9)

CNC Programming for Figure 9.3

O10021; This block represents a program number, which is starting from alphabet **O,** after thealphabet **O** drawing number will write as a program number. This number can be 4/5 digits. When alphabet **O** and numbers meet together, it becomes a CNC program number

G54; This is called **Work Zero [origin (0, 0)]. Work Zero is the reference point of all coordinates and cutting tools and it is** located on the center of the work piece face. Generally reference or origin point takes on the front face of the work piece.

G50 S1600; When we use G50 command, it controlled maximum spindle speed of the spindle. When we use G50 and G96 command, it means spindle rpm will vary between maximum and minimum rpm.

T0202; When we use T0202 command it means, CNC machine will select the tool from the 02 number station with same number 02 tool geometry.

G00 X100 Z80; When this programming block will execute (work), it means the tool will take position at coordinates X100 Z80 in rapid motion. It is an imaginary position of the cutting tool (Figure 9.4).

G96 M03 S200; When this programming block will execute, the work piece will rotate in clockwise direction and rpm will

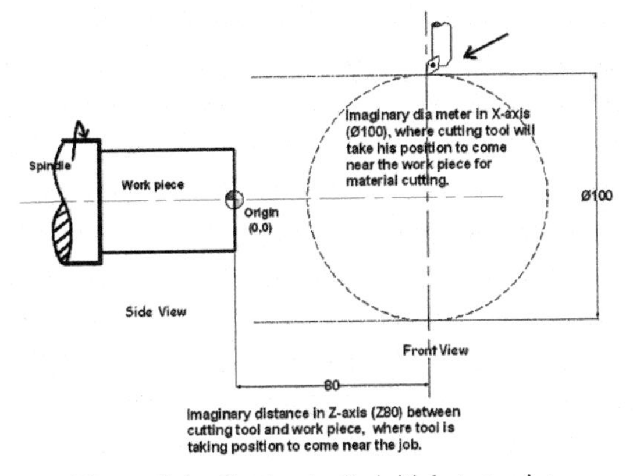

Figure 9.4 Cutting tool's initial start point.

vary between smaller and bigger diameter and rpm will be change between 1600 to 200 as per diameter. (Figure 9.4)

G00 X37.4 Z1.0; In Figure 9.5, the tool will come near the work piece and take position at the coordinates X37.4Z1.0 in rapid motion.

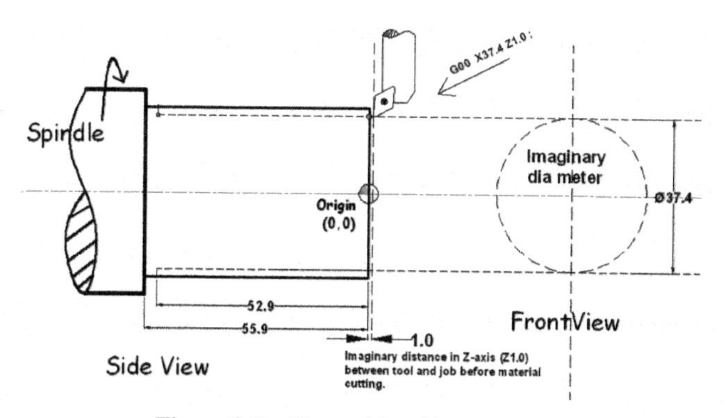

Figure 9.5 The tool is taking position.

In Figure 9.6, the cutting tool is continuously cutting the material and going towards the coordinates X37.4 and Z-52.9 in linear motion (G01) with feed.

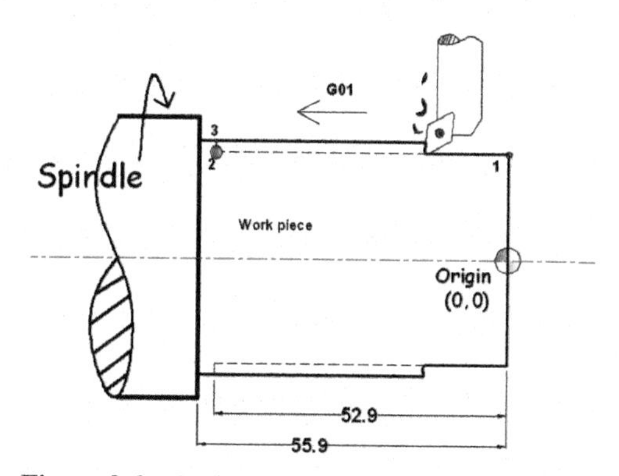

Figure 9.6 Cutting tool is removing the material.

In Figure 9.7, the cutting tool is still continuously cutting the material in linear motion G01 with feed (f) and going towards the coordinates X37.4 and Z-52.9.

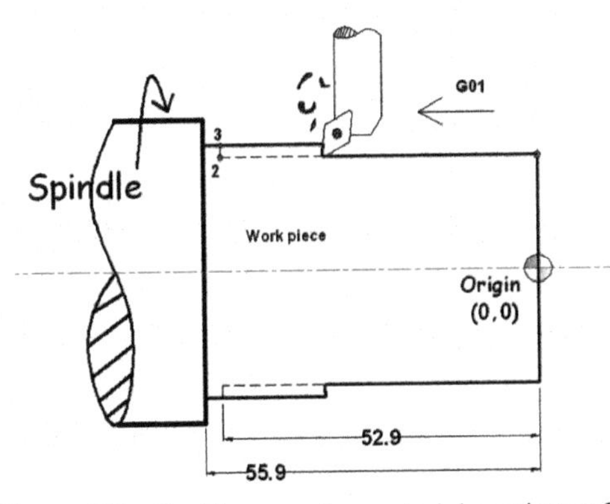

Figure 9.7 Tool is removing material continuously.

G01 X37.4 Z-52.9 F0.12; In Figure 9.8, the cutting tool reached at the coordinates X37.4 and Z-52.9 after removing the material as a straight turning operation in linear motion (G01) with feed (F).

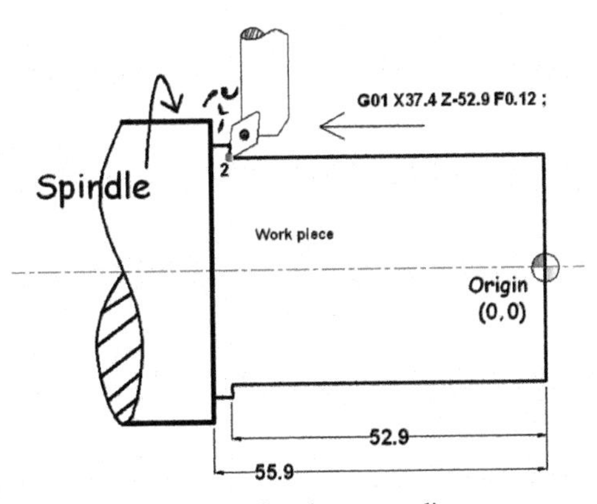

Figure 9.8 Tool reached at the given coordinates as per program.

G01 X42.5 Z-52.9 F0.12; In Figure 9.9, after straight turning, now the cutting tool is making a step during material removal. The tool is also retracting in X-axis at a diameter of 38.5 mm in linear motion (G01) with feed (F).

G00 X42.5 Z1.0; In Figure 9.10, after retracting in X-axis cutting tool will go away (retract) from the work piece and take safe distance in Z+ (positive) axis as a Z1.0 mm.

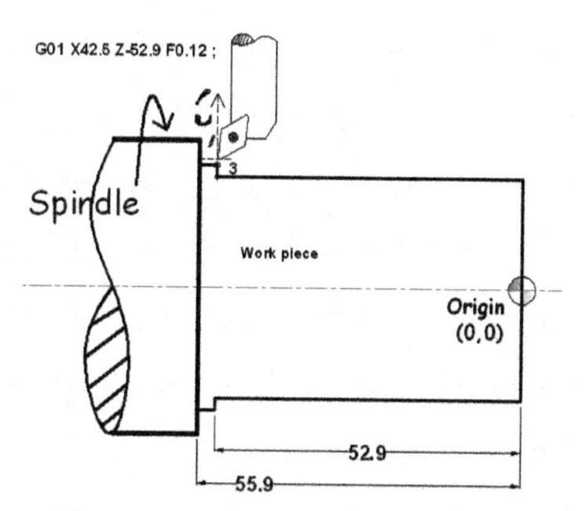

Figure 9.9 Tool is retracting in X-axis.

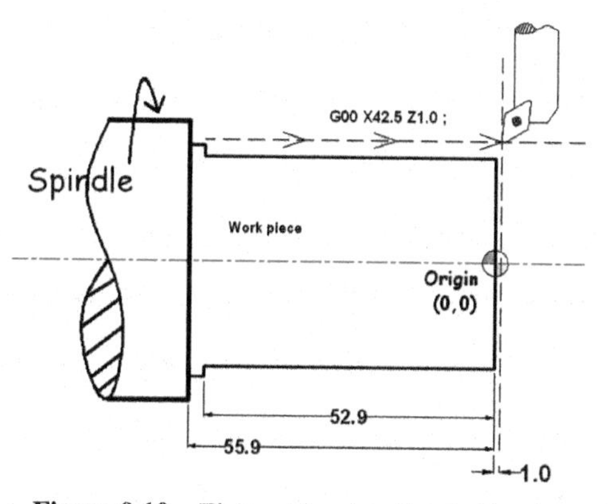

Figure 9.10 The tool is retracting in Z-axis.

M09; After cutting the material and retraction in Z-axis coolant motor will stop.

M05; Now spindle motor will also stop.

G00 X100 Z80; In Figure 9.11, after the spindle stops, the cutting tool is taking safe position rapidly from the work piece at the coordinates X100 & Z80. Due to this safe position the operator can change the next work piece for production without any interruption and maintain the distance between the work piece and cutting tool.

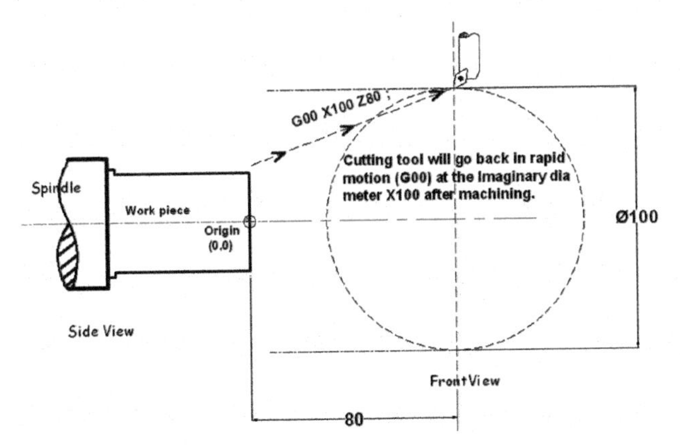

Figure 9.11 Cutting tool is retracting in safe distance.

M30; After execution of this programming line (block), the CNC program will stop and reset. Now cursor will reach the first programming line.

9.1.1.1 Drilling operation in step by step with CNC programming

Generally in the manufacturing industry, the drilling operation is performed on the work piece in two ways:

- **Continuous drilling operation**
  During the continuous drilling operation, the drilling tool starts removing the material until the drill tool does not reach at the end point of the required drill length as per drawing. It is possible only then when the material is to be soft or drill hole length is small.
- **Peck drilling operation**
  Here, in this operation, drill tool will not drill the hole thru in single cut it will drill the hole in separate – separate small segments. This depth of cut can be changed depending on the material, material size, drill hole length and tool size.

  We can say drill tool will make complete drill hole in small segments using by the small depth of cut.

Note:

Always, constant spindle speed (rpm) is used in the drilling operation.

Figure 9.12 shows the machining drawing.

Figure 9.13 shows the raw material drawing.

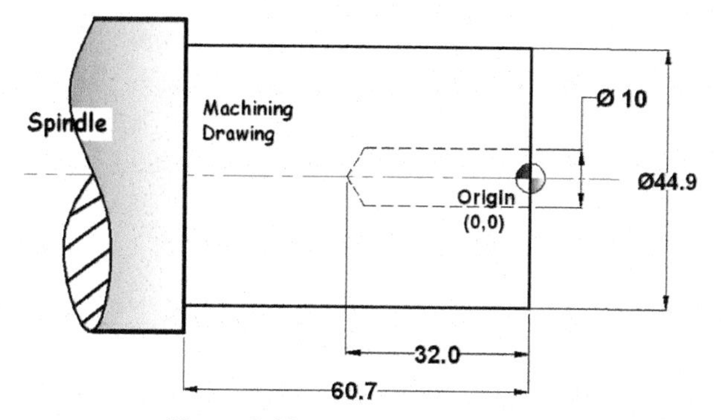

Figure 9.12 Machining drawing.

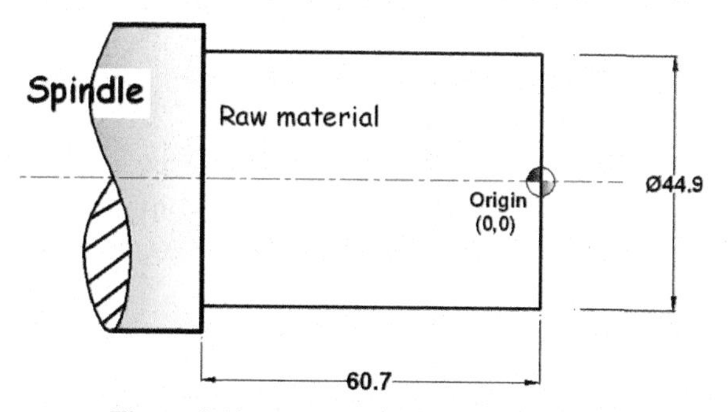

Figure 9.13 Drawing of the raw material.

O10082; This is called program number. In this, the first letter is **O** and another five digits are numbers and last sign (**;**) is called EOB (end of the block).

G54; This is called **Work Zero [origin (0,0)]. Work Zero is the reference point of all coordinates and cutting tools and it is** located on the center of the work piece face. Generally reference or origin point takes on the front face of the work piece.

T0505; When we use T0505 command it means the CNC machine will select the tool from the 02 number station with the same number of (02) tool geometry.

G00 X100 Z100; G00 command is used for rapid motion (non-cutting motion). Here Cutting tool T0505 will take position rapidly in X-axis at diameter 100 mm and Z-axis at distance 100 mm of the respect of origin point. The tool will use these coordinates to come near the job for cutting. See Figure 9.14.

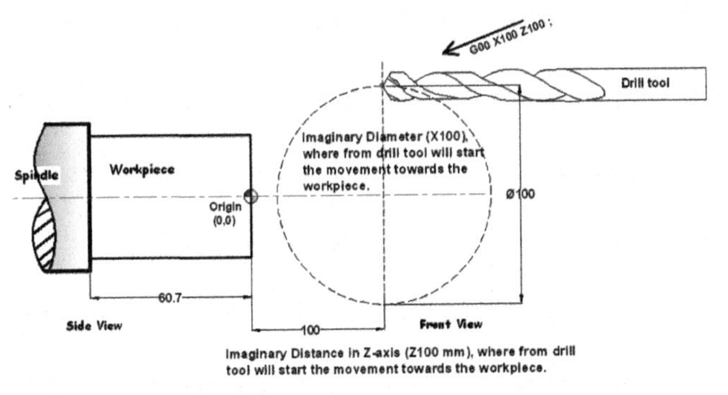

Figure 9.14 Drill tool is taking initial position.

G97 M03 S500; In Figure 9.15, **G97** command is used for constant spindle speed. By using this command spindle will rotate in clock wise direction in 500 rpm. Now spindle is rotating.

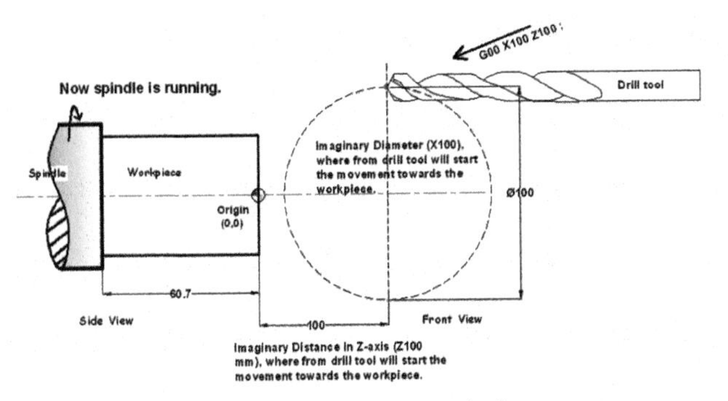

Figure 9.15 Rotating spindle.

G00 X30 Z50.0; In Figure 9.16, drilling tool is coming near the work piece together in X-axis (30 mm diameter) and Z-axis direction (Z50.0) in rapid motion (G00).

G00 X0.0 Z5.0; In Figure 9.17, the drilling tool is moving towards the work piece center and reached at the work piece center X0.0 in X-axis and reached at Z5.0 (in the Z direction) in rapid motion (G00).

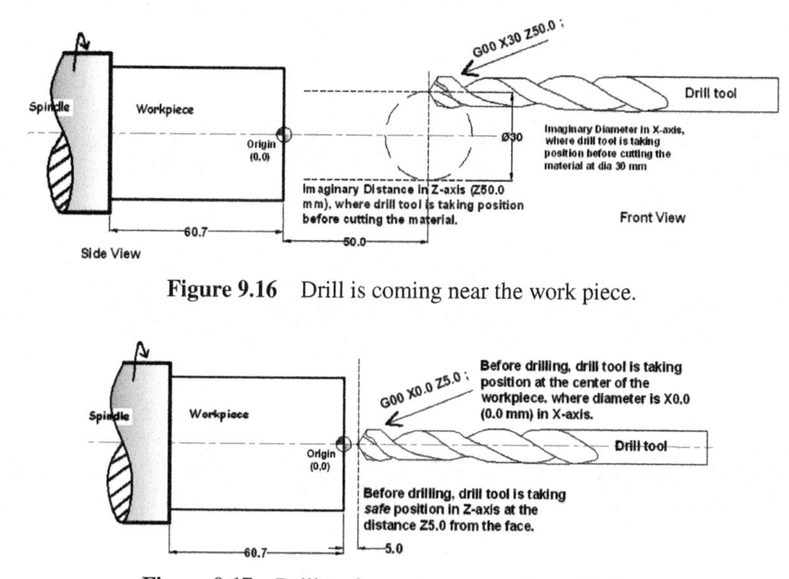

Figure 9.16 Drill is coming near the work piece.

Figure 9.17 Drill tool came very near the work piece.

G00 X0.0 Z1.0; In Figure 9.18, before cutting the material drill is taking position at co-ordinates X0.0 and Z 1.0, nearest the work piece in rapid motion G00 but does not touch.

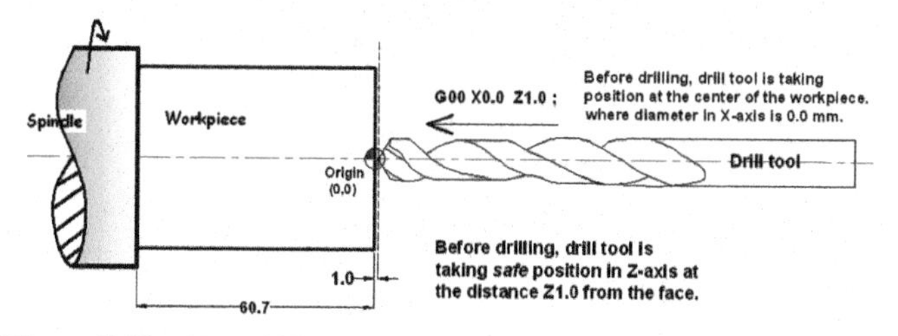

Figure 9.18 Now drill came nearest the job before drilling operation.

Figure 9.19 shows zoomed figure.

G01 X0.0 Z-5.0 F0.06; In Figure 9.20, the drilling tool is drilling (removing the material) the work piece very slowly and going 5.0 mm in depth inside the work piece at the center of the work piece X 0.0 in linear motion G01 with very slow feed F 0.06.

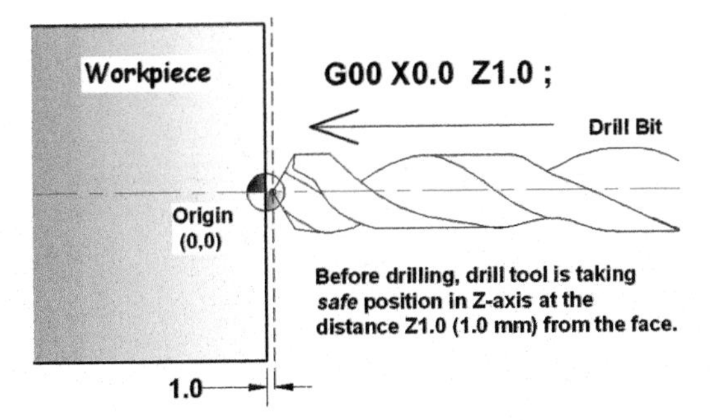

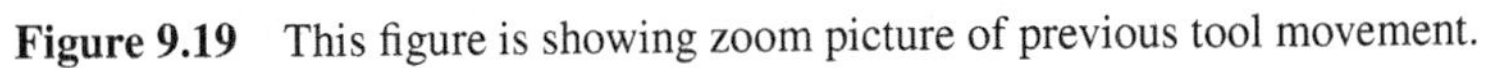
Figure 9.19 This figure is showing zoom picture of previous tool movement.

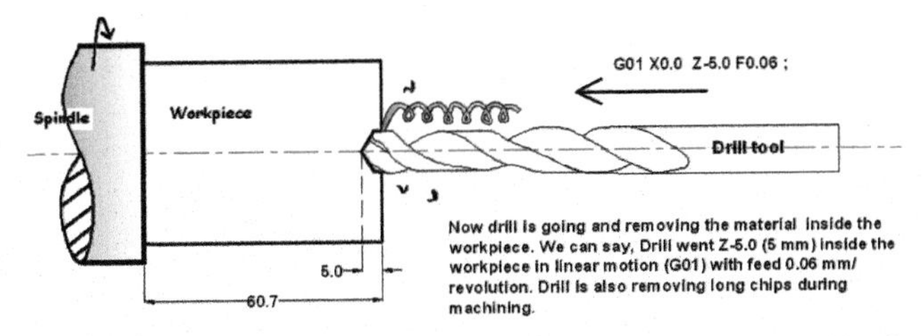

Figure 9.20 Material is removed by drill tool in linear motion with feed.

G00 X0.0 Z5.0; In Figure 9.21, after the first 5.0 mm drilling in depth, drill tool is retracting and reached at the co-ordinates X0.0 and Z1.0 in rapid motion.

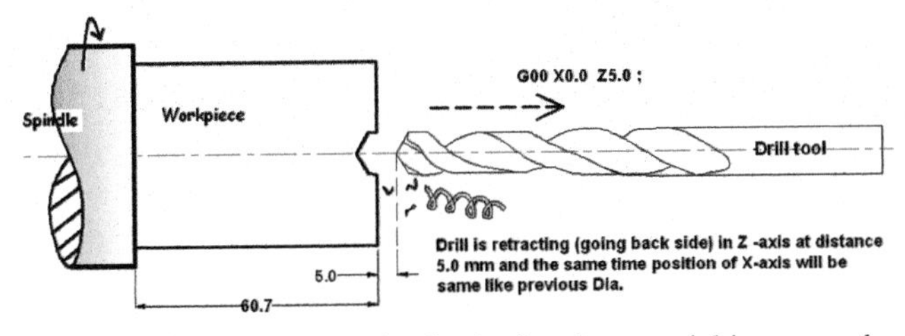

Figure 9.21 Drill is coming back after the material is removed.

Due to retraction of the drilling tool in Z-axis, coolant goes properly inside the hole and scrap (chips) does not stick with drill tool nose.

G00 X0.0 Z-4.0; In Figure 9.22, drill tool is moving fastly 4.0 mm inside the hole in rapid motion and stopped 1.0 mm before from the total drill hole depth (5 mm) which is done previously.

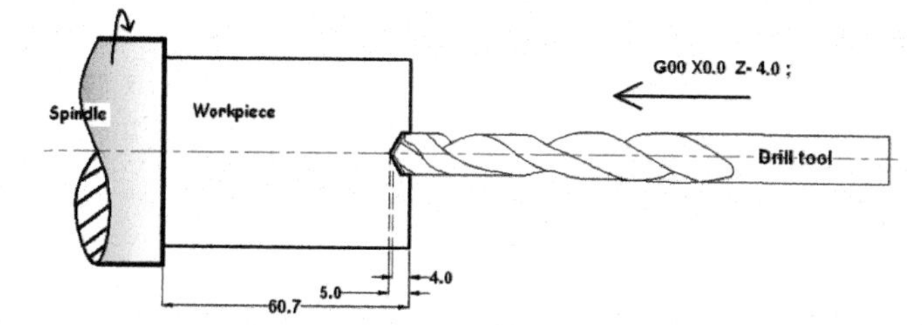

Figure 9.22 Again for next cut, drill tool is going rapidly inside the job but it will not touch the drill surface only taking position.

Note:

"It is taking position inside the hole before the next drilling operation. During this process drill tool will not touch the material inside the hole because on this depth, material already removed in previous drilling operation". The drill is just taking position before next drilling.

Before drilling operation, first we take following step (Figures 9.22 and 9.23)

- Drill tool will go rapidly in the forward direction inside the work piece because that area already drilled (Z-5.0) during the previous drilling operation.
- So this time drill will go only 4.0 mm (Z-4.0) inside the work piece rapidly and take the immediate position before drilling operation.
- But during this movement, material will not cut because it is already cut in the previous operation till Z-5.0 (5 mm) in depth.

Figure 9.23 is zoomed figure.

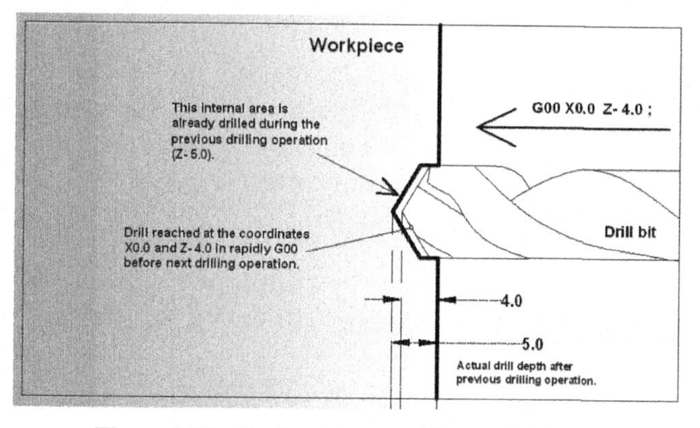

Figure 9.23 Tool movement with zoomed figure.

G01 X0.0 Z-10.0 F0.06; In Figure 9.24, now again drill is drilling (removing) the work piece in linear motion G01 with feed F 0.06 and increasing the total hole length till 10.0 mm depth.

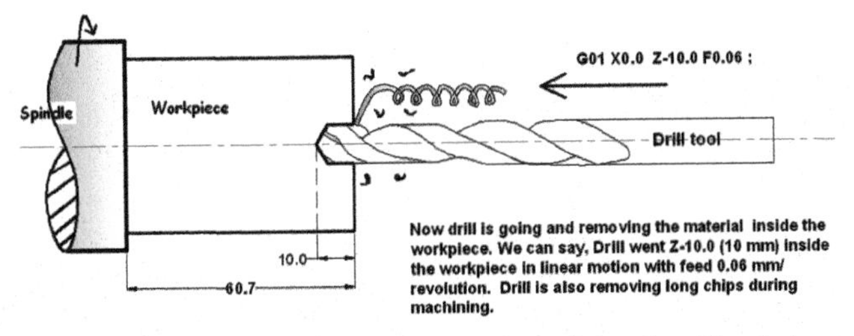

Figure 9.24 Now drill is cutting the material and making hole.

G00 X0.0 Z5.0; In Figure 9.25, after 5 mm more drilling (total 10 mm) drilling, drill tool is going back in Z-axis for proper coolant and removing the scrap inside the drill hole.

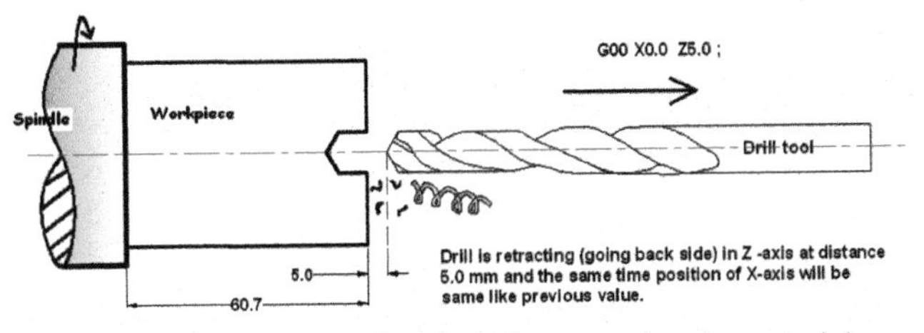

Figure 9.25 Drill is retracting after removing the material.

G00 X0.0 Z-9.0; In Figure 9.26, the drill tool is moving rapidly 9.0 mm inside the hole in rapid motion and stopped 1.0 mm before from the total drill hole depth (10 mm) which is drilled.

Before drilling operation, first we take the following step (Figure 9.26)

- Drill tool will go rapidly in the forward direction inside the work piece because that area already drilled (Z-10.0) during the previous drilling operation.
- So this time drill will go only 9.0 mm (Z-9.0) inside the work piece rapidly and take the immediate position before drilling operation.

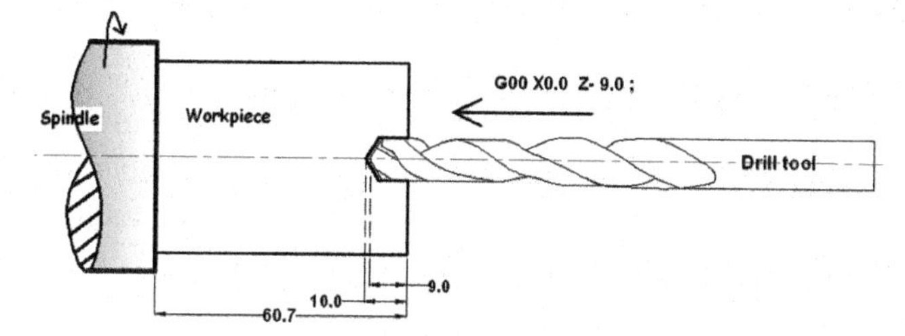

Figure 9.26 Drill tool is going rapidly inside the job but it will not touch with drilling surface.

- But during this movement, the material will not cut because it is already cut in the previous operation till Z-10.0 (10 mm) in depth.

G01 X0.0 Z-15.0 F0.06; In Figure 9.27, now again the drill is drilling (removing) the work piece in linear motion G01 with feed F 0.06 and increasing the hole length till 15.0 mm depth.

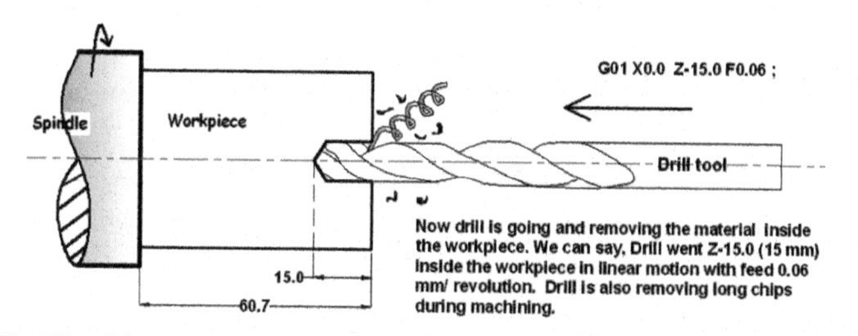

Figure 9.27 The drill is now going forward direction and removing the material with slow feed.

G00 X0.0 Z5.0; In Figure 9.28, after 5 mm more drilling (tota15 mm), drill tool is going back in Z-axis

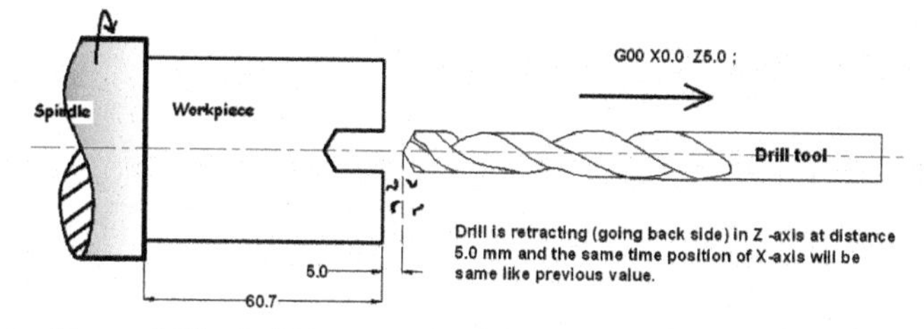

Figure 9.28 Drill is retracting again after cutting the material.

for proper coolant and removing the scrap inside the drill hole.

G00 X0.0 Z-14.0; In Figure 9.29, the drill tool is moving rapidly 14.0 mm inside the hole in rapid motion and stopped 1.0 mm before from the total drill hole depth (15 mm) which is drilled.

Before drilling operation, first we take the following step (Figure 9.29)

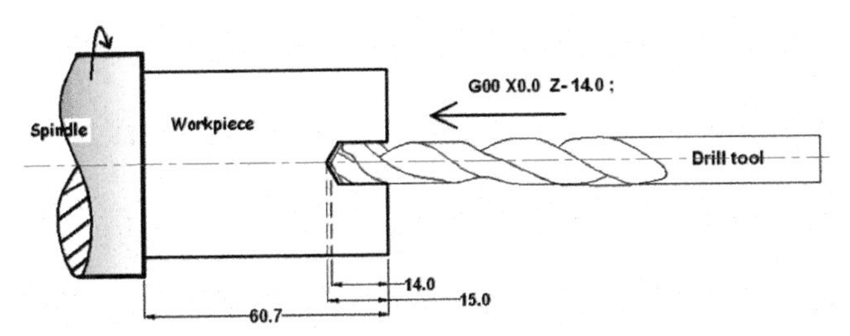

Figure 9.29 Drill is taking position inside the job for next cut.

- Drill tool will go rapidly in the forward direction inside the work piece because that area already drilled (Z-15.0) during the previous drilling operation.
- So this time drill will go only 14.0 mm (Z-14.0) inside the work piece rapidly and take the immediate position before drilling operation.
- But during this movement, the material will not cut because it is already cut in the previous operation till Z-15.0 (15 mm) in depth.

G01 X0.0 Z-20.0 F0.06; In Figure 9.30, now again the drill is drilling (removing) the work piece in linear motion G01 with feed F 0.06 and increasing the hole length till 20.0 mm depth.

G01 X0.0 Z-20.0 F0.06 ;
Spindle
Workpiece
Drill tool
Now drill is going and removing the material inside the workpiece. We can say, Drill went Z-20.0 (20 mm) inside the workpiece in linear motion with feed 0.06 mm/revolution. Drill is also removing long chips during machining.
20.0
60.7

Figure 9.30 The drill is removing material with feed.

G00 X0.0 Z5.0; In Figure 9.31, after 5 mm more (total 20.0 mm) drilling, drill tool is going back in Z-axis for proper coolant and removing the scrap inside the drill hole.

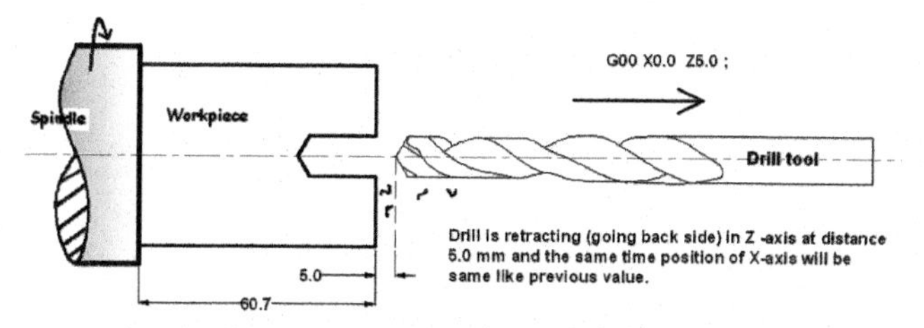

Figure 9.31 Dill is going back after removing the material.

G00 X0.0 Z-19.0; In Figure 9.32, the drill tool is moving rapidly 19.0 mm inside the hole in rapid motion and stopped 1.0 mm before from the total drill hole depth (20 mm) which is drilled.

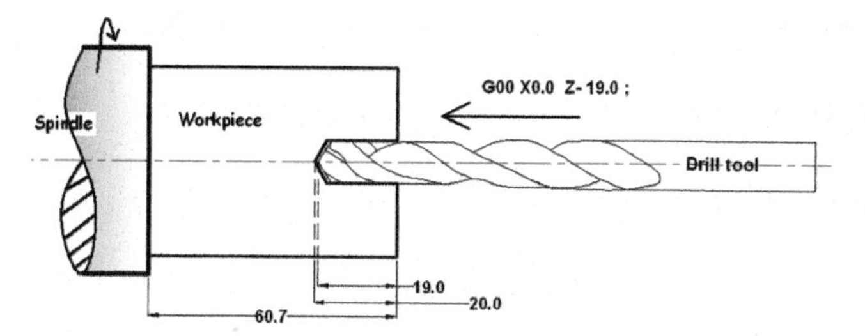

Figure 9.32 Drill is taking position inside the work piece but it will not touch the drilling face.

Before drilling operation, first we take the following step (Figure 9.32)

- Drill tool will go rapidly in the forward direction inside the work piece because that area already drilled (Z-20.0) during the previous drilling operation.
- So this time drill will go only 19.0 mm (Z-19.0) inside the work piece rapidly and take the immediate position before drilling operation.
- But during this movement, the material will not cut because it is already cut in the previous operation till Z-20.0 (20 mm) in depth.

G01 X0.0 Z-25.0 F0.06; In Figure 9.33, now again the drill is drilling (removing) the work piece in linear motion G01 with feed F 0.06 and increasing the hole length till 25.0 mm depth.

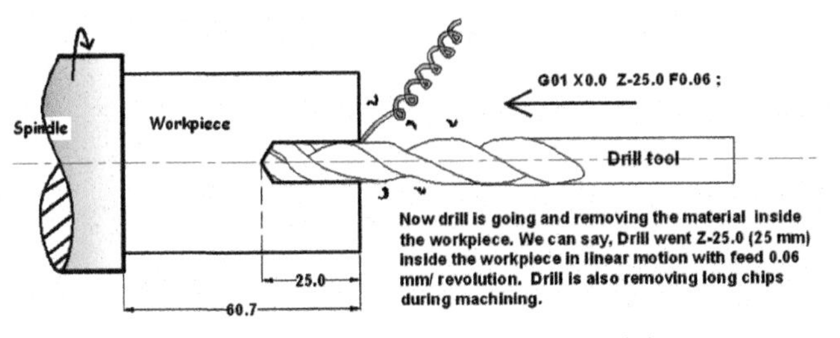

Figure 9.33 Drill is removing material.

G00 X0.0 Z5.0; In Figure 9.34, after 5 mm more (total 25.0 mm) drilling, drill tool is going back in Z–axis for proper coolant and removing the scrap inside the drill hole.

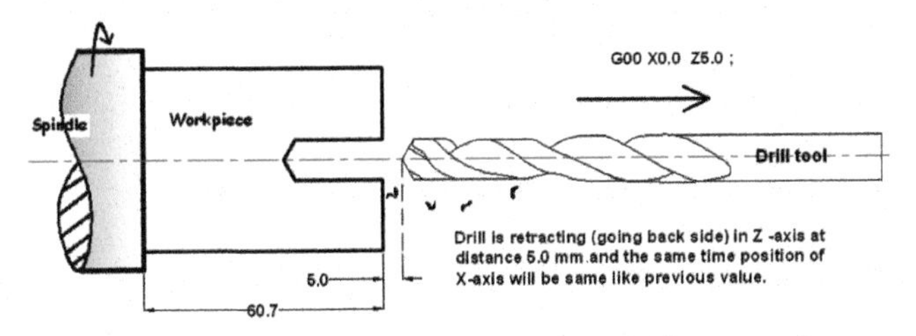

Figure 9.34 Now drill is retracting after drilling operation.

G00 X0.0 Z-24.0; In Figure 9.35, the drill tool is moving rapidly 24.0 mm inside the hole in rapid motion and stopped 1.0 mm before from the total drill hole depth (25 mm) which is drilled.

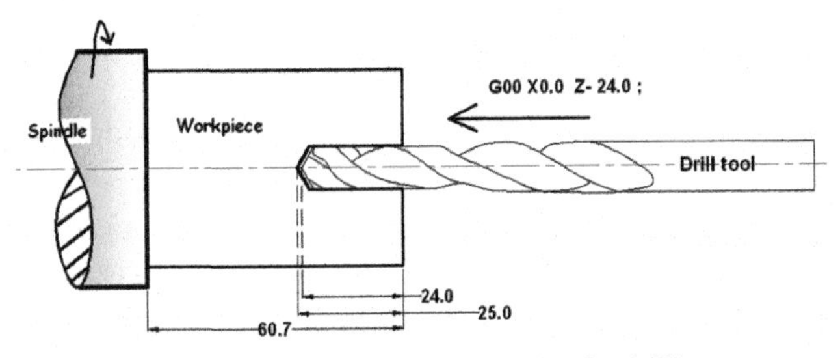

Figure 9.35 Drill is taking position for drilling.

Before drilling operation, first we take the following step (Figures 9.35 and 9.36)

- Drill tool will go rapidly in the forward direction inside the work piece because that area already drilled (Z-25.0) during the previous drilling operation.
- So this time drill will go only 24.0 mm (Z-24.0) inside the work piece rapidly and take the immediate position before drilling operation.
- But during this movement, the material will not cut because it is already cut in the previous operation till Z-25.0 (25 mm) in depth.

Figure 9.36 shows Zoomed figure.

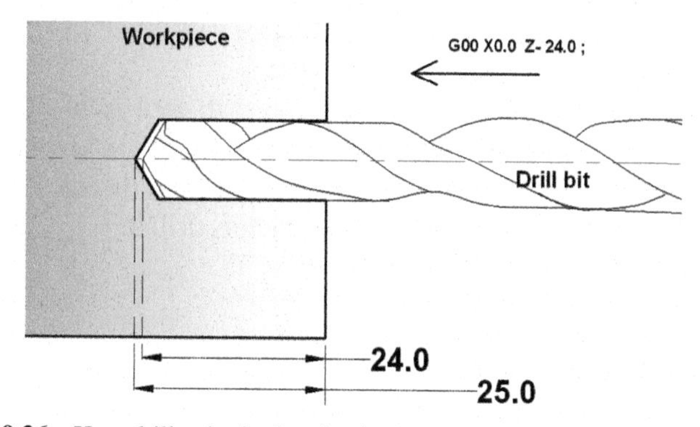

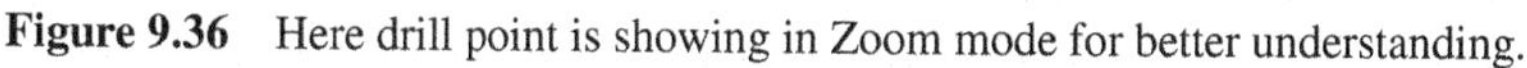
Figure 9.36 Here drill point is showing in Zoom mode for better understanding.

G01 X0.0 Z-30.0 F0.06; In Figure 9.37, now again the drill is drilling (removing) the work piece in linear motion G01 with feed F 0.06 and increasing the hole length till 30.0 mm depth.

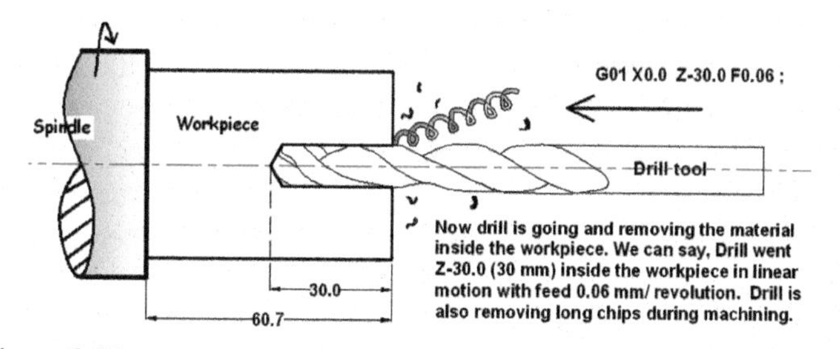

Figure 9.37 Material is removing by drill tool in linear motion with feed.

G00 X0.0 Z5.0; In Figure 9.38, after 5 mm more (total 30.0 mm) drilling, drill tool is going back in Z-axis for proper coolant and removing the scrap inside the drill hole.

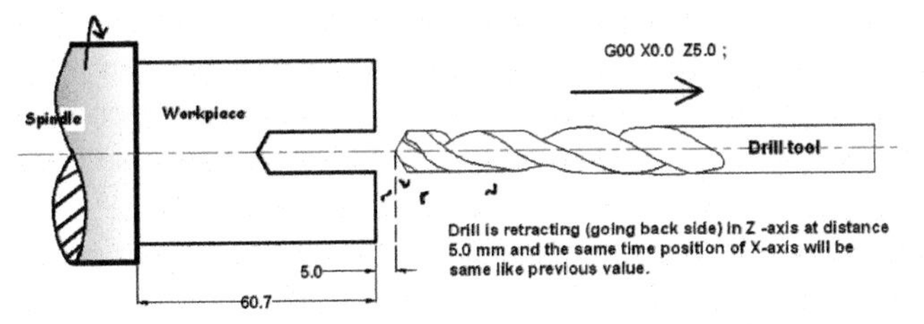

Figure 9.38 Drill is retracting after making the hole as per the CNC programming line.

G00 X0.0 Z-29.0; In Figure 9.39, the drill tool is moving fastly 29.0 mm inside the hole in rapid motion and stopped 1.0 mm before from the total drill hole depth (30 mm) which is drilled.

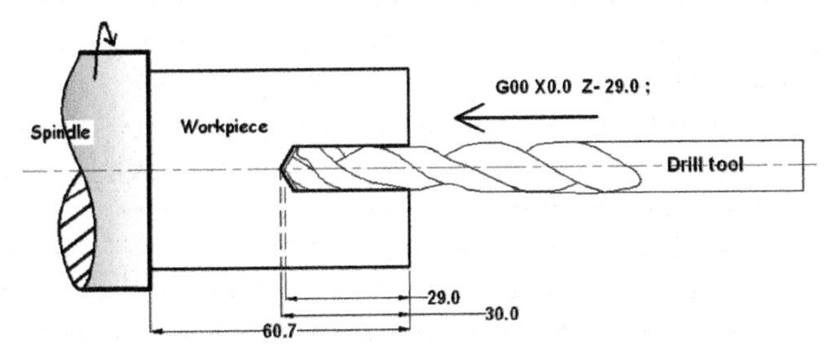

Figure 9.39 Drill is taking position for making the drill hole.

Before drilling operation, first we take the following step (Figure 9.39)

- Drill tool will go rapidly in the forward direction inside the work piece because that area already drilled (Z-30.0) during the previous drilling operation.
- So this time drill will go only 29.0 mm (Z-29.0) inside the work piece rapidly and take the immediate position before drilling operation.
- But during this movement, the material will not cut because it is already cut in the previous operation till Z-30.0 (30 mm) in depth.

G01 X0.0 Z-32.0 F0.06; In Figure 9.40, now again the drill is drilling (removing) the work piece in linear motion G01

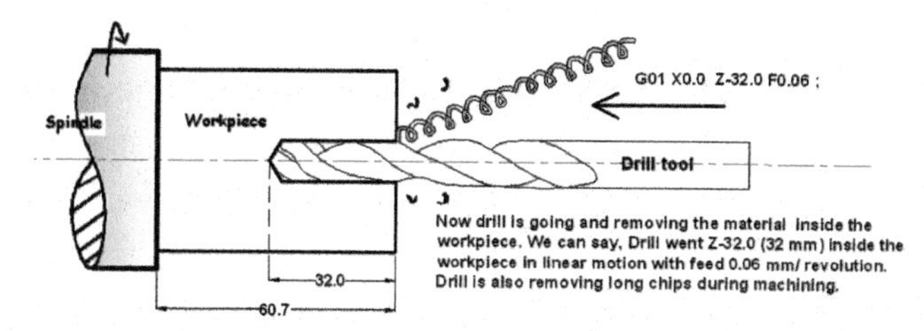

Figure 9.40 Drill tool is removing the material at the length Z-32.

with feed F 0.06 and increasing the hole length till 32.0 mm depth.

G00 X0.0 Z5.0; In Figure 9.41, finally drill tool is retracting from the work piece in Z-axis direction after complete drilling operation.

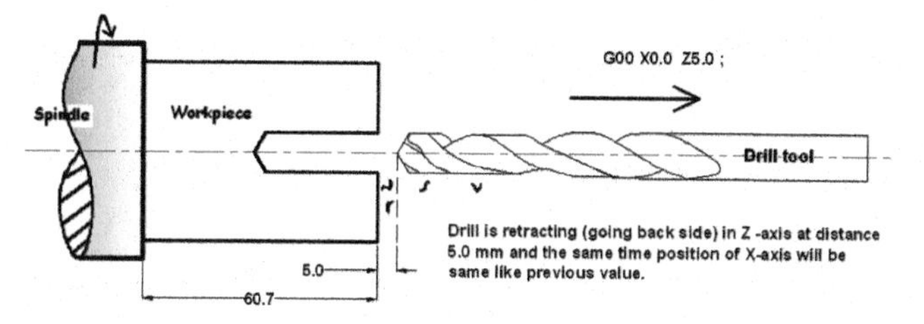

Figure 9.41 Drill is coming back after material cutting at length Z 5.

M09; After tool retracting, coolant motor will stop.

M05; When M05 will execute. Spindle motor will stop.

G00 X100.0 Z100.0; In Figure 9.42, the cutting tool is taking the safe position after drilling operation in X-axis and

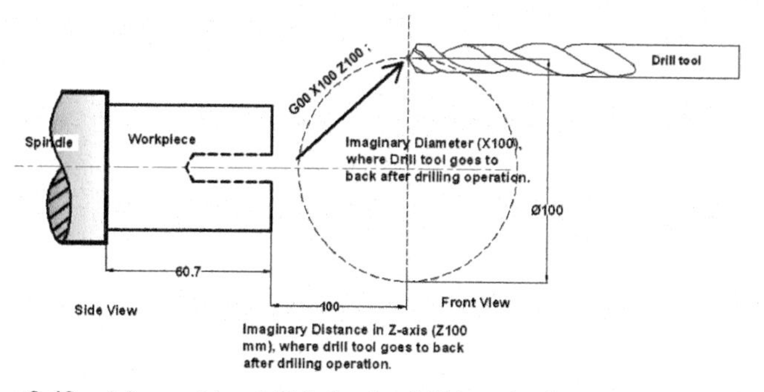

Figure 9.42 After making drill hole, the drill is going back at his initial point.

Z-axis direction. It is imaginary coordinates of the cutting tool, where cutting tool stays.

M30; After executing of this programming block, CNC program will stop and reset. Now cursor will reach on the first programming line.

9.1.2 Step Turning Operation with CNC Programming

9.1.2.1 What is the rough cutting during machining operation?

Before the machining operation, if we found more extra raw material on the desired surface as per drawing. Then first we cut (remove) the extra raw material in several steps from the work piece surface after that we take final cut on the work piece surface as per drawing. Those several cuts are called rough cutting during the machining operation.

The rough cutting operation performed on the desired work piece as per drawing. In the given below figures, the cutting tool will cut the material step by step as per drawing. First of all, the cutting tool will start the machining, step by step from bigger diameter (example: Ø32) to smaller diameter (example: Ø22). It means, first, bigger diameter will machine than next diameter will machine back to back again next diameter will machine this process will run continue until cutting tool will not achieve the last diameter (smaller diameter, example: Ø22) as per drawing.

If any step diameter/profile diameter has more extra material (an extra material more than the depth of cut of cutting tool) then cutting tool will machine that extra material in more than one cut, it depends on the extra raw material (material stock).

But the most important thing is during rough machining of every step, cutting tool will leave very less material on each final dimensional surface for finishing cut. So that after finishing cut work piece gets required surface finish as per recommended drawing.

9.1.2.2 Machining example of rough cutting

Figure 9.43 shows Machining drawing.

Figure 9.44 shows Raw material drawing.

In Figure 9.45, given points (1 to 9) are coordinates points. Cutting tool will follow these coordinates points and remove the raw material from the round shaft as per given drawing during the machining process.

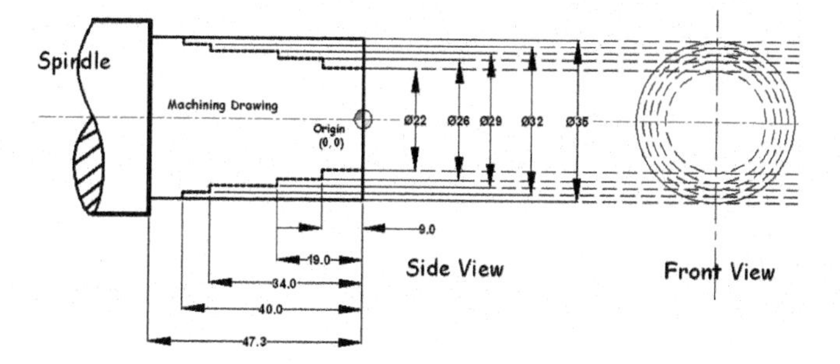

Figure 9.43 Machining drawing.

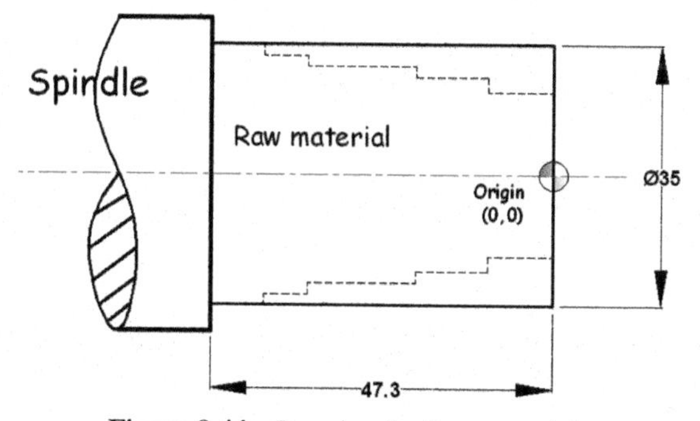

Figure 9.44 Drawing for Raw material.

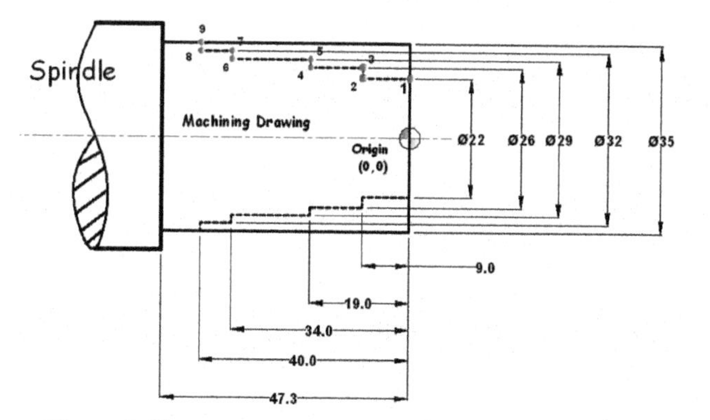

Figure 9.45 Machining drawing with coordinates points.

Coordinates of the above given points.

	(X, Z)		(X, Z)
1.	(22, 0)	6.	(29, −34)
2.	(22, −9.0)	7.	(32, −34)
3.	(26, −9.0)	8.	(32, −40)
4.	(26, −19)	9.	(35, −40)
5.	(29, −19)		

CNC program for above drawing

O10042; This is called program number. In this, the first letter is **O** and another five are numbers (digits) and last sign (;) is called EOB (end of the block).

G54; This is called work zero. When we take **G54** command on the center of the face of the round work piece, that time coordinates of the center of that cylindrical job are (0, 0). It means all cutting tool will assume that point as an origin (0, 0). and important think is when we make the CNC program in absolute mode G90. Origin (0,0) will take as a reference point for all coordinates in the CNC program.

G50 S2500; G50 command controls the maximum spindle speed. In this program, spindle speed will not exceed more than 2500 rpm.

T0101; This programming line is saying to machine that turret (tool post) will change the cutting tool and take the tool number one (1) for machining. In this line, First two digits of T**01**01 are representing the tool station number, where from tool is coming for machining and second two digits of T01**01** are representing the tool offset, tool offset means location number of tool data, where the tool data (tool geometry) is stored, finally we can say tool offset is the brain of cutting tool.

G00 X70 Z50; G00 command is used for rapid motion (non-cutting motion). Here Cutting tool T0101 will take position rapidly in X-axis at diameter 70 mm and Z-axis at distance 50 mm from the origin (0, 0).

The tool will use these coordinates to come near the job for cutting. See Figure 9.46.

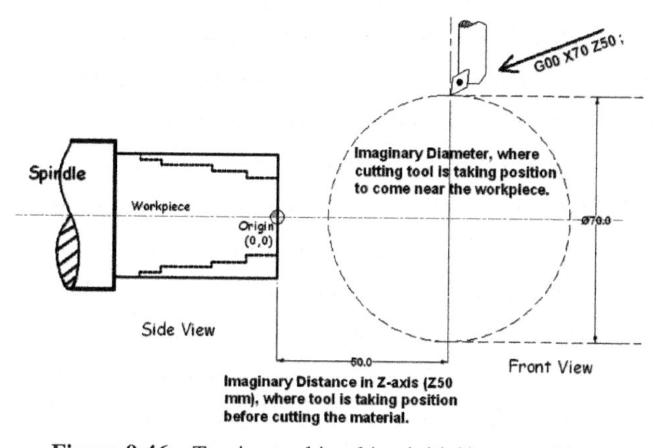

Figure 9.46 Turning tool is taking initial/start position.

G96 M03 S200; G96 command is used for constant surface speed on, it controls the spindle rpm (rotating of work piece), when the cutting tool moves on different diameters. According to this programming line, the spindle will rotate in clockwise direction (M03) and S200 is minimum rpm. It means when the cutting tool is reached on maximum diameter that time spindle speed will be 200 rpm (revolution per minute).

On the other hand, we can say **rpm is inversely proportional to work piece diameter in CNC turning machine**. It means when the tool goes towards the big diameter that time spindle rpm will decrease and when the tool goes towards the small diameter that time spindle rpm will increase. But this minimum and maximum rpm will depend as per given rpm in G96 and G50 in the CNC program.

G00 X32 Z1.0; In Figure 9.47, before cutting the material cutting tool is taking position nearest the job, together in X-axis and Z-axis direction at Ø32 mm and length 1.0 mm from the work zero (origin) in rapid motion (G00).

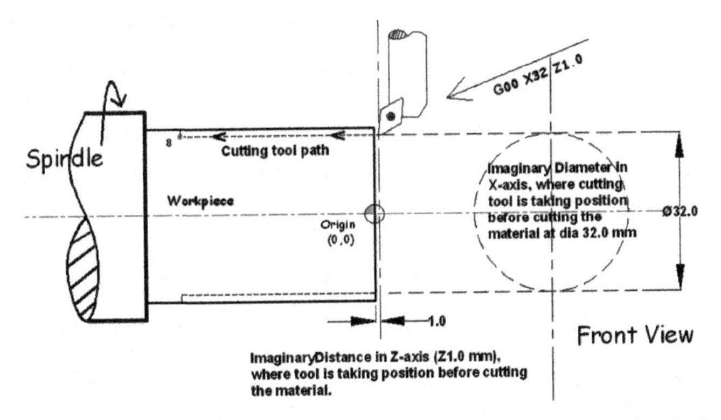

Figure 9.47 Before cutting the material turning tool came nearest the work piece.

M08; Coolant motor starts before cutting the material.

G01 X32 Z-40 F0.15; In Figure 9.48, now the tool is going and removing the material in Z-axis direction (Z-40) in linear motion (G01) with feed F 0.15 at the diameter 32.0 (X32.0).

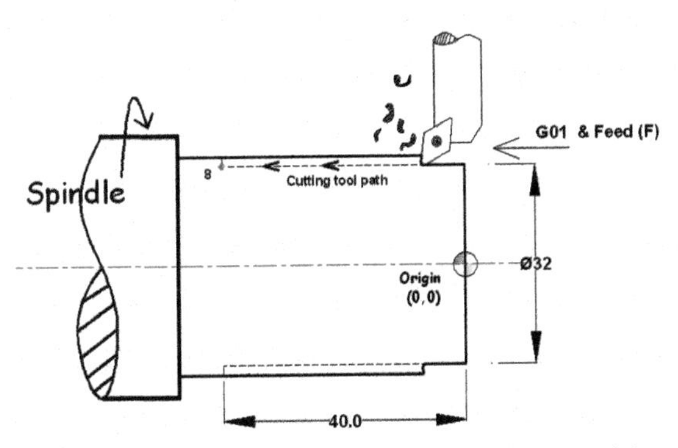

Figure 9.48 Now turning tool is cutting the material.

In Figure 9.49, the cutting tool is continuously going and removing the material in Z-axis direction (Z-40) at the previous diameter.

In Figure 9.50, cutting tool reached at the coordinates X32 and Z-40 in linear motion (G01) with feed 0.15. It means first, step turning operation completed by the tool.

G01 X36 Z-40 F0.15; In Figure 9.51, after straight turning operation. Now tool will make the step during material removing and also retracting in X-axis direction.

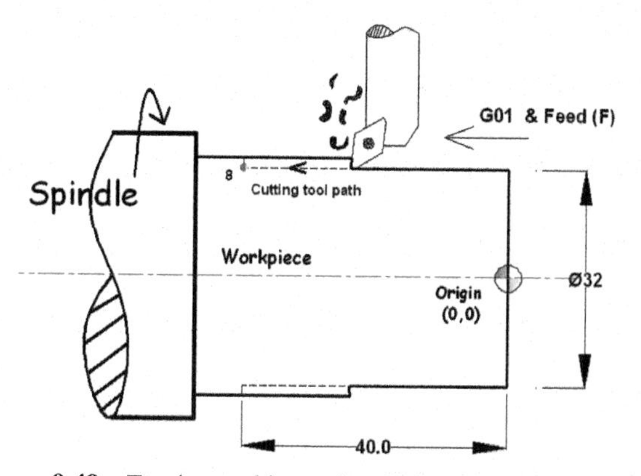

Figure 9.49 Turning tool is continuously cutting the material.

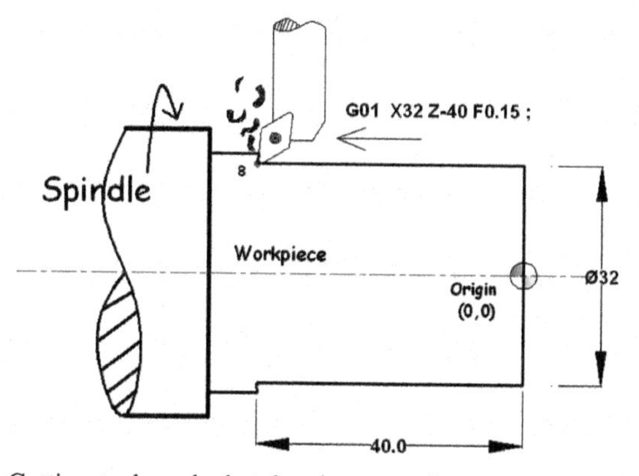

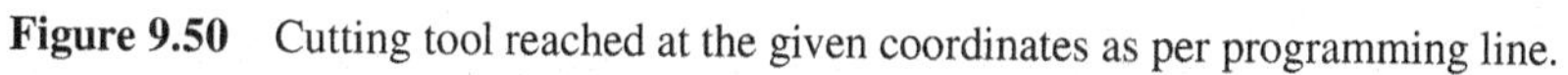

Figure 9.50 Cutting tool reached at the given coordinates as per programming line.

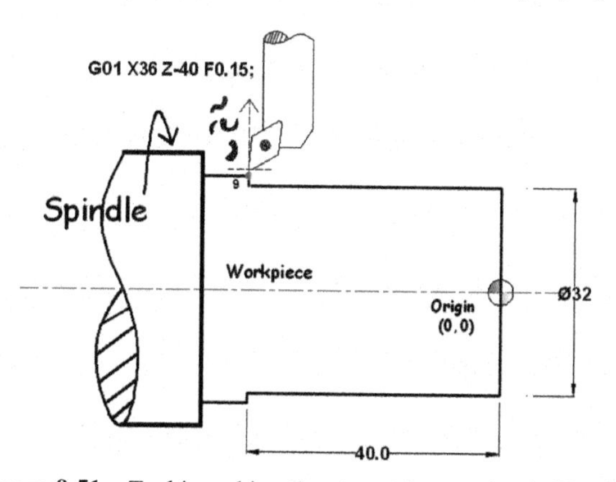

Figure 9.51 Tool is making the step and retracting in X-axis.

Tool will take safe positions at diameter X34 in linear motion (G01) during retraction.

G00 X36 Z1.0; In Figure 9.52, the cutting tool is retracting rapidly in Z-axis the direction at distance Z1.0 mm from the work zero (0, 0) at the same time tool position in X-axis will be same as previous.

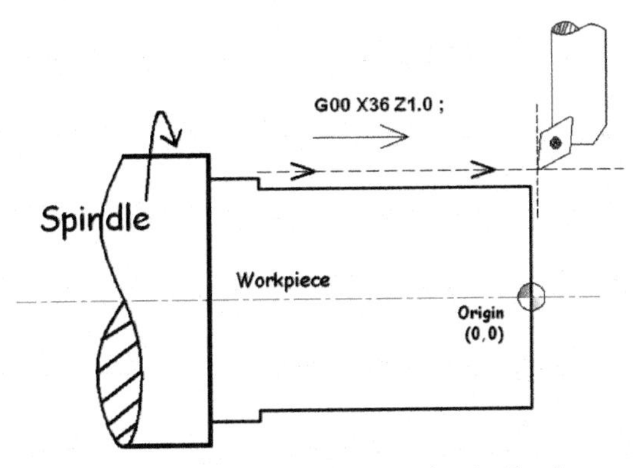

Figure 9.52 Now tool is retracting in Z-axis.

G00 X29 Z1.0; In Figure 9.53, before cutting the material in the next step turning the operation, the tool is **taking position** in X-axis direction (X29) in rapid motion (G00) and this time Z-axis position will be same as previous.

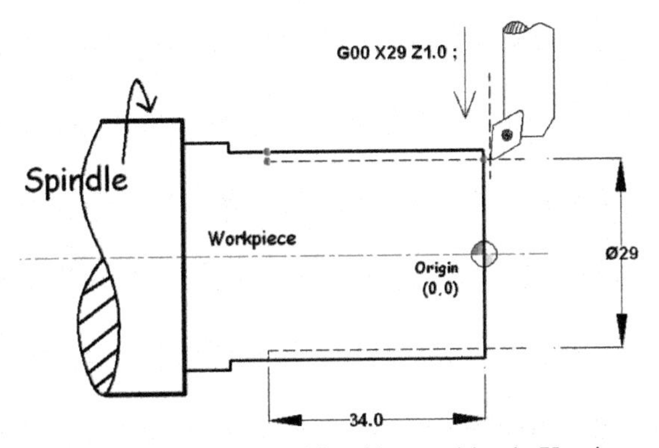

Figure 9.53 Again tool is taking position in X-axis.

G01 X29 Z-34 F0.15; In Figure 9.54, now Cutting tool is continuously going and removing the material in Z-axis (Z-34) direction in linear motion (G01) with feed (f) 0.15 at same previous diameter 29.0 mm.

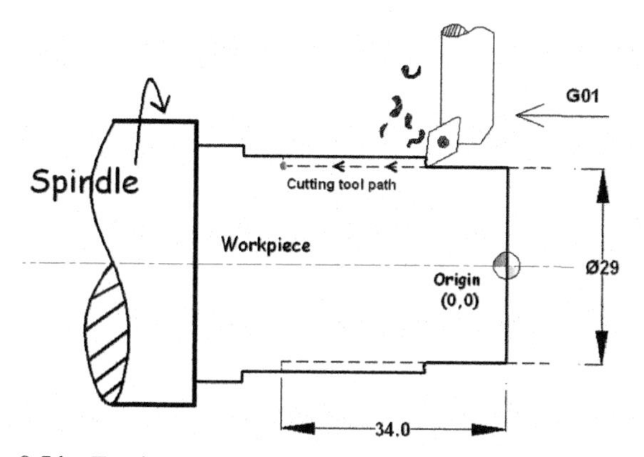

Figure 9.54 Turning tool is removing material in the continuous mode.

In Figure 9.55, now cutting tool reached at the coordinates X-29 and Z-34 as per drawing in linear motion (G01) with feed 0.15.

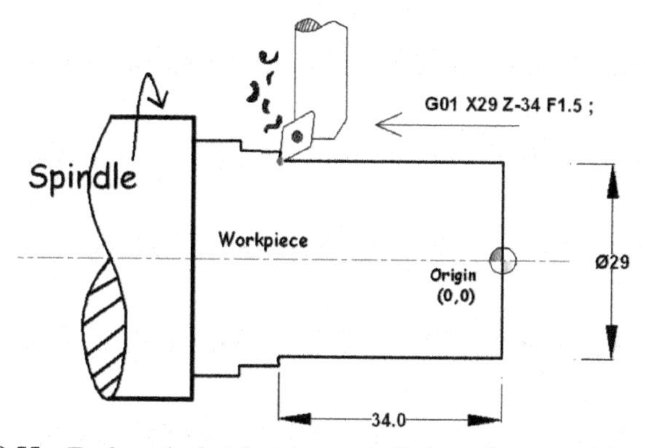

Figure 9.55 Tool reached at the given coordinates after material removing.

G01 X33 Z-34 F0.15; In Figure 9.56, after straight turning operation. Now tool will make the step during material removing and also retracting in X-axis direction. The tool will take safe position at diameter X33 in linear motion (G01) during retraction.

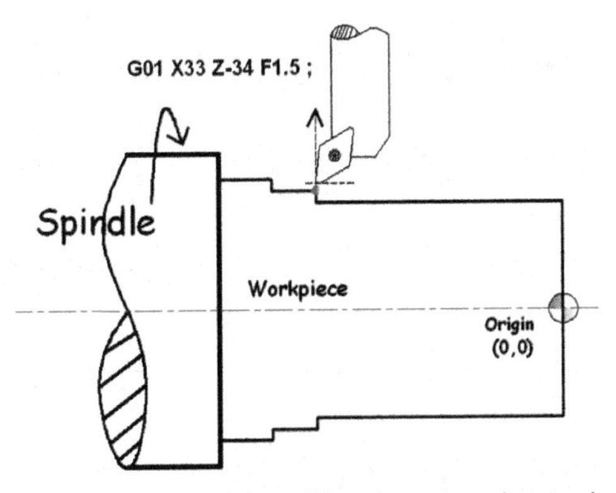

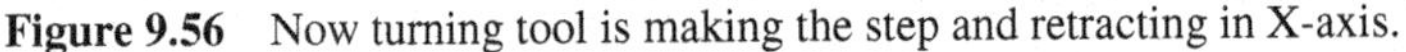

Figure 9.56 Now turning tool is making the step and retracting in X-axis.

G00 X33 Z1.0; In Figure 9.57, after retracting in X-axis, now toll will retract in Z-axis (Z1.0 mm) direction in rapid motion (G00) during this, tool position in X-axis will be the same like the previous position.

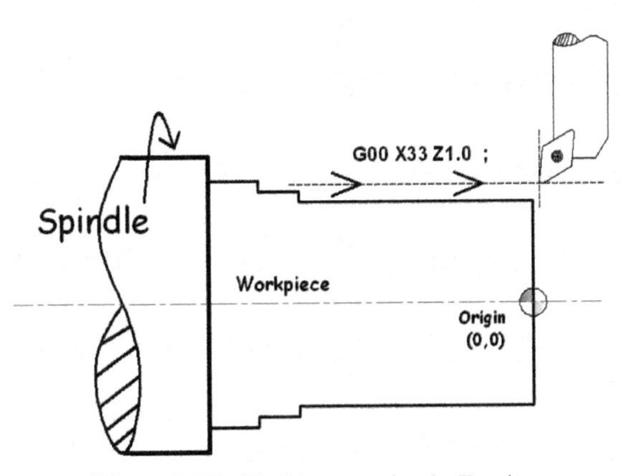

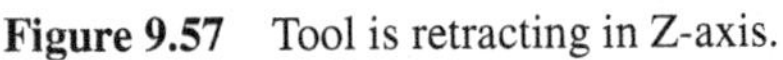

Figure 9.57 Tool is retracting in Z-axis.

G00 X26 Z1.0; In Figure 9.58, before cutting the material in the next step turning operation, the tool is **taking position** in X-axis direction (X26) in rapid motion (G00) and this time Z-axis position will be same as previous.

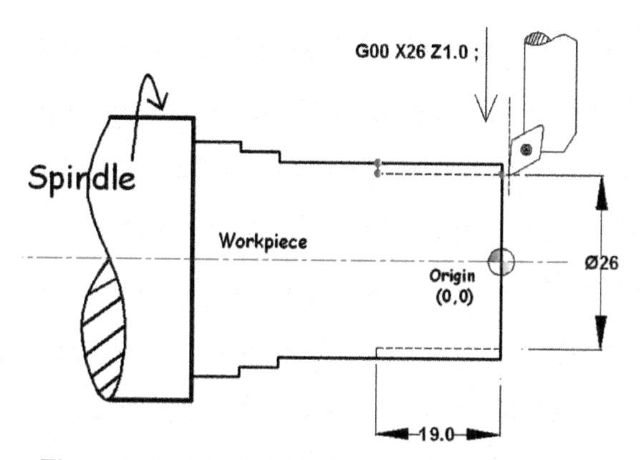

Figure 9.58 Tool is taking new position for cutting.

G01 X26 Z-19.0 F0.15; In Figure 9.59, now cutting tool is continuously going and removing the material in Z-axis (Z-19) direction in linear motion (G01) with feed (f) 0.15 at same previous diameter 26.0 mm.

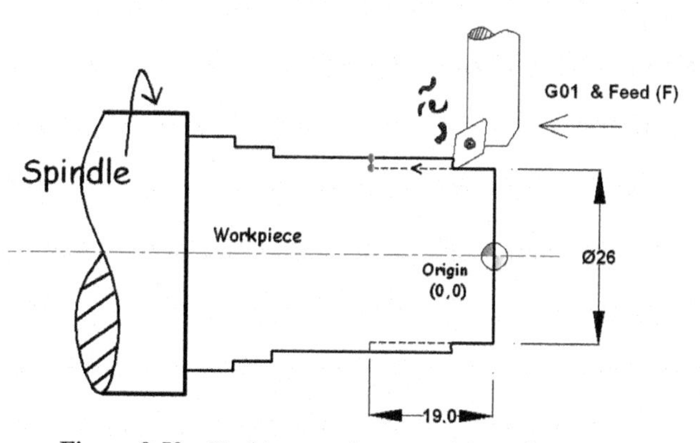

Figure 9.59 Tool is removing material continuously.

In Figure 9.60, cutting tool removed the material and reached at the coordinates X26 and Z-19 in linear motion (G01) with feed 0.15 mm/revolution.

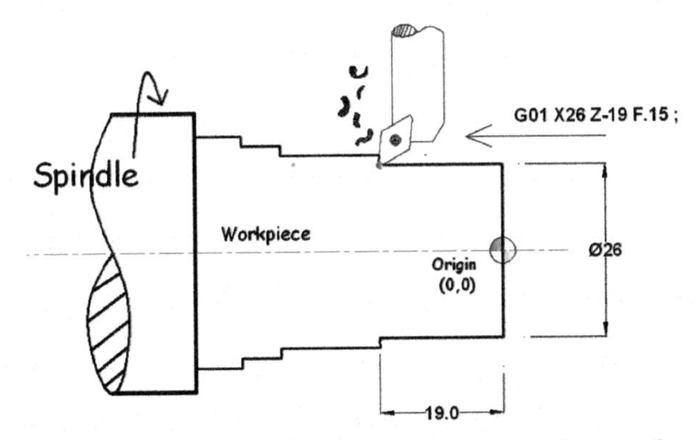

Figure 9.60 Cutting tool reached at the given programming distance after removing the material.

G01 X30 Z-19.0 F0.15; In Figure 9.61, after straight turning operation. Now tool will make the step during material removing and also retracting in X-axis direction. The tool will take safe position at diameter X30 in linear motion (G01) during retraction.

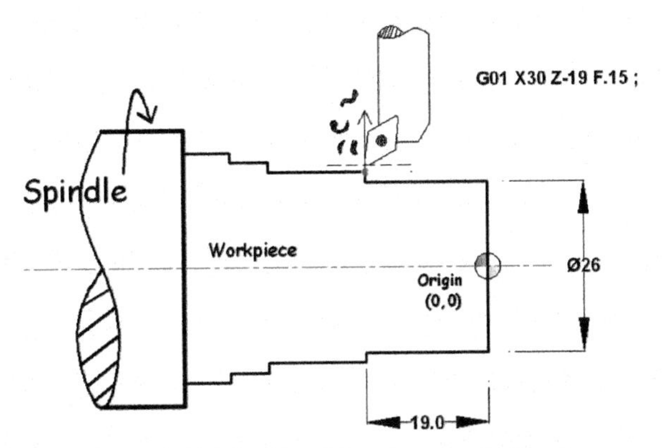

Figure 9.61 Tool is taking clearance in X-axis.

G00 X30 Z1.0; In Figure 9.62, cutting tool is retracting in Z-axis direction at distance Z1.0 mm at same previous diameter X30.0 mm.

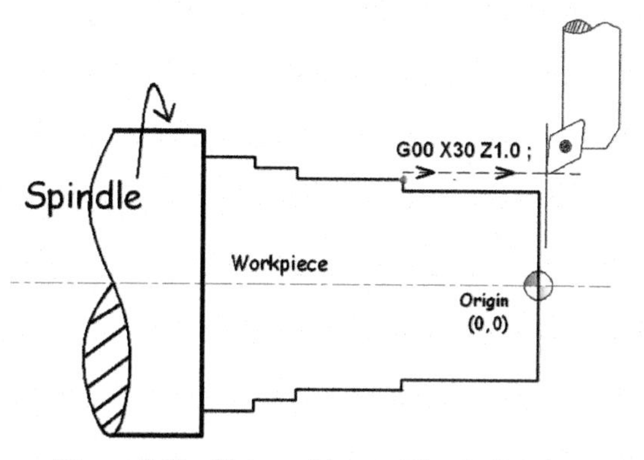

Figure 9.62 Now tool is retracting in Z-axis.

G00 X22 Z1.0; In Figure 9.63, the cutting tool is taking position for the last step turning operation in X-axis direction at diameter 22.0 mm at the same time position of the cutting tool in Z-axis will be same as the previous position.

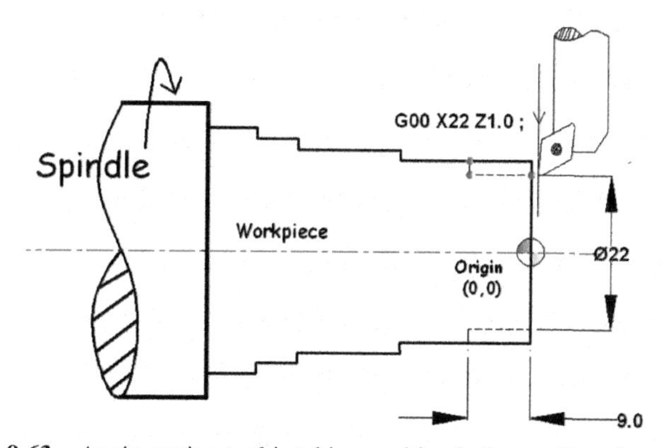

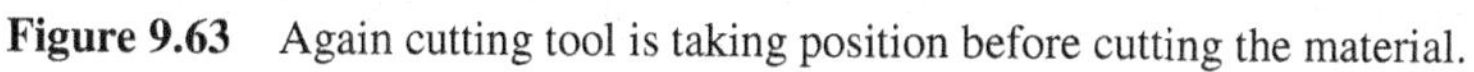

Figure 9.63 Again cutting tool is taking position before cutting the material.

G01 X22 Z-9.0 F0.15; In Figure 9.64, cutting tool starts the cut in Z-axis direction in linear motion with feed F0.15 at diameter X22.

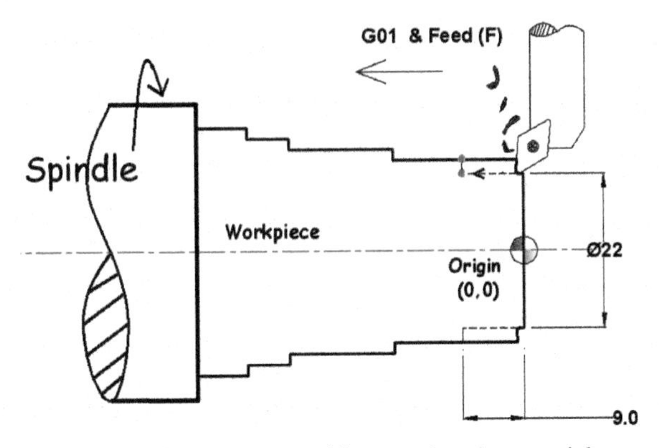

Figure 9.64 Cutting tool is removing the material.

In Figure 9.65, the cutting tool is continuously removing the material in Z-axis direction in linear motion with feed F0.15 at diameter X22.

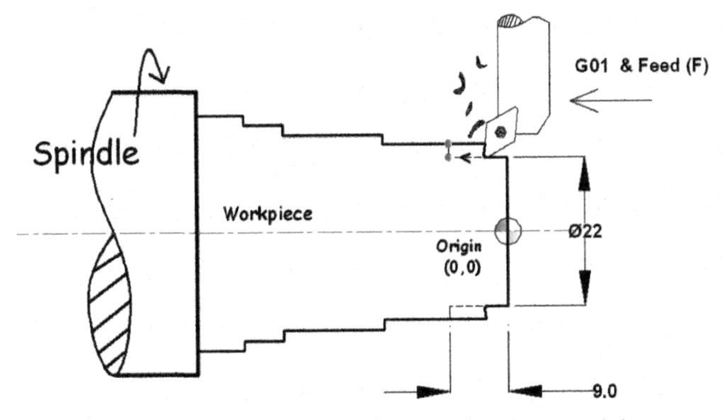

Figure 9.65 Cutting tool is removing the material.

In Figure 9.66, cutting tool removed the material and reached at the coordinates X22 and Z-9.0 in linear motion with feed F 0.15.

G01 X27 Z-9.0 F0.15; In Figure 9.67, After straight turning operation. Now tool will make the step during material removing and also retracting in X-axis direction. The tool will take safe position at diameter X27 in linear motion (G01) during retraction.

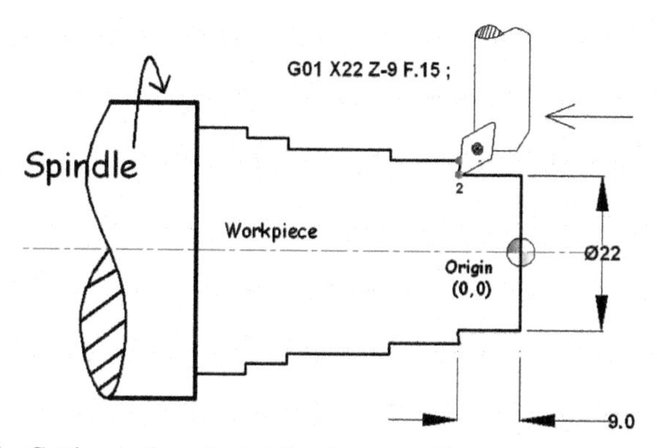

Figure 9.66 Cutting tool reached at the given coordinates as per programming line.

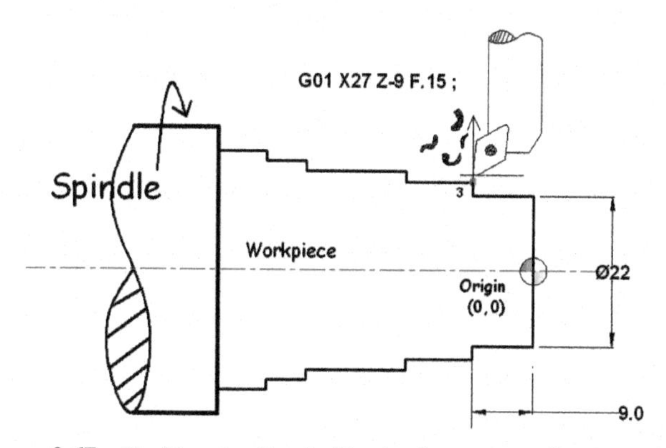

Figure 9.67 Tool is retracting in X-axis after step turning operation.

G00 X27 Z1.0; In Figure 9.68, Tool is retracting rapidly in Z-axis direction at Z1.0 mm, at the same time position of the cutting tool in X-axis same like the previous diameter.

M09; After complete machining of step turning operation, coolant motor will stop.

M05; And now spindle motor will stop.

G00 X70 Z50; In Figure 9.69, now tool is retracting in X and Z-axis in rapid motion and took safe distance from the work piece.

M30; M30 command is used for program stop and reset. After complete machining of the work piece CNC program will be stop and reset.

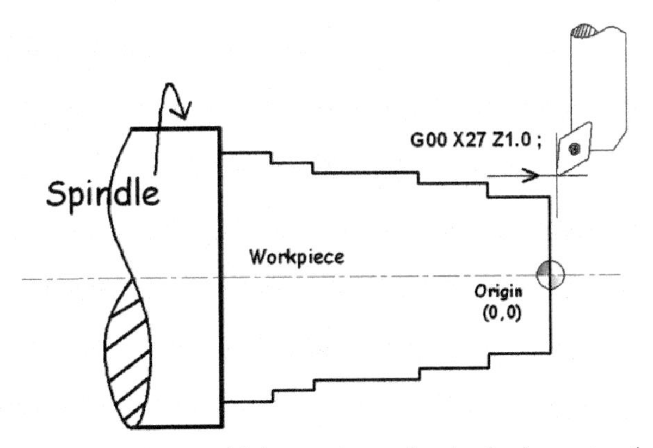

Figure 9.68 After complete machining cutting tool going back or retracting in Z-axis.

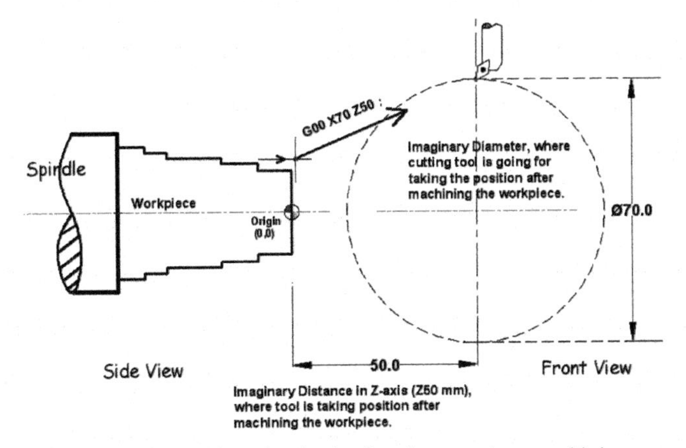

Figure 9.69 Turning tool is going back after the complete machining operation.

9.1.2.3 Graph of the cutting tool movement during rough cutting operation

In Figure 9.70, two lines are showing continuous lines and dotted lines, both are the path of the cutting tool. Where the cutting tool moves.

I. **Dotted Line** – These dotted lines with arrow indicate the rapid motion (G00) of the cutting tool. It is non-cutting motion of the tool.

II. **Continuous Line** – This continuous line with an arrow indicates the material cutting path of the tool. Whenever tool follow this path, the tool will cut the material very slow in linear motion (G01) with feed.

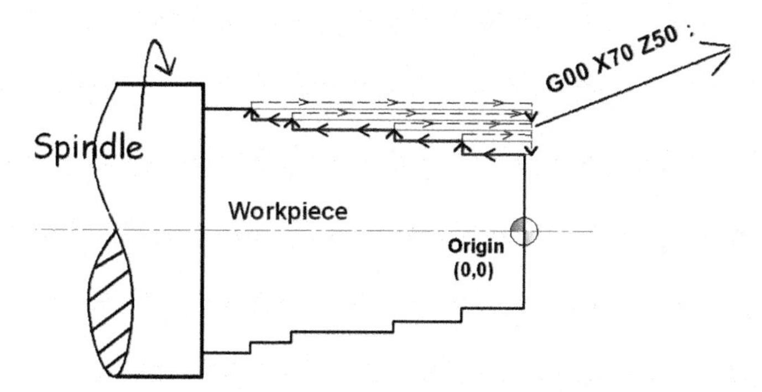

Figure 9.70 Above figure is showing tool cutting path of the cutting tool.

9.1.3 Facing Operation in Step by Step with CNC Programming

Figure 9.71 is a raw material drawing.

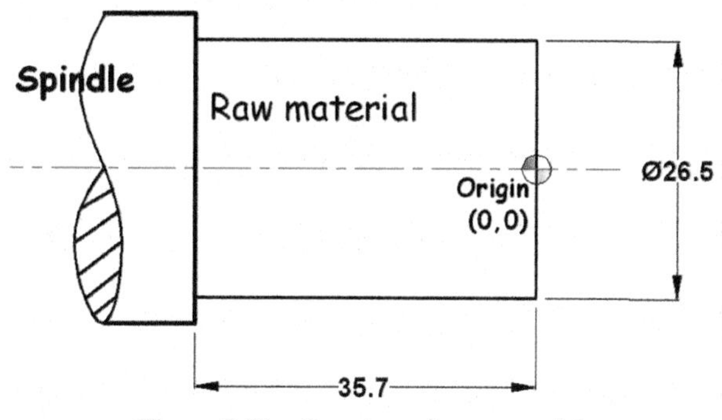

Figure 9.71 Drawing of raw material.

Figure 9.72 shows machining drawing.

O1003; This is a first line of the CNC program. It is called the program number.

G54; This is called work zero. When we take **G54** command on the center of the face of the round work piece, that time coordinates of the center of that cylindrical job are (0, 0). It means all cutting tool will assume that point as an origin (0, 0). And important thing is when we make the CNC program in absolute mode G90. Origin (0,0) will take as a reference point for all coordinates in the CNC program.

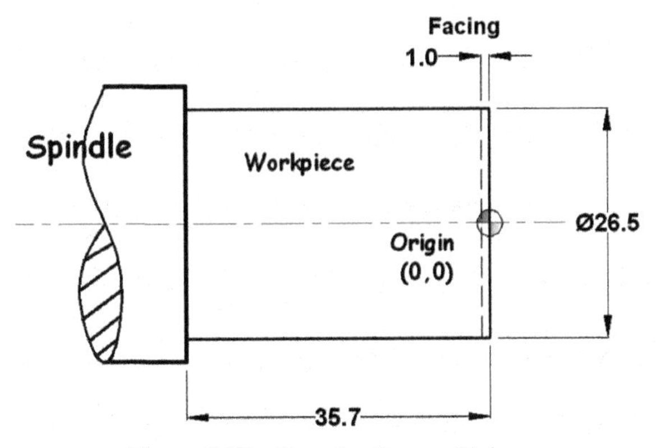

Figure 9.72 Drawing for machining.

G50 S1800; G50 command controls the maximum spindle speed. In this program, spindle speed will not exceed more than 1800 rpm.

T0303; This programming line is saying to machine that turret (tool post) will change the cutting tool and take the tool number one (03) for machining. In this line, First two digits of T**03**03 are representing the tool station number, where from tool is coming for machining and second two digits of T03**03** are representing the tool offset, tool offset means location number of tool data, where the tool data (tool geometry) is stored, finally we can say tool offset is the brain of cutting tool.

G00 X90 Z100; G00 command is used for rapid motion (non-cutting motion). Here Cutting tool T0303 will take position rapidly in X-axis at diameter 90 mm and Z-axis at distance 100 mm from the origin (0, 0). The tool will use these coordinates to come near the job for cutting. See Figure 9.73.

G96 M03 S180; G96 command is used for constant surface speed on, it controls the spindle rpm (rotating of work piece), when the cutting tool moves on different diameters. According to this programming

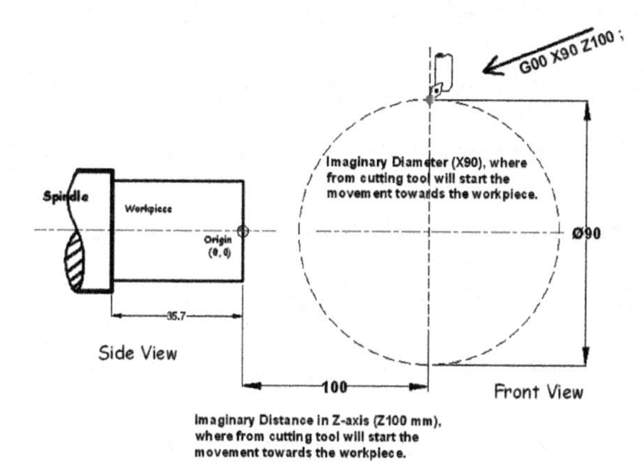

Figure 9.73 Turning tool is taking initial position.

line, the spindle will rotate in clockwise direction (M03) and S180 is minimum rpm. It means when the cutting tool is reached on maximum diameter that time spindle speed will be 180 rpm (revolution per minute).

On the other hand, we can say **rpm is inversely proportional to work piece diameter in CNC turning machine**. It means when the tool goes towards the big diameter that time spindle rpm will decrease and when the tool goes towards the small diameter that time spindle rpm will increase. But this minimum and maximum rpm will depend as per given rpm in G96 and G50 in the CNC program. See Figure 9.74.

G00 X30 Z1.0; In Figure 9.75, the tool is taking position nearest the job, together in X-axis and Z-axis direction at Ø30 mm and length 1.0 mm from the work zero (origin) in rapid motion (G00).

M08; Coolant motor start before cutting the material.

G00 X30 Z-1.0; In Figure 9.76, before cutting the material, the cutting tool is taking the position above the work piece surface at diameter 30 mm at the same time tool takes the position

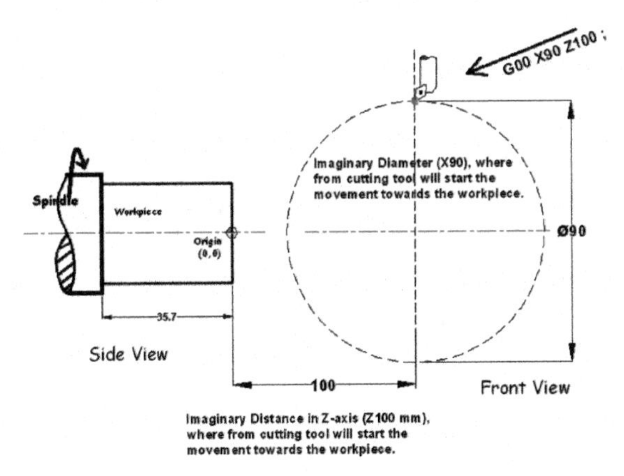

Figure 9.74 Now spindle is rotating.

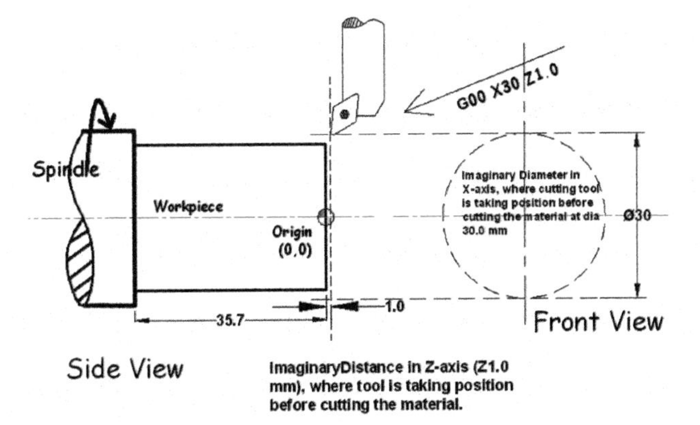

Figure 9.75 Turning tool came near over the work piece.

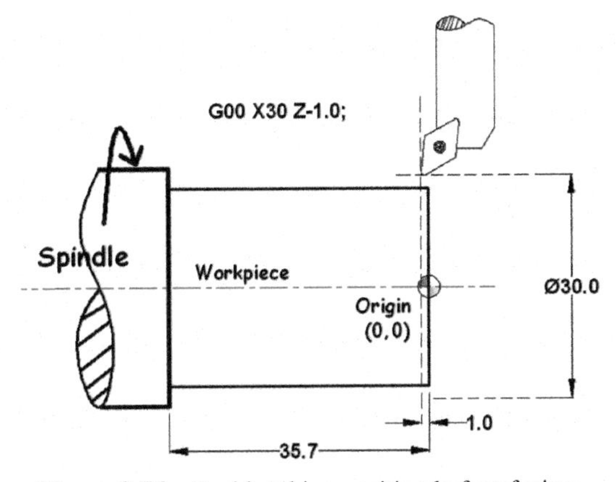

Figure 9.76 Tool is taking position before facing.

inside the work piece face in Z– (negative) axis but the tool will not touch with the work piece surface.

G01 X0.0 Z-1.0 F0.1; In Figure 9.77, cutting tool is removing the material from the front face of the work piece in linear motion G01with feed (F 0.10) and the tool is going towards the X-axis direction at diameter X0.0 (work piece center).

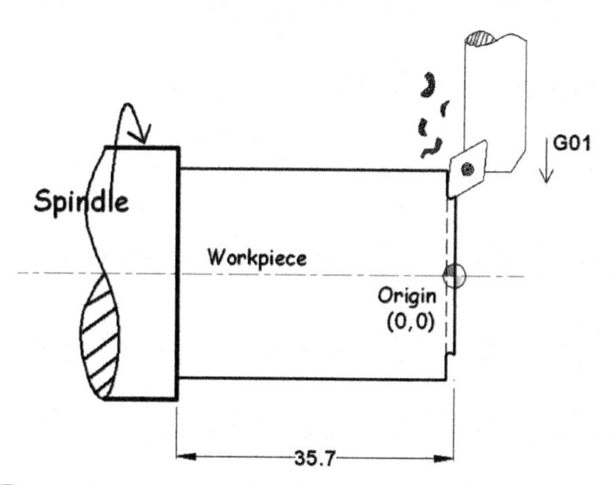

Figure 9.77 Cutting tool is going towards the center and removing the material.

In Figure 9.78, the tool is continuously cutting the material towards the center of the work piece.

In Figure 9.79, cutting tool removed the material and reached at the work piece center. After removing the material (1.0 mm) from the work piece face, now length will remain 34.7 mm.

G00 X0 Z1.0; In Figure 9.80, after facing operation tool will retract from the the face of the work piece in rapid motion G00 in Z -axis (Z1.0) positive direction.

M09; After retracting in Z-axis, coolant motor will stop (off).

M05; After retracting in Z-axis spindle motor will stop. It means spindle will not rotate.

G00 X90 Z100; Now, finally the tool will retract and reach at coordinates X90 Z100 in rapid motion. We can say this position is also the starting position of the tool (before moving) and

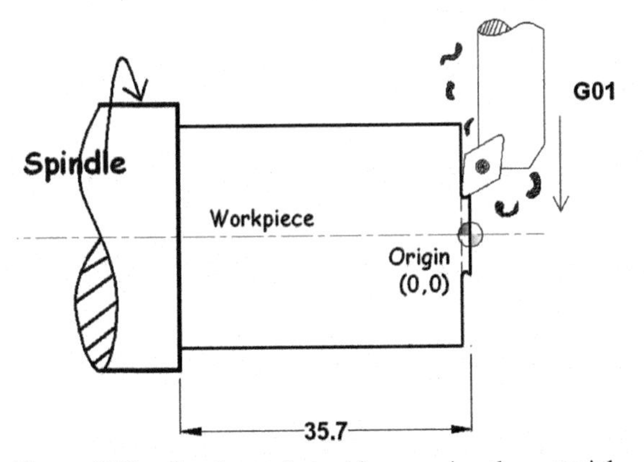

Figure 9.78 Continuously tool is removing the material.

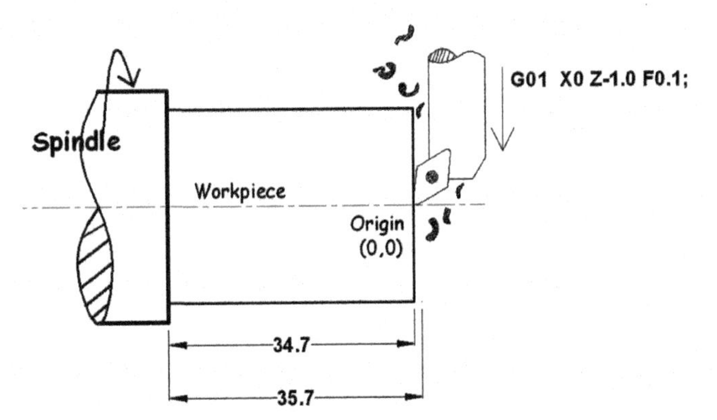

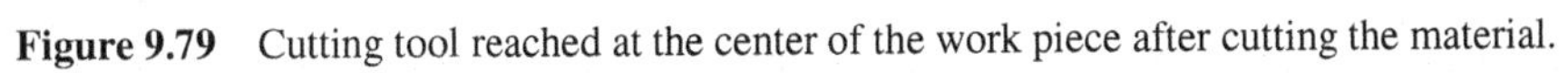

Figure 9.79 Cutting tool reached at the center of the work piece after cutting the material.

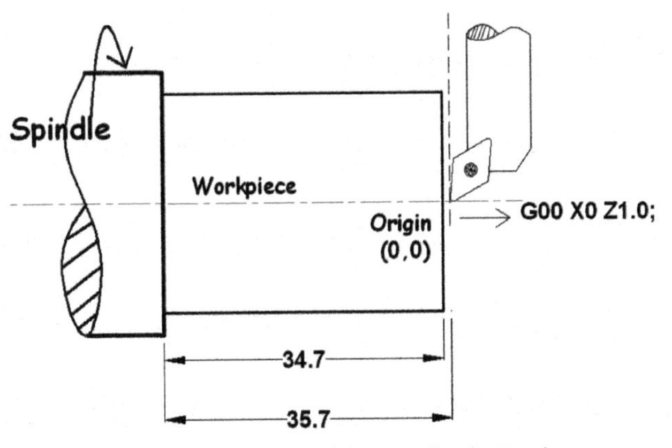

Figure 9.80 Now tool is retracting in Z-axis.

last position (after material cutting) of the tool, where the cutting tool will stay finally. See Figure 9.81.

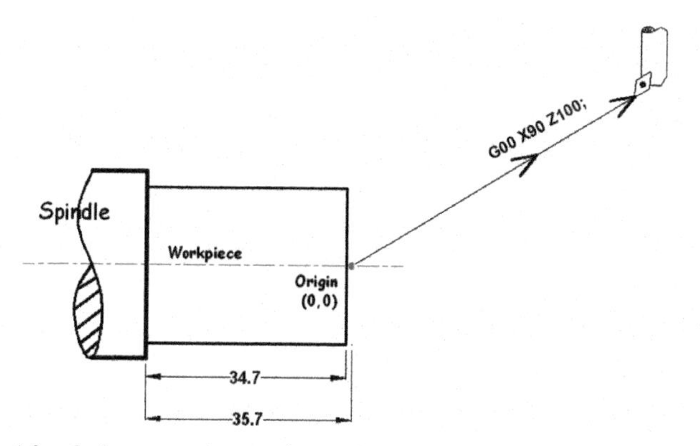

Figure 9.81 After facing operation cutting tool retracted and reached at the starting point of the tool.

M30; And finally CNC program will be stop and reset. It means to program and CNC machine will be ready again for the next production.

9.1.4 Multiple Facing Operation with CNC Programming

Figure 9.82 is a raw material drawing.

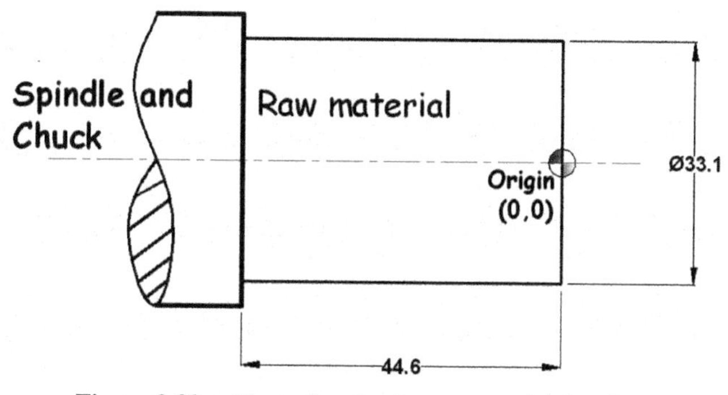

Figure 9.82 Above drawing is raw material drawing.

Figure 9.83 shows machining drawing.

O1004; This is a first line of the CNC program. This is called program number.

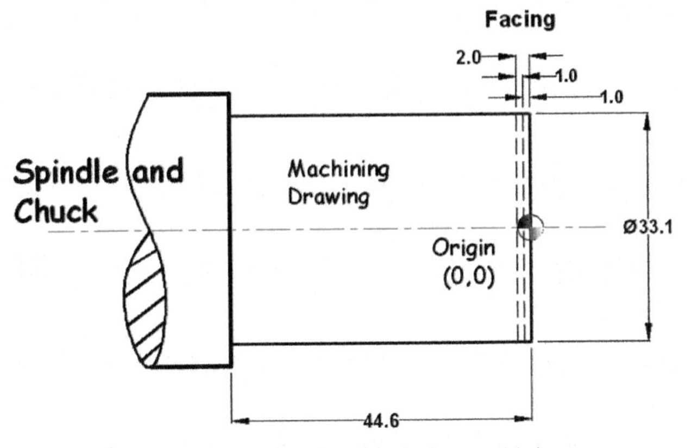

Figure 9.83 This drawing is for machining.

G54; This is called work zero. When we take **G54** command on the center of the face of the round work piece, that time coordinates of the center of that cylindrical job are (0, 0). It means all cutting tool will assume that point as an origin (0, 0) and important think is when we make the CNC program in absolute mode G90. Origin (0, 0) will take as a reference point for all coordinates in the CNC program.

G50 S1600; G50 command controls the maximum spindle speed. In this program, spindle speed will not exceed more than 1600 rpm.

T0303; This programming line is saying to machine that turret (tool post) will change the cutting tool and take the tool number three (3) for machining. In this line, First two digits of T**03**03 are representing the tool station number, where from tool is coming for machining and second two digits of T03**03** are representing the tool offset, tool offset means location number of tool data, where the tool data (tool geometry) is stored, finally we can say tool offset is the brain of cutting tool.

G00 X90 Z100; In Figure 9.84, G00 command is used for rapid motion (non-cutting motion). Here Cutting tool

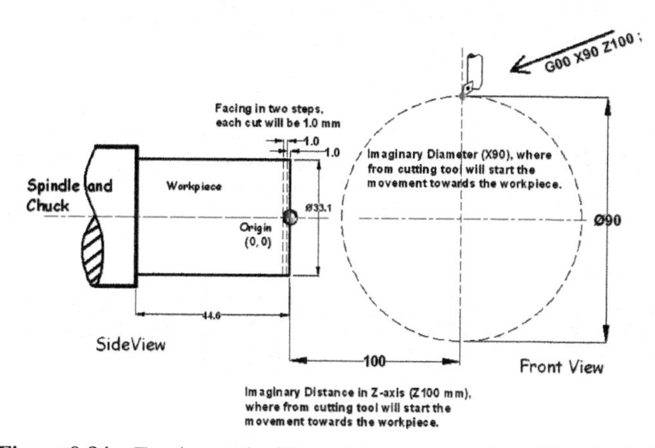

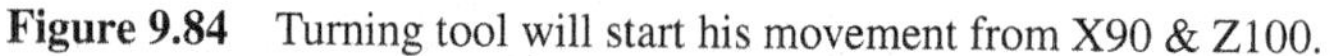
Figure 9.84 Turning tool will start his movement from X90 & Z100.

T0303 will take position rapidly in X-axis at diameter 90 mm and Z-axis at distance 100 mm from the origin (0, 0). The tool will use these coordinates to come near the job for cutting.

G96 M03 S180; In Figure 9.85, now spindle is rotating and vary between 225 rpm to 1600 rpm. It is possible because we are using G50 and G96 command in this program.

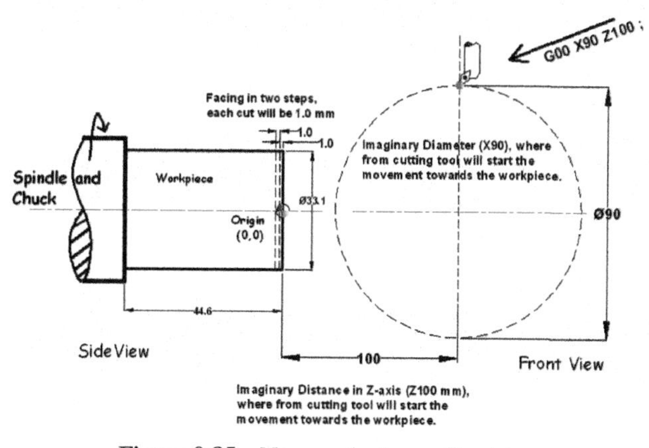

Figure 9.85 Now work piece will rotate.

G00 X35 Z1.0; In Figure 9.86, the tool is taking position nearest the job, together in X-axis and Z-axis direction at Ø35 mm and length 1.0 mm from the work zero (origin) in rapid motion (G00).

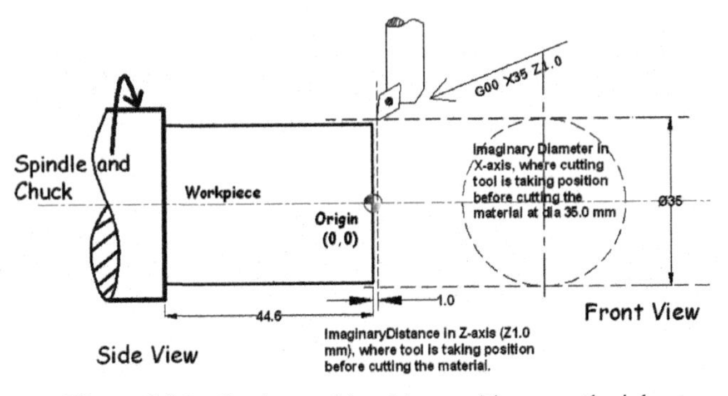

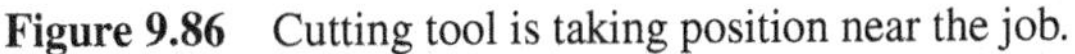

Figure 9.86 Cutting tool is taking position near the job.

M08; Coolant motor start before cutting the material.

G00 X35 Z-1.0; In Figure 9.87, before cutting the material, the cutting tool is taking position above the work piece surface at diameter 35 mm at the same time tool takes the position inside the work piece face in Z– (negative) axis but the tool will not touch with the work piece surface.

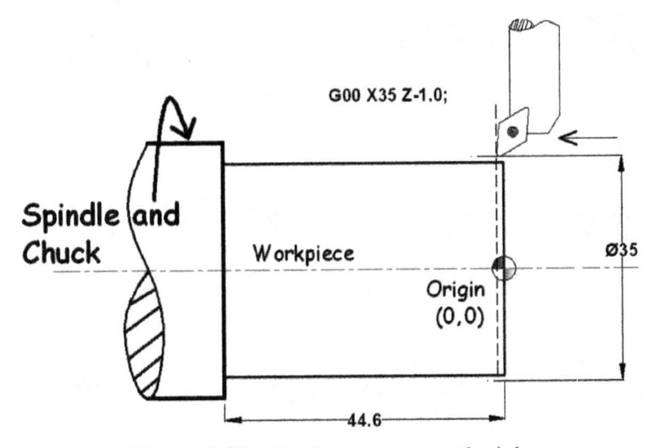

Figure 9.87 Tool came nearest the job.

G01 X0.0 Z-1.0 F0.1; In Figure 9.88, cutting Tool is removing the material from the front face of the work piece in linear motion G01 with feed F 0.10 and tool is going towards the X-axis direction at diameter X0.0 (work piece center).

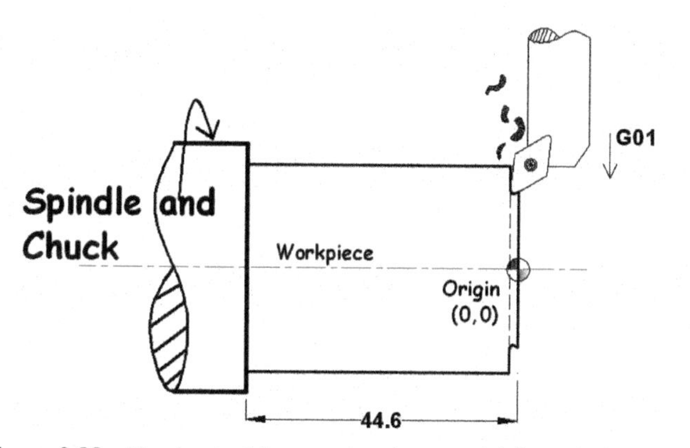

Figure 9.88 Turning tool is removing the material from the job face.

In Figure 9.89, the cutting tool is removing the material continuously towards the center.

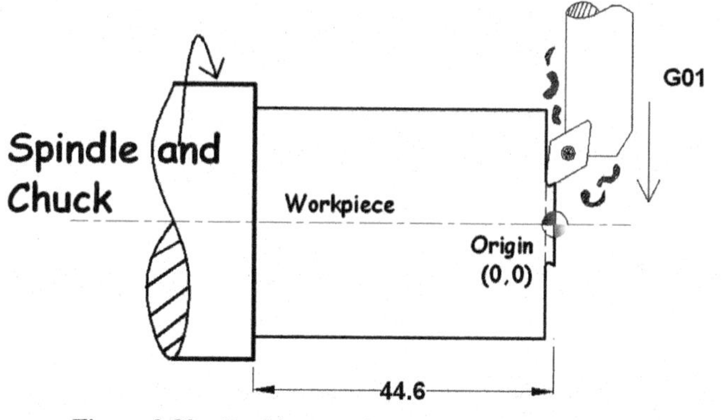

Figure 9.89 Tool is removing material continuously.

In Figure 9.90, cutting tool removed the material and reached at the work piece center.

After the first facing of the job, the remaining length of the job is 43.6 mm.

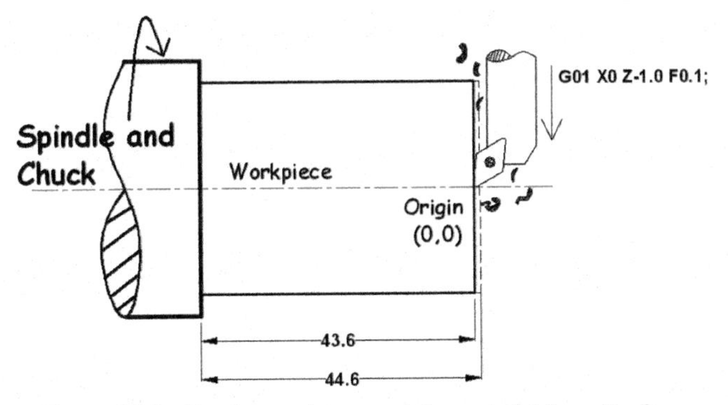

Figure 9.90 Turning tool removed the material from the face.

G00 X0.0 Z0.0; In Figure 9.91, the tool is retracting in Z-axis direction and reached at Z-axis direction at Z0 in rapid motion (G00).

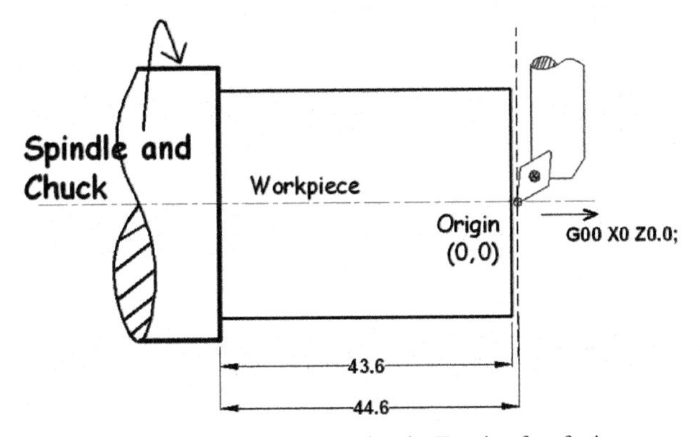

Figure 9.91 Tool is retracting in Z-axis after facing.

G00 X35 Z0.0; In Figure 9.92, the tool is retracting in X-axis direction at diameter 35.0 mm but Z value will be same like previous value.

G00 X35 Z-2.0; In Figure 9.93, before taking 2nd cut on the face of the material, cutting tool is taking position above the work piece surface at diameter 35 mm at the same time tool takes the position inside the work

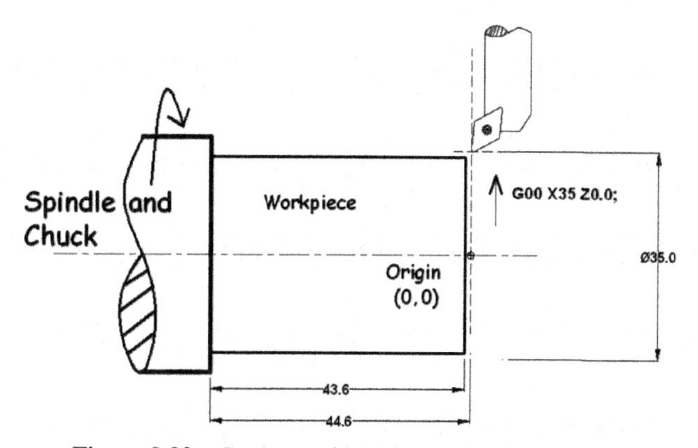

Figure 9.92 Cutting tool is going upside for next cut.

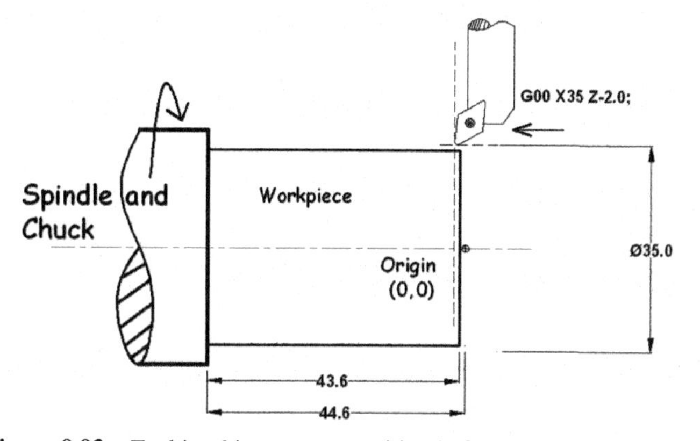

Figure 9.93 Tool is taking nearest position before cutting the material.

piece face in Z– (negative) axis but the tool will not touch with the work piece surface.

G01 X0.0 Z-2.0 F0.1; In Figure 9.94, cutting tool is removing the material from the front face of the work piece in linear motion G01 with feed F 0.10 and the tool is going towards the X-axis direction at diameter X0.0 (work piece center).

In Figure 9.95, the cutting tool is removing the material continuously towards the center.

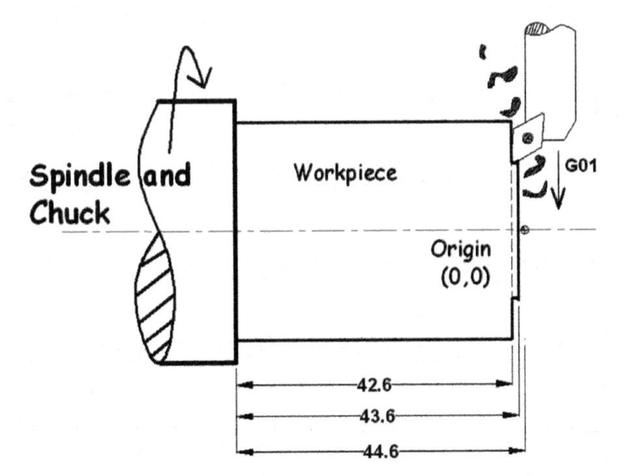

Figure 9.94 Tool is removing the material and going towards the center of the job.

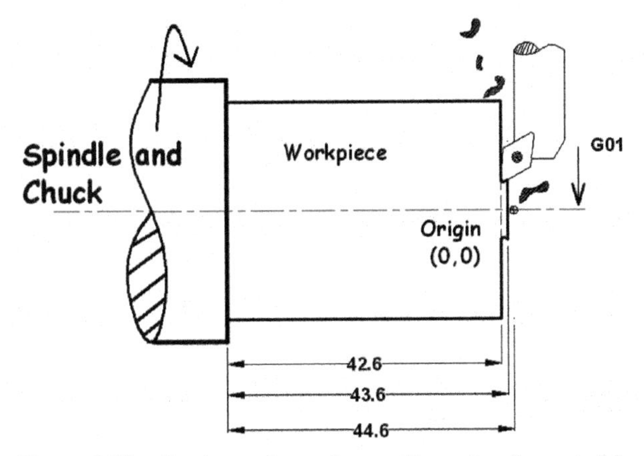

Figure 9.95 Continuously turning tool is cutting the material.

In Figure 9.96, cutting tool removed the material and reached at the work piece center.

After 2nd cut as a facing, remaining length of the job is 42.6 mm.

G00 X0.0 Z0.0;	In Figure 9.97, the tool is retracting in Z-axis direction and reached at Z-axis direction at Z0 in rapid motion (G00).
M09;	After completion of facing operation coolant motor will stop (off).
M05;	Now spindle motor will stop.

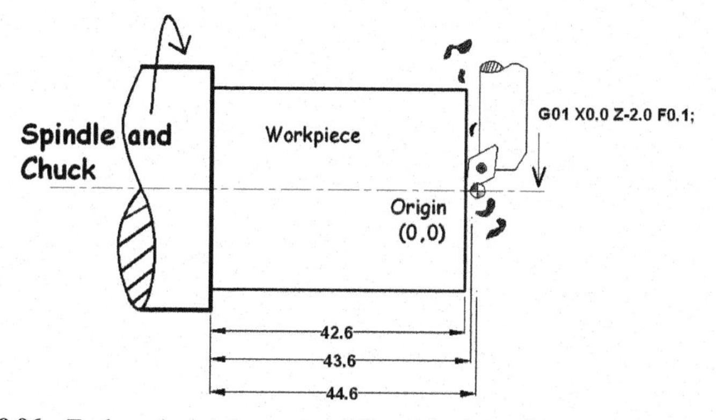

Figure 9.96 Tool reached at the center of the work piece after removing the material.

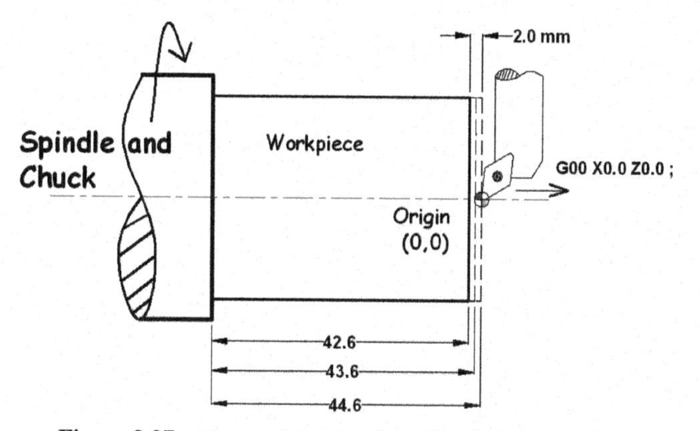

Figure 9.97 Now tool is retracting after facing operation.

G00 X90 Z100; In Figure 9.98, now finally tool will retract and reach at co-ordinates X90 Z100 in rapid motion. We can say this position is the starting position of the tool (before moving) and last position (after material cutting) of the tool, where cutting tool will stay finally.

M30; CNC program will be stop and reset. Reset means this CNC program will ready again for production.

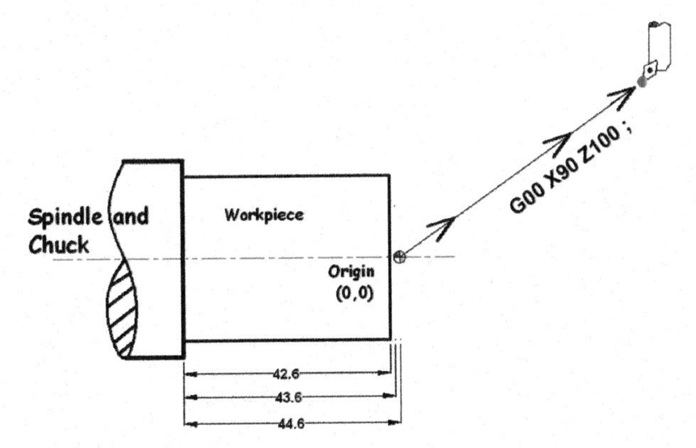

Figure 9.98 Now cutting tool is going back after facing operation as per drawing.

9.1.5 Chamfering Operation in Step by Step with CNC Programming

Figure 9.99 shows machining drawing.

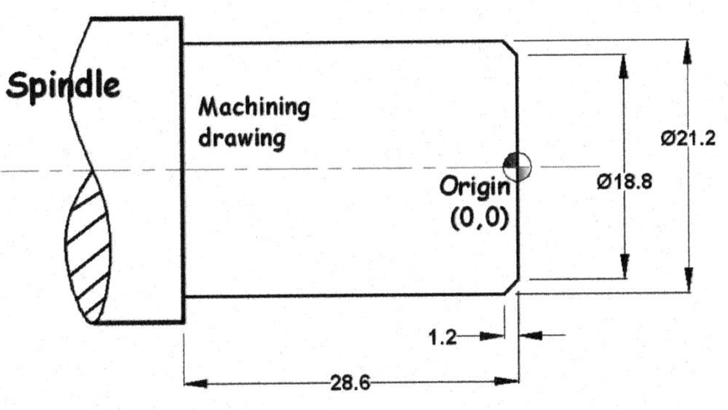

Figure 9.99 Above drawing is for machining.

Figure 9.100 is a raw material drawing.

O1024; This is a first line of the CNC program. This is called program number.

G54; This is called work zero. When we take **G54** command on the center of the face of the round work piece, that time coordinates of the center of that cylindrical job are (0, 0). It means all cutting tool will assume that point as an origin (0, 0) and important think is when we make

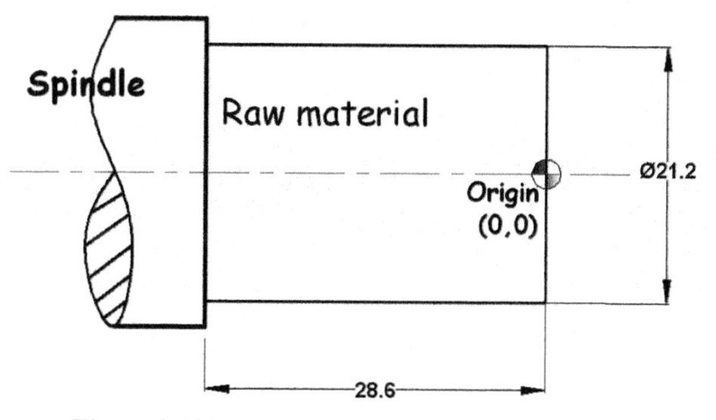

Figure 9.100 This is a drawing for raw material.

the CNC program in absolute mode G90. Origin (0, 0) will take as a reference point for all coordinates in the CNC program.

G50 S2000; G50 command controls the maximum spindle speed. In this program, spindle speed will not exceed more than 2000 rpm.

T0101; This programming line is saying to machine that turret (tool post) will change the cutting tool and take the tool number one (1) for machining. In this line, First two digits of T**01**01 are representing the tool station number, where from tool is coming for machining and second two digits of T01**01** are representing the tool offset, tool offset means location number of tool data, where the tool data (tool geometry) is stored, finally we can say tool offset is the brain of cutting tool.

G00 X90 Z100; G00 command is used for rapid motion (non-cutting motion). Here, cutting tool T0101 will take position rapidly in X-axis at diameter 90 mm and Z-axis at distance 100 mm from the origin (0, 0). The tool will use these coordinates to come near the job for cutting.

Figure 9.101 shows the rapid motion of the tool.

G96 M03 S230; In Figure 9.102, after executing this programming line, the spindle is rotating in clock wise direction at constant surface speed.

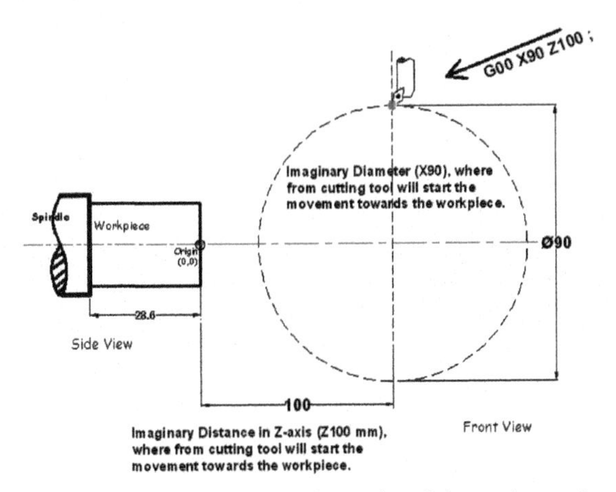

Figure 9.101 This is a starting point of the cutting tool.

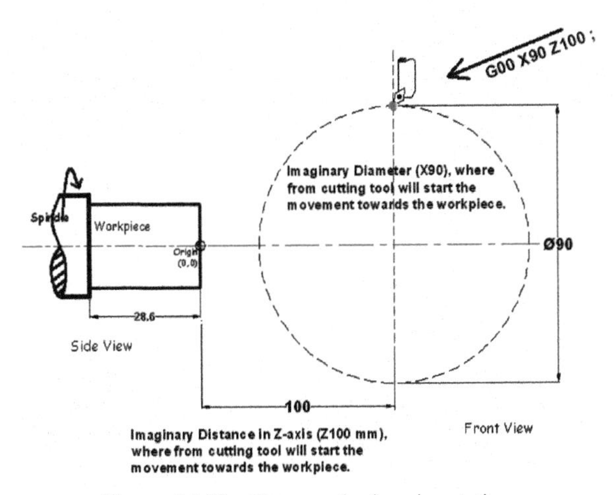

Figure 9.102 Here work piece is rotating.

G00 X17.8 Z1.0; In Figure 9.103, before making the chamfer, cutting tool came rapidly near the Job at diameter 17.8 mm in X-axis and 1.0 mm in Z-axis.

Note:

It is very important that cutting tool must take pre-position in X-axis and Z-axis before making the correct chamfer/taper/radius profile as per drawing. This **pre-position of the cutting will be 1.0 mm less in diameter (it depends on cutting tool direction) from actual chamfer diameter and 0.5 mm away from the job face in Z-axis.** So that cutting tool can make a correct

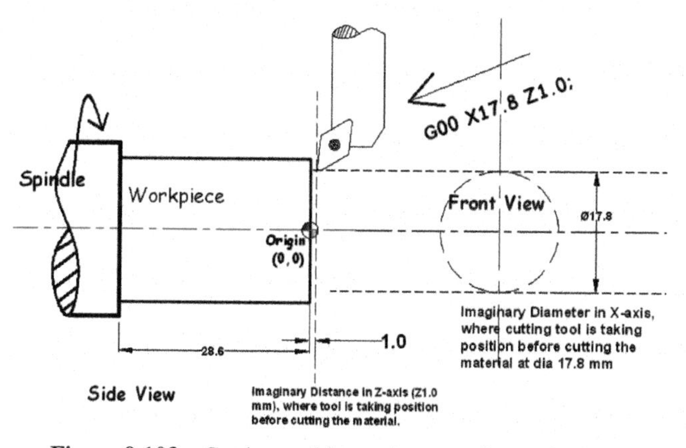

Figure 9.103 Cutting tool is coming near the work piece.

chamfer profile and get correct dimensions of starting and ending point as per given drawing.

In Figure 9.104, cutting tool tip position with the zoomed figure.

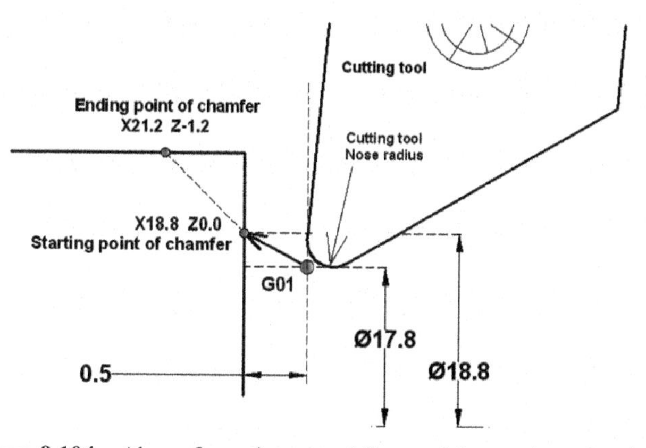

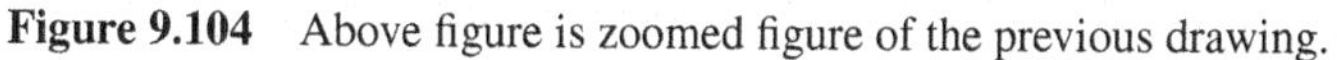

Figure 9.104 Above figure is zoomed figure of the previous drawing.

The figure is showing cutting tool position. cutting tool position before reaching at the starting point of the chamfer. Due to this condition, we get correct chamfer dimensions after machining at starting and ending point.

M08; Coolant motor start before cutting the material.

G01 X18.8 Z0.0 F0.12; In Figure 9.105, just before cutting, cutting tool reached at coordinates X18.8 in X-axis direction and Z0.0 (touched on the job face) in Z-axis direction in linear motion G01 with feed 0.12 mm/revolution.

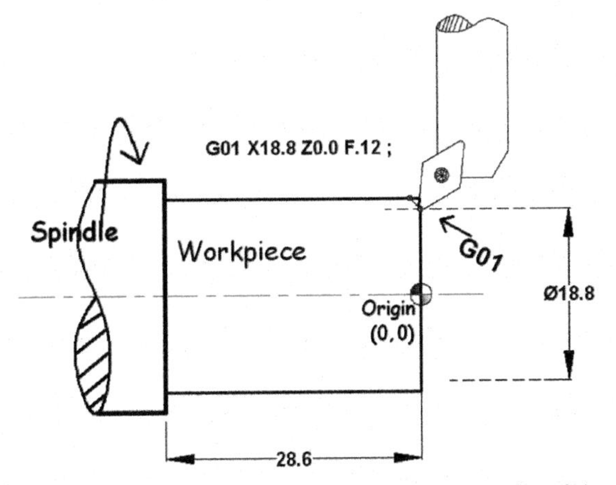

Figure 9.105 Cutting tool is taking position at the start point of the chamfer.

G01 X21.2 Z-1.2 F0.12; In Figure 9.106, the cutting tool is removing the material continuously and going in taper towards the end point of the chamfer.

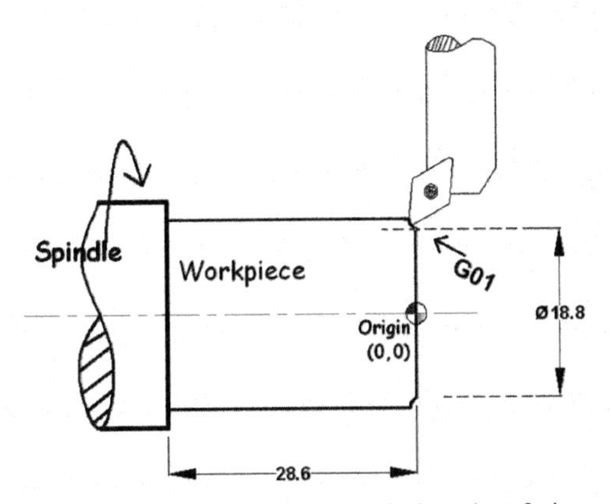

Figure 9.106 Tool is removing material during chamfering operation.

In Figure 9.107 cutting tool reached at the ending point of the chamfer (X21.2 Z-1.2) in linear motion with feed 0.12 mm/revolution.

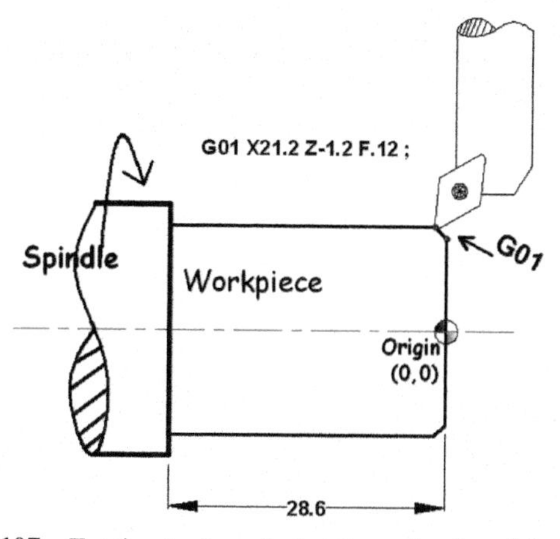

Figure 9.107 Turning tool reached at the end point of the chamfer.

G01 X22.2 Z-1.7 F0.12; In Figure 9.108, the cutting tool is retracting from the job in linear motion G01 with slow feed at diameter 22.2 mm in X-axis and −0.5 mm [total length −1.7 (1.2 + 0.5) mm in negative direction] in Z-axis after chamfering. On the other hand, we can say the cutting tool is retracting at 45° angle in linear motion **towards**

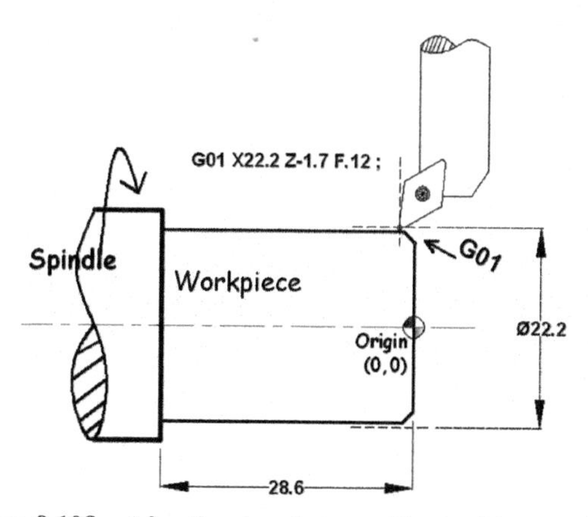

Figure 9.108 After the chamfering cutting tool is retracting.

to chamfer direction and reached at safe distance (X22.2, Z-1.7) far from the last chamfer point.

In Figure 9.109, cutting tool tip position with the zoomed figure.

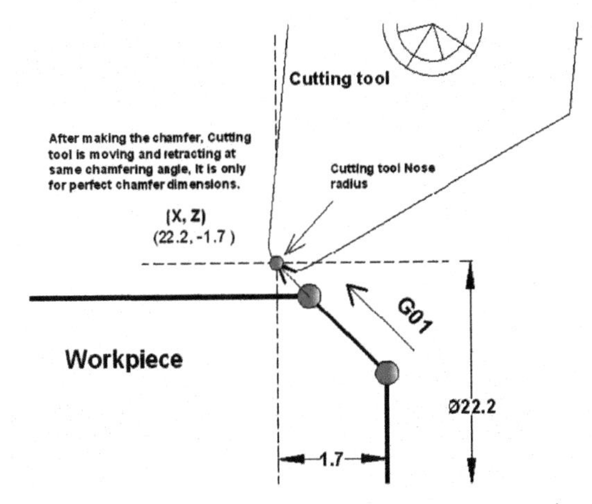

Figure 9.109 Above figure is Zoomed figure of the previous step.

After chamfering operation cutting tool position during safe retraction at coordinates X22.2 and Z-1.7.

G00 X22.2 Z1.0; In Figure 9.110, the cutting tool is retracting in X-axis and Z-axis (going back) but the diameter is same as the previous diameter.

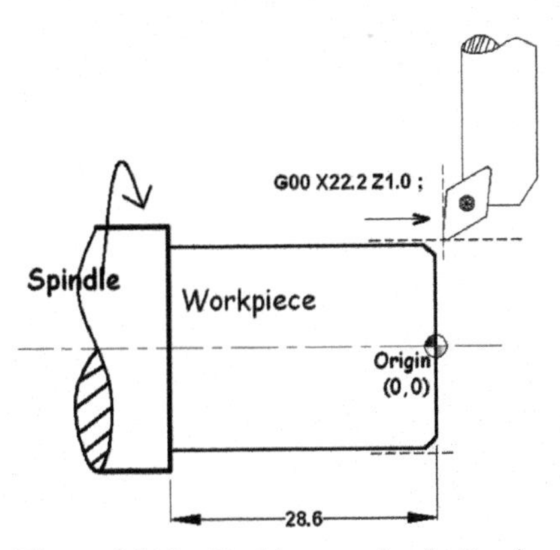

Figure 9.110 Tool is retracting in Z-axis.

M09; After retracting of the cutting tool, coolant motor stops. It means coolant will not flow on the tool tip.

M05; When M05 command executes, it means spindle motor will stop.

G00 X90 Z100; In Figure 9.111, finally cutting tool retracted from the job and reached at safe distance.

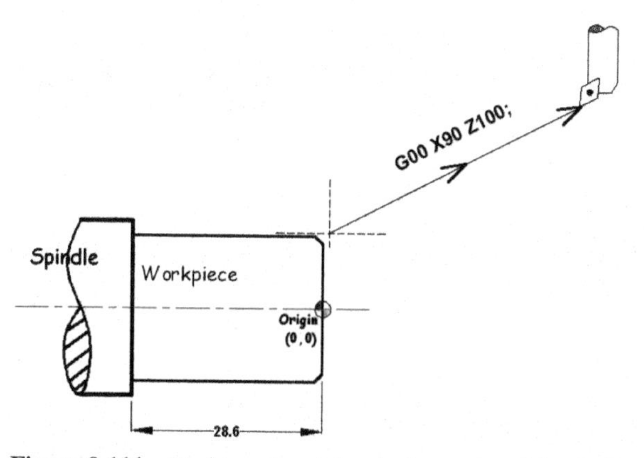

Figure 9.111 Tool is going at the starting point of the tool.

M30; When M30 command executes, it means CNC program will stop and reset.

Note:

If the chamfer size is big and you want to make or machine that big chamfer. You should take more than one cut to make that chamfer. It depends on the total depth of cut and material material hardness. Only follow the above procedure. You can take 0.5 mm/1.0 mm/2.0 mm depth of cut to make that chamfer. It depends on cutting tool size.

9.1.6 Corner Radius Operation with CNC Programming

Figure 9.112 shows machining drawing.
Figure 9.113 is a raw material drawing.

O1005; This is a first line of the CNC program. This is called program number.

G54; This is called work zero. When we take **G54** command on the center of the face of

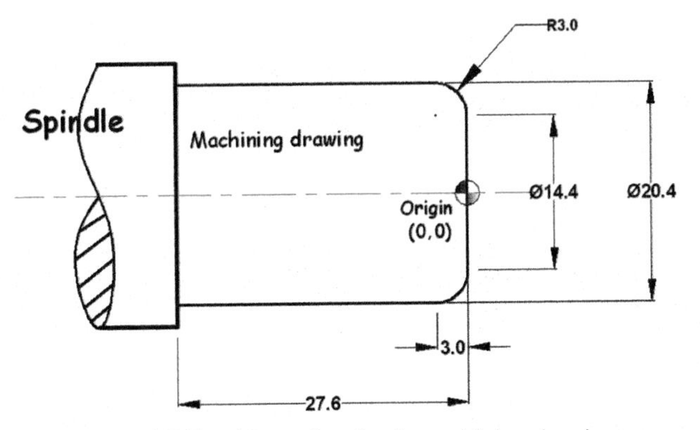

Figure 9.112 Above drawing is machining drawing.

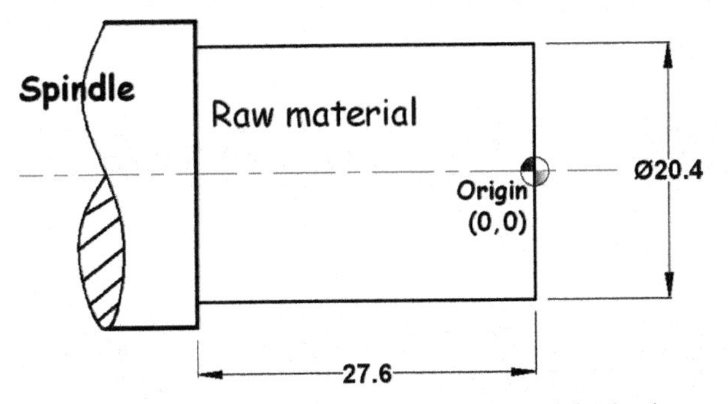

Figure 9.113 Above drawing is for raw material selection.

the round the work piece, that time coordinates of the center of that cylindrical job are (0, 0). It means all cutting tool will assume that point as an origin (0, 0) and important think is when we make the CNC program in absolute mode G90. Origin (0, 0) will take as a reference point for all coordinates in the CNC program.

G50 S2400; G50 command controls the maximum spindle speed. In this program, spindle speed will not exceed more than 2400 rpm.

T0101; This programming line is saying to machine that turret (tool post) will change the cutting tool and take the tool number one (1) for machining. In this line, First two digits of T**01**01 are representing the tool station number, where from tool is coming for machining and second two digits of T01**01** are representing the tool offset, tool offset means location number of tool data, where the tool data (tool geometry) is stored, finally we can say tool offset is the brain of cutting tool.

G00 X90 Z100; In Figure 9.114, G00 command is used for rapid motion (non-cutting motion). Here Cutting tool T0101 will take position rapidly in X-axis at diameter 90 mm and Z-axis at distance 100 mm from the origin (0, 0). The tool will use these coordinates to come near the job for cutting.

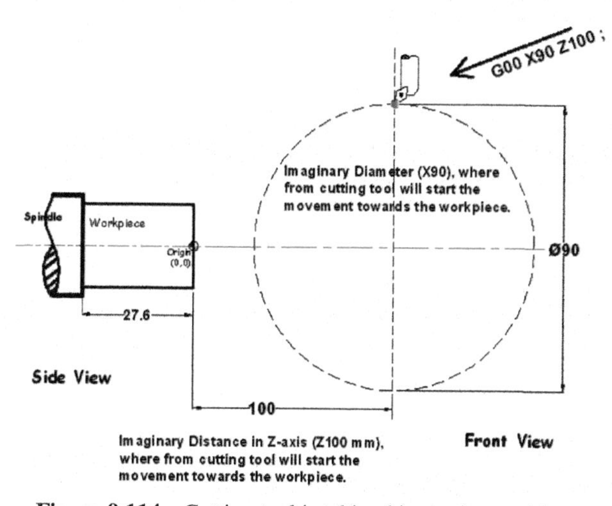

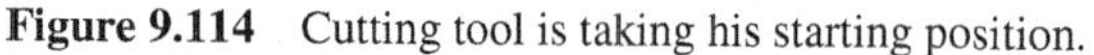
Figure 9.114 Cutting tool is taking his starting position.

G96 M03 S210; After executing this programming line, in Figure 9.115, now spindle is rotating in clock wise direction on constant surface speed.

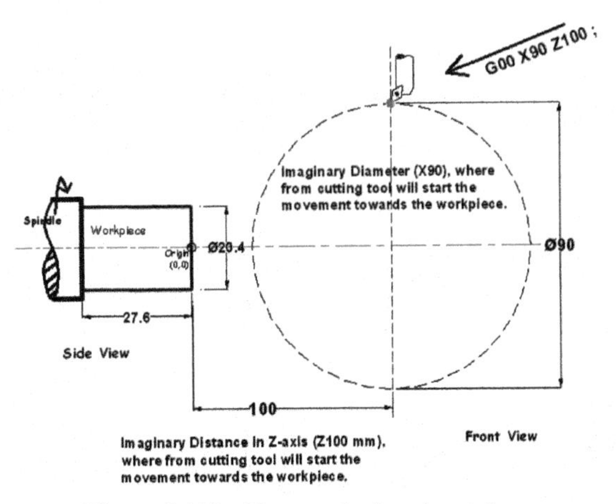

Figure 9.115 Here work piece is rotating.

G00 X17.4 Z1.0; In Figure 9.116, the cutting tool is taking position in rapid motion near the job at coordinates X17.4 and Z1.0 before making the rough radius of **R1.5**.

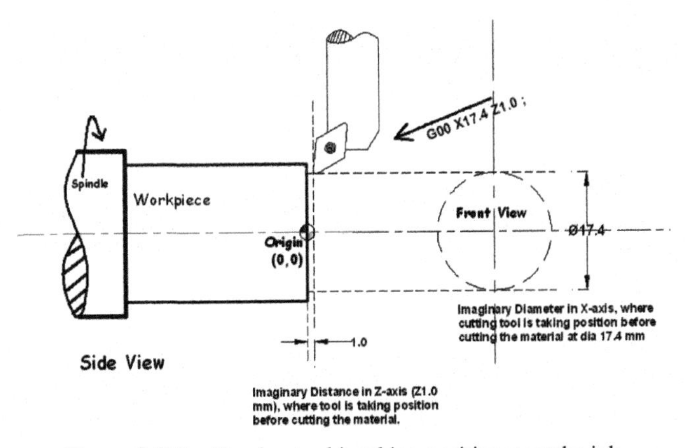

Figure 9.116 Turning tool is taking position near the job.

G01 X17.4 Z0 F0.12; In Figure 9.117, cutting tool came slowly and **touched on the rough Starting point of the radius** in Linear motion G01 with feed 0.12 mm/revolution.

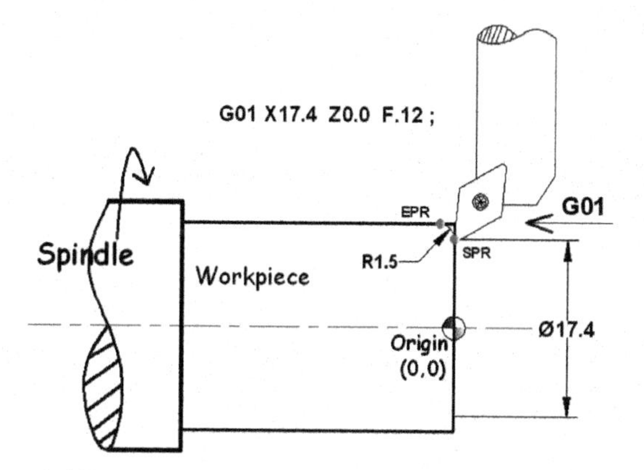

Figure 9.117 Here tool reached at the starting point of the radius.

G03 X20.4 Z-1.5 R1.5 F0.12; In Figure 9.118, now the tool is removing the material continuously in counter clockwise motion (G03) and going towards the **ending point of the radius** at coordinates X20.4 Z-1.5. In general terms, we can say counter clockwise direction (G03) is also called **anti clockwise direction**.

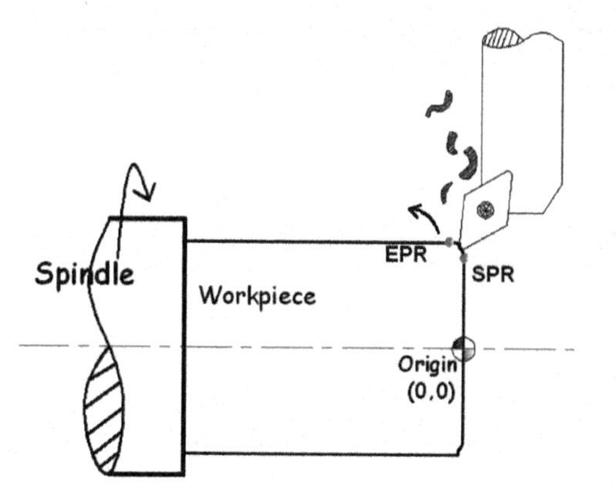

Figure 9.118 Tool is removing the material and making the radius.

In Figure 9.119, cutting tool removed the material and **reached at the end point of the radius** at diameter 20.4 mm in X-axis and simultaneously tool reached at the distance 1.5 mm (Z-1.5) in Z-axis with radius 1.5 mm in counter clock wise direction (G03).

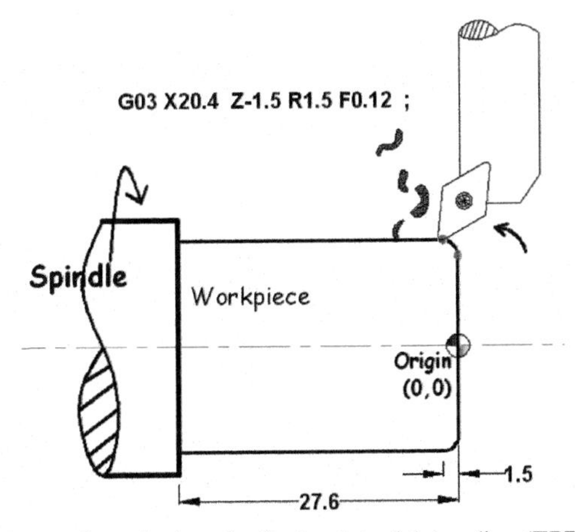

Figure 9.119 Cutting tool reached at the End point of the radius (EPR) after removing the material.

G00 X20.4 Z1.0; In Figure 9.120, after **making rough corner radius of 1.5 R,** now cutting tool retracts in Z-axis on same previous diameter in rapid motion.

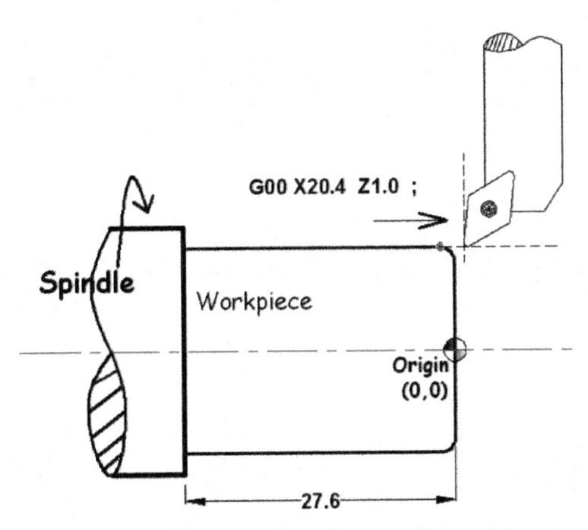

Figure 9.120 Now after making rough radius profile tool is retracting in Z-axis.

G00 X13.4 Z1.0; In Figure 9.121, for final radius profile, the cutting tool is taking position in X-axis at diameter X13.4 and Z1.0 in rapid motion G00.

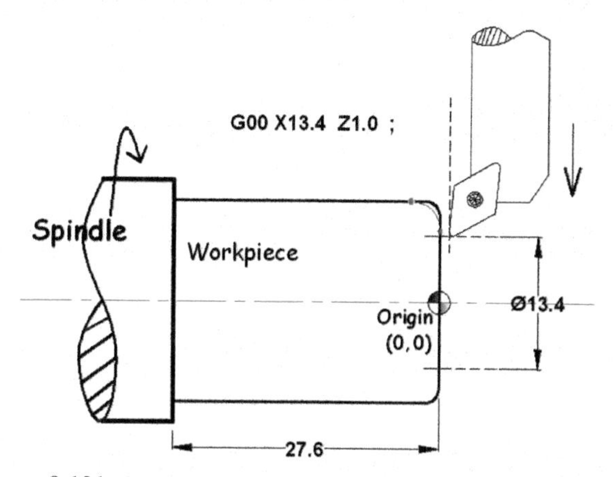

Figure 9.121 Now tool is taking position for final radius cutting.

Note:

Always cutting tool will take position very carefully to cut the profile in the form of chamfer/taper or radius on the face of the work piece.

Coordinates of ***Starting point of radius*** (SPR) is **X14.4 mm** and **Z0.0** (where from tool start the cut).

Coordinates of the ***Ending point of radius*** (EPR) is **X20.4 and Z-3.0** (where tool reached at last after radius profile).

In Figure 9.122, cutting tool tip position with the zoomed figure.

G01 X14.4 Z0.0 F0.12; In Figure 9.123, the cutting tool is moving slowly towards the **The starting point of radius** (SPR) in linear motion (G01) with feed 0.12 mm/revolution. In Figure 9.123, cutting tool tip position with zoomed figure.

Figure 9.124, cutting tool took the position on the starting point of the radius for making radius profile.

Figure 9.125, cutting tool tip position with the zoomed figure.

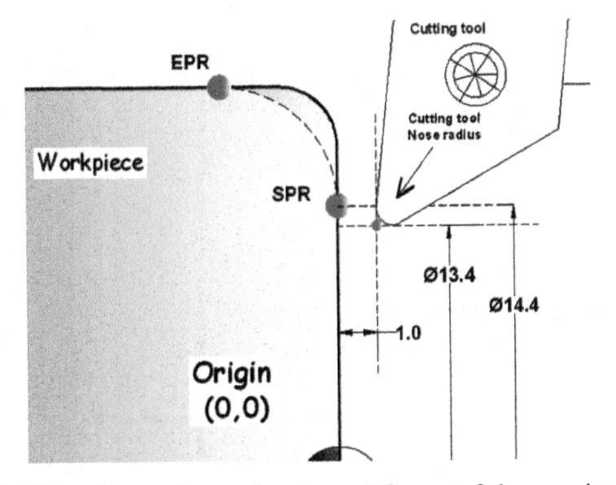

Figure 9.122 Above figure is zoomed figure of the previous figure.

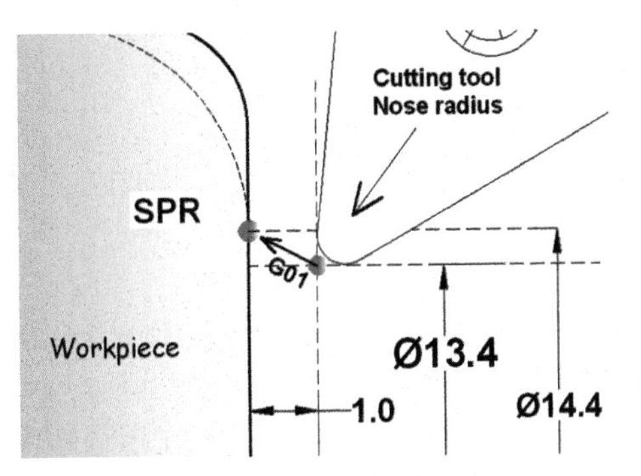

Figure 9.123 In this zoomed figure, the tool will take position for final radius cutting.

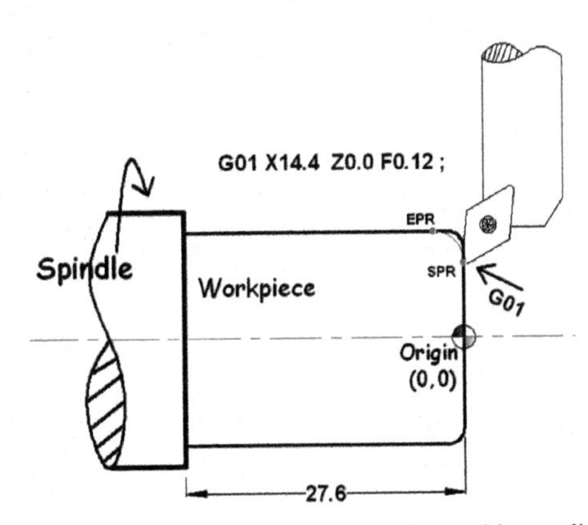

Figure 9.124 Here tool took the position for making radius profile.

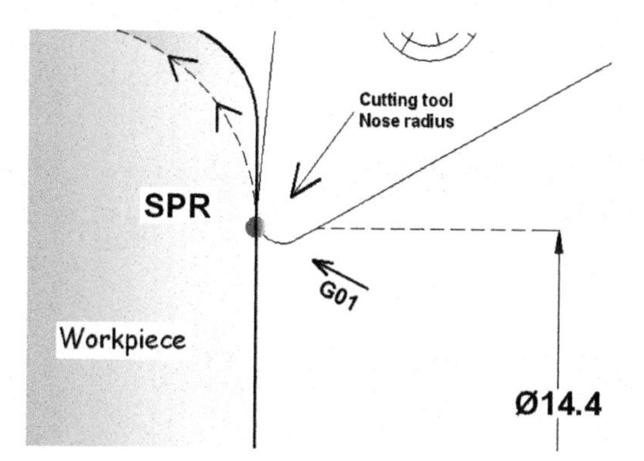

Figure 9.125 It is a zoomed figure of the previous figure.

Before cutting the material cutting tool reached and touched on the **starting point of radius** (SPR) at coordinates **X14.4** and **Z0.0** in linear motion with feed.

G03 X20.4 Z-3.0 R3.0 F0.12; In Figure 9.126, the cutting tool is removing the material continuously in counter clock wise direction (Anti clock wise direction) and **going towards end point of the radius** (EPR) **X20.4** and **Z-3.0** with feed 1.2 mm/Rev.

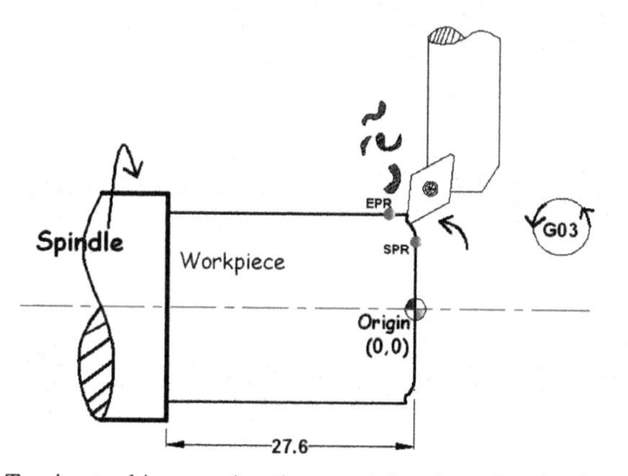

Figure 9.126 Turning tool is removing the material and making the final radius profile.

In Figure 9.127, cutting tool removed the material and reached at coordinates **X20.4** and **Z-3.0** with radius **3.0 R**.

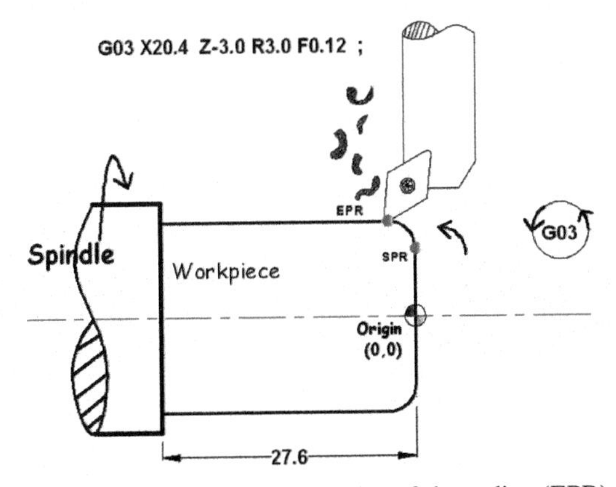

Figure 9.127 Cutting tool reached at the End point of the radius (EPR) after removing the material.

G00 X20.4 Z1.0; In Figure 9.128, after removing the material, cutting tool will retract in Z-axis 1.0 mm at same diameter X2 0.4 mm.

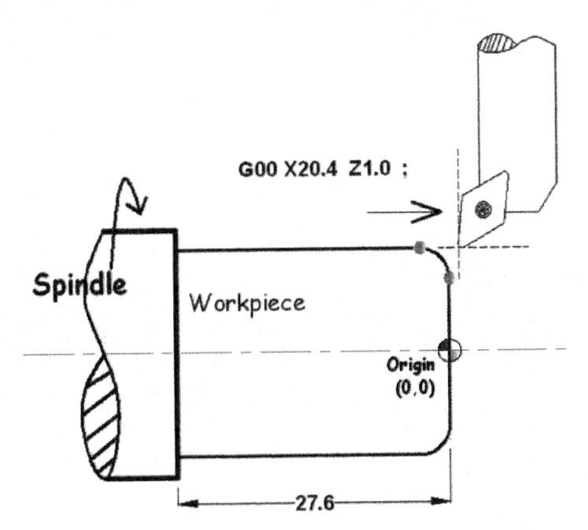

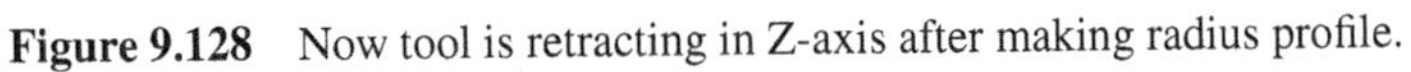

Figure 9.128 Now tool is retracting in Z-axis after making radius profile.

M09; After retracting in Z-axis, now coolant motor will be stopped.

M05; Now spindle motor will stop. It means spindle will not rotate.

G00 X90 Z100; Finally tool will go back in safe distance at imaginary diameter X90 and distance Z100. See Figure 9.129.

M30; Now CNC program will be stop and reset.

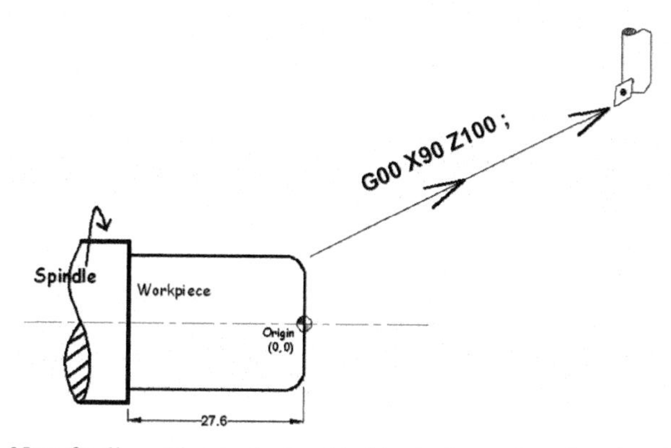

Figure 9.129 Now finally cutting tool retract to his starting point after complete machining.

Note:

i. In Figure 9.130, dotted lines are showing rapid motion (G00) of the cutting tool.
ii. In Figure 9.130, continuous lines are showing cutting tool path of the cutting tool.

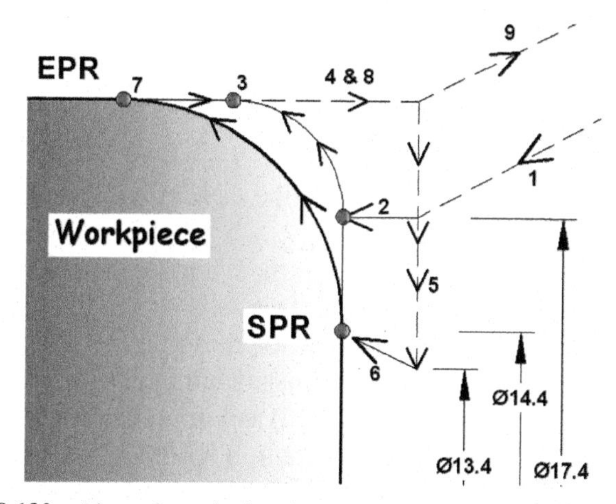

Figure 9.130 Above figure is showing cutting tool path of the radius profile.

9.1.7 Taper Turning Operation in Step by Step with CNC Programming

Figure 9.131 shows machining drawing.

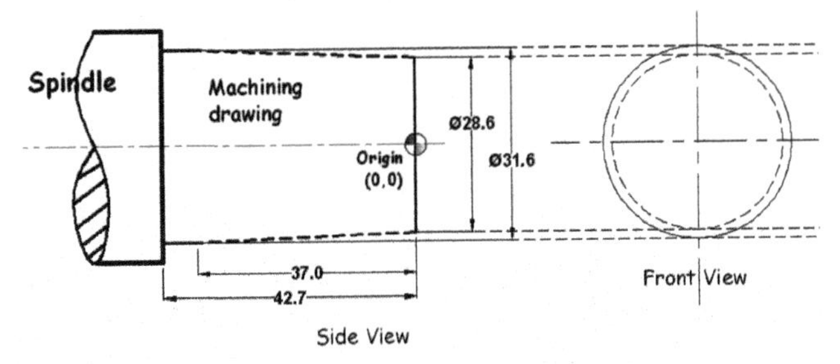

Figure 9.131 Machining drawing.

Figure 9.132 is a raw material drawing.

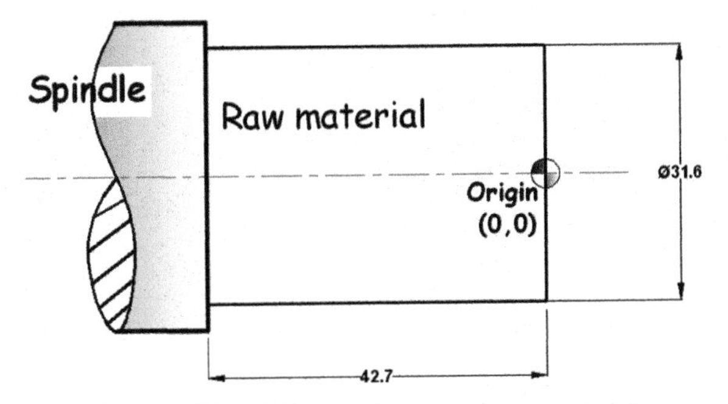

Figure 9.132 This is a drawing of raw material.

O1006; This is a first line of the CNC program. This is called program number.

G54; This is called work zero. When we take **G54** command on the center of the face of the round the work piece, that time coordinates of the center of that cylindrical job are (0, 0). It means all cutting tool will assume that point as an origin (0, 0) and important think is when we make the CNC program in absolute mode G90. Origin (0, 0) will take as a reference point for all coordinates in the CNC program.

G50 S2300; G50 command controls the maximum spindle speed. In this program, spindle speed will not exceed more than 2300 rpm.

T0101; This programming line is saying to machine that turret (tool post) will change the cutting tool and take the tool number one (1) for machining. In this line, First two digits of T**01**01 are representing the tool station number, where from tool is coming for machining and second two digits of T01**01** are representing the tool offset, tool offset means location number of tool data, where the tool data (tool geometry) is stored, finally we can say tool offset is the brain of cutting tool.

G00 X100 Z100; In Figure 9.133, G00 command is used for rapid motion (non-cutting motion). Here, cutting tool T0101 will take position rapidly in X-axis at diameter 100 mm and Z-axis at distance 100 mm from the origin (0, 0). The tool will use these coordinates to come near the job for cutting.

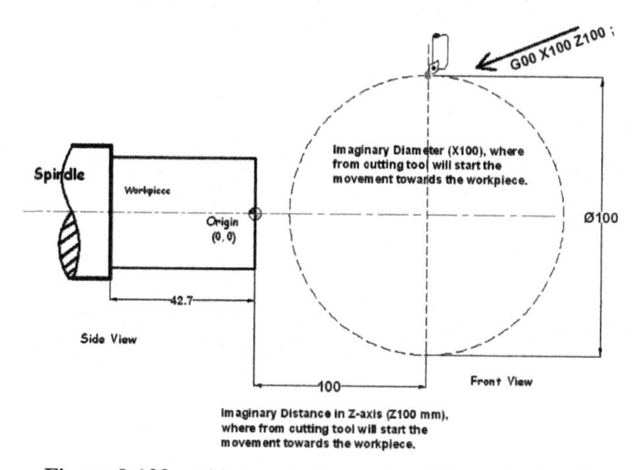

Figure 9.133 This is a starting point of the cutting tool.

G96 M03 S220; After executing this programming line. In Figure 9.134, **spindle** is rotating in clockwise direction at constant surface speed.

G00 X27.6 Z1.0; In Figure 9.135, the cutting tool is taking position in rapid motion near the job at coordinates X27.6 and Z1.0.

In Figure 9.136, cutting tool tip position with the zoomed figure.

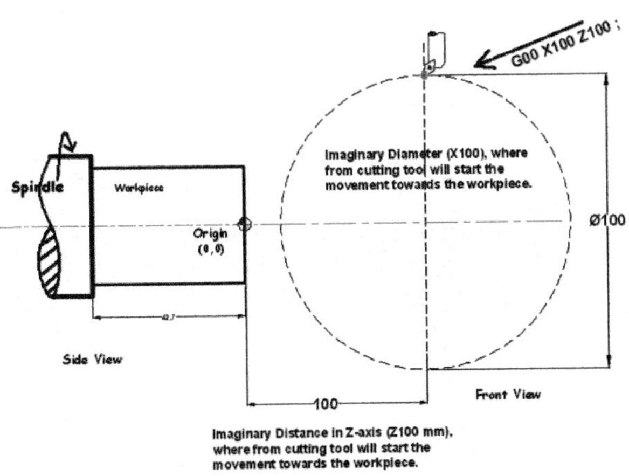

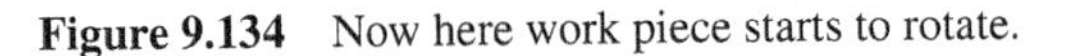
Figure 9.134 Now here work piece starts to rotate.

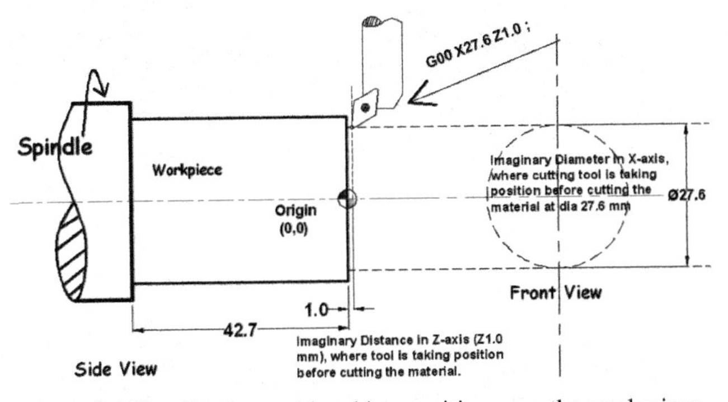

Figure 9.135 Cutting tool is taking position near the work piece.

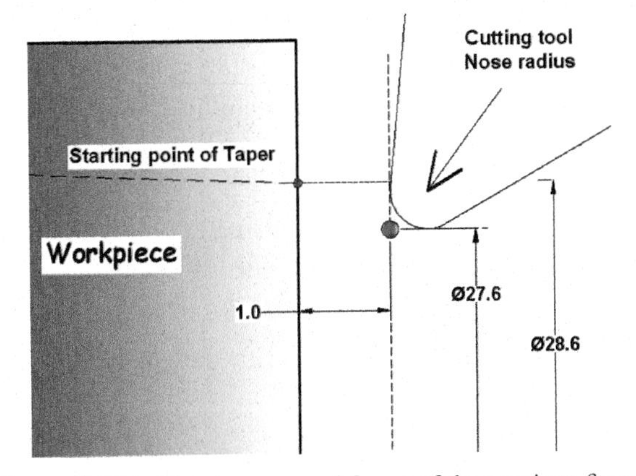

Figure 9.136 This is a zoomed figure of the previous figure.

G01 X28.6 Z0.0 F0.12; The Cutting tool is moving very slowly towards the **Starting Point of Taper** (SPT) (**X28.6 Z0.0**) in linear motion (G01) with feed 0.12. See Figure 9.137.

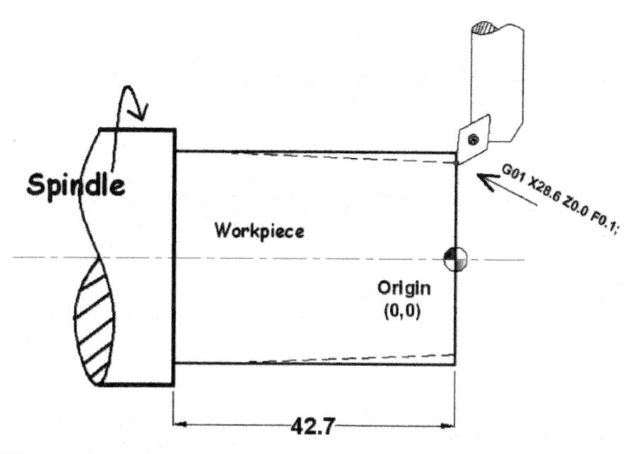

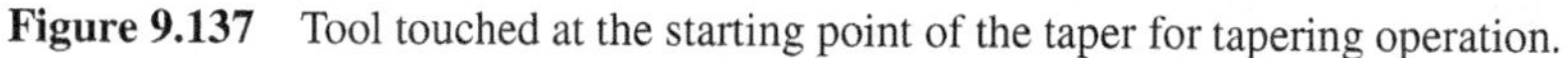

Figure 9.137 Tool touched at the starting point of the taper for tapering operation.

In Figure 9.138, cutting tool tip position with the zoomed figure.

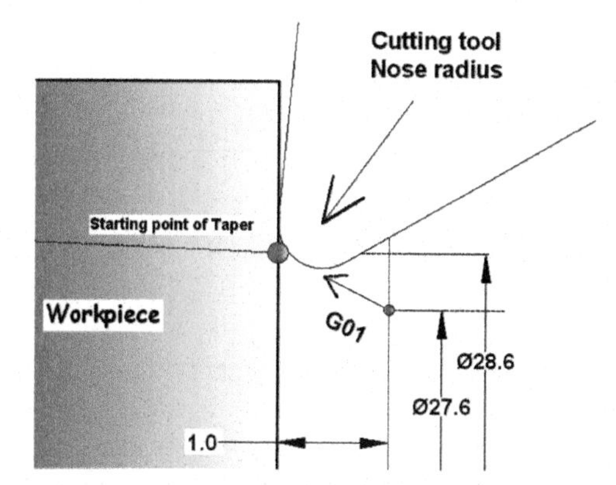

Figure 9.138 This is a zoomed figure of the previous figure.

For taper turning operation, cutting tool touched very slowly at the **starting point of the taper (SPT)** in linear motion with feed 0.12.

G01 X31.6 Z-37.0 F0.12; In Figure 9.139, know the cutting tool is removing the material and **making taper** in

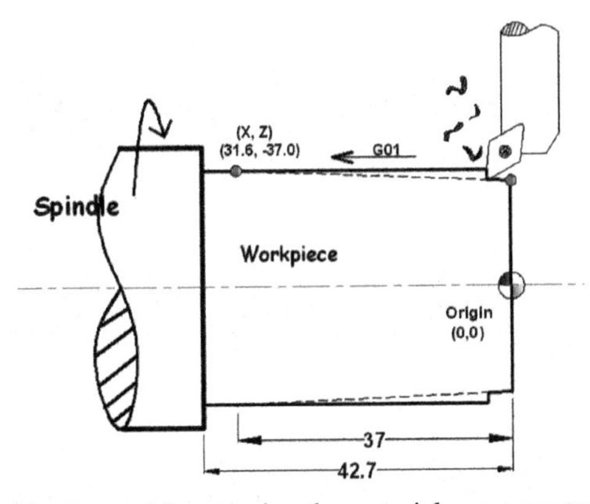

Figure 9.139 Turning tool is removing the material as a taper turning operation.

linear motion (G01) with feed 0.12 and also going continuously towards the coordinates X31.6 in X-axis and Z-37.0 in Z-axis.

In Figure 9.140, cutting tool continuously removing the material and making taper profile in linear motion with feed 0.12.

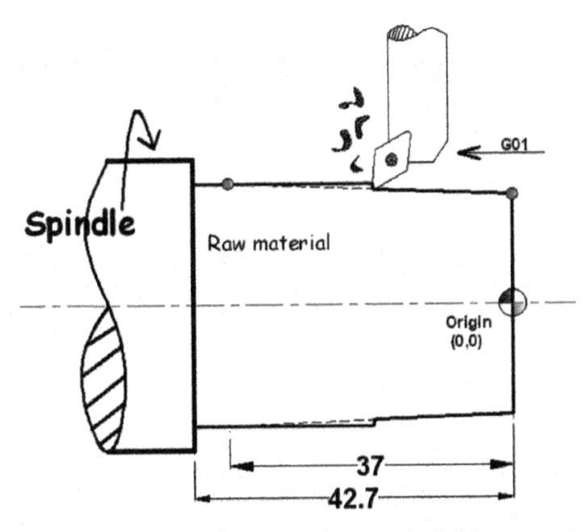

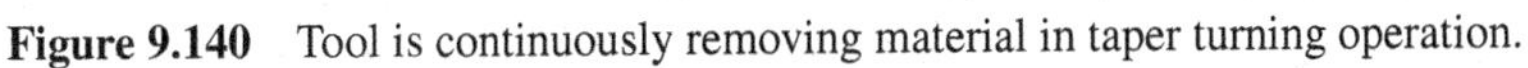

Figure 9.140 Tool is continuously removing material in taper turning operation.

In Figure 9.141, the cutting tool removed the material and reached at coordinates **X31.6 Z-37.0** in linear motion and machined the taper turning profile.

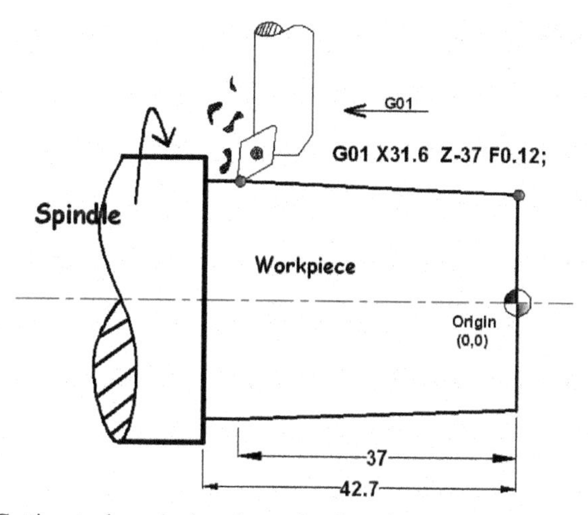

Figure 9.141 Cutting tool reached at the end point of the taper (EPT) after removing the material.

G00 X31.6 Z1.0; In Figure 9.142, after making taper profile, the tool is retracting in Z-axis (1.0) at same previous diameter (Ø31.6) in rapid motion.

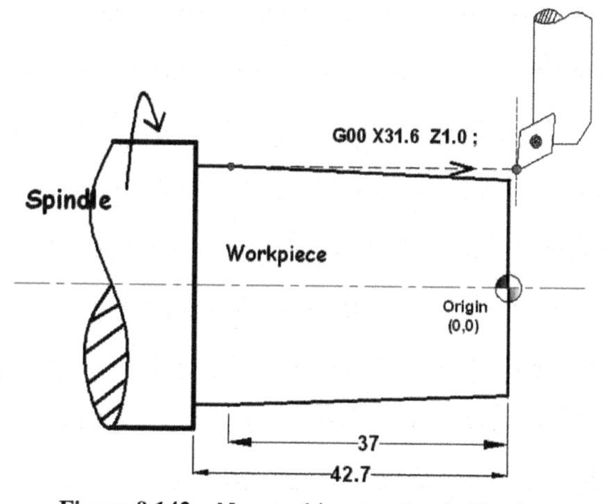

Figure 9.142 Now tool is retracting in Z-axis.

M09; After retracting the tool in Z-axis, when this line will execute. Coolant motor will be stopped.

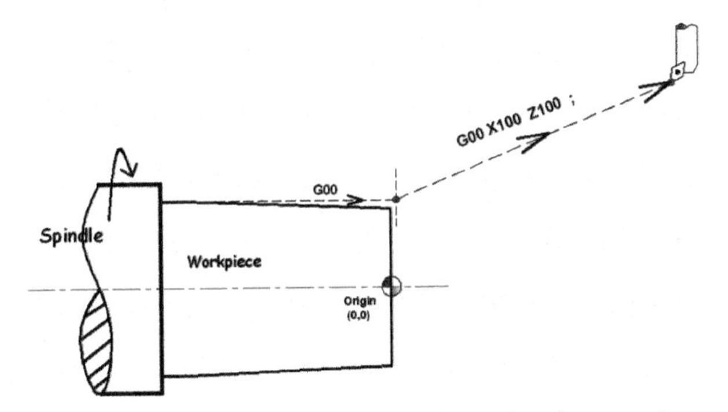

Figure 9.143 Now cutting tool is going his starting point after complete machining.

M05; After complete machining, know the rotation of the spindle will be stopped. On the other hand, we can say rotating work piece will stop.

G00 X100 Z100; At last cutting tool will retract and keep safe distance from the work piece. See Figure 9.143.

M30; When this programming line will execute, above CNC program will stop and reset.

Following Figures 9.144 and 9.145 are representing the graph of the cutting tool path. On the other hand, we can say that the below figures are showing cutting and non-cutting movement of the tool, movement from Start to End.

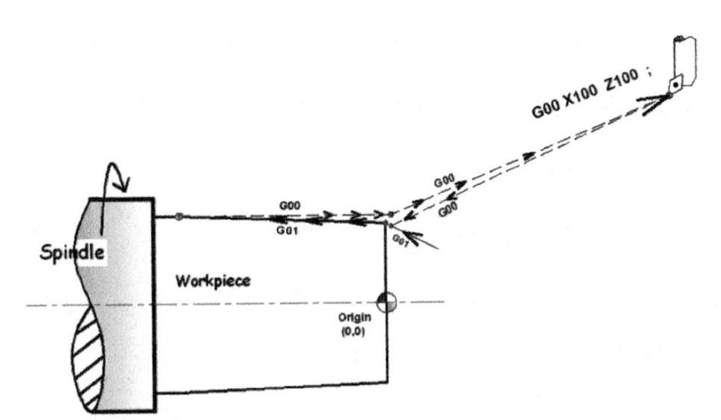

Figure 9.144 Above figure is showing cutting tool path.

Note:

- Dotted lines are showing rapid motion (G00) of the cutting tool. It means the tool is going toward the work piece for cutting the material or retract from the work piece after cutting the material (Figure 9.144).
- Continuous lines are showing linear motion (G01), it means the tool is removing the material for making taper profile (Figure 9.144).

Figure 9.145 is a zoomed figure of cutting tool path.

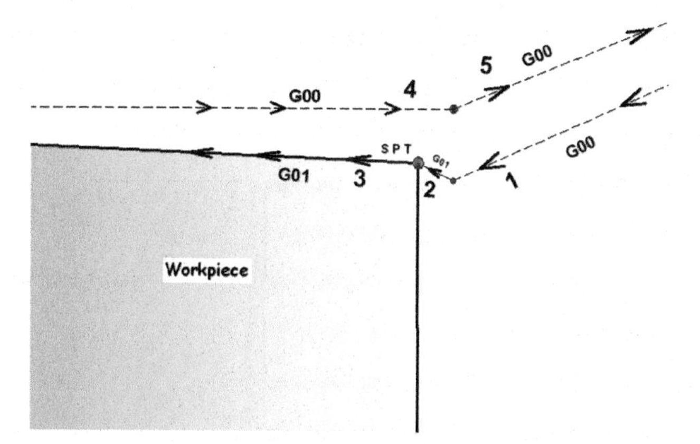

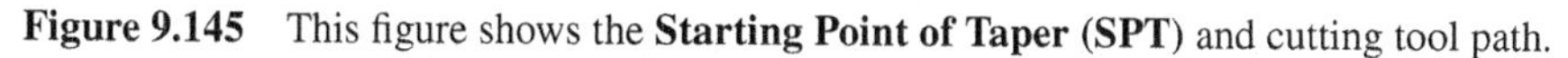

Figure 9.145 This figure shows the **Starting Point of Taper (SPT)** and cutting tool path.

9.1.8 Threading Parameters

What is thread in lathe/turning machine?

A thread is a uniform helical groove or it is a process on lathe machine in which the threading tool produces a helical ridge of the uniform section on the work piece surface.

How to show thread parameters in the drawing?

You can see some examples of identification method of thread parameters. See Figures 9.146 and 9.147.

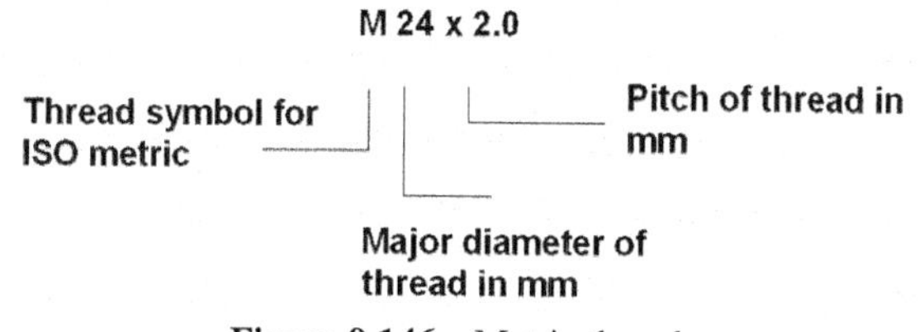

Figure 9.146 Metric thread.

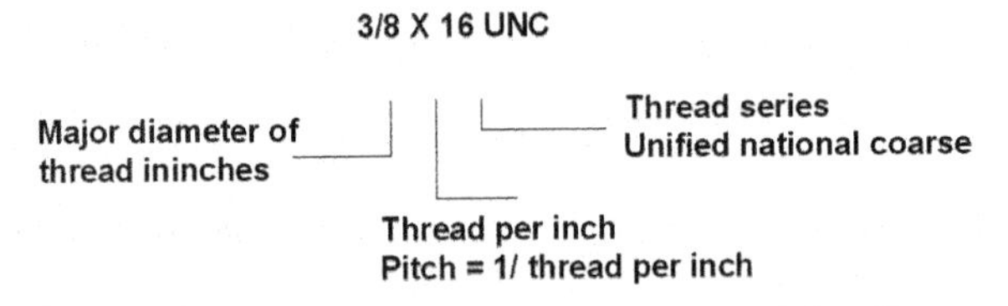

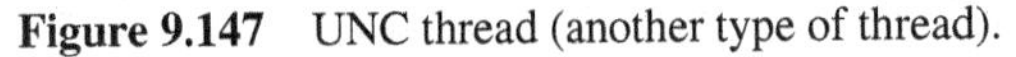
Figure 9.147 UNC thread (another type of thread).

9.1.8.1 Threading operation with CNC programming

Figure 9.148 shows the machining drawing.

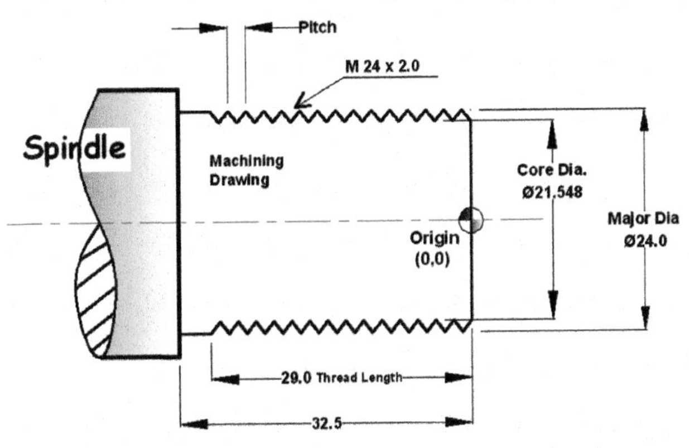

Figure 9.148 Above drawing is a machining drawing.

Figure 9.149 shows the raw material drawing.

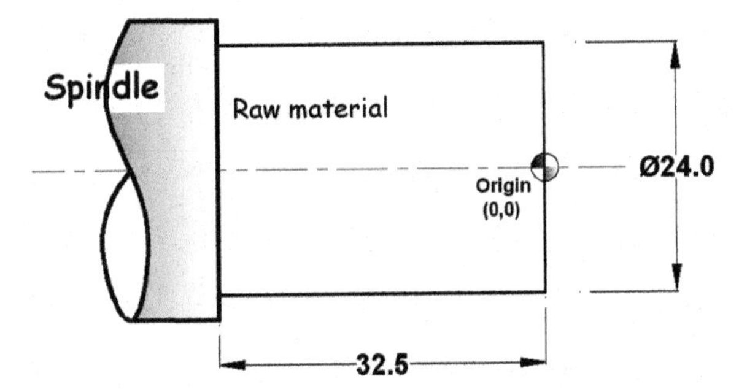

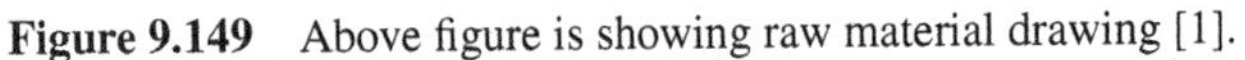
Figure 9.149 Above figure is showing raw material drawing [1].

Note:

In every cut of the thread, cutting tool will start his movement diametrically 10 mm above from the thread crest and 5.0 mm far from the work piece face.

Figure 9.150 shows the threading tool image. This threading tool is used in external threading operation.

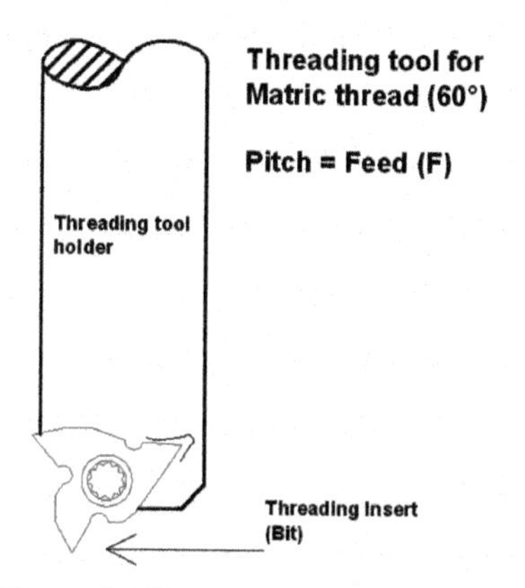

Figure 9.150 External threading tool.

For Metric Thread: If the thread is metric thread and we do not have one of them from pitch or thread height, than we can use following formula to find out pitch/thread height.

$$\textbf{Thread Height} = 0.613 \times \text{Thread Pitch}$$

For Unified Thread: If we do not have one of them from pitch or thread height, we can use the following formula to find out the pitch/thread height.

$$\textbf{Thread Height} = 0.7500 \times \text{Thread Pitch}$$
$$\textbf{Core Diameter} = \text{Major Diameter} - 2 \times \text{Thread Height}$$
$$\textbf{Pitch} = 1/\text{Number of Threads per inch (TPI)}$$

O1507; This is a first line of the CNC program. This is called program number.

G54; This is called work zero. When we take **G54** command on the center of the face of the round the work piece, that time coordinates of the center of that cylindrical job are (0, 0). It means all cutting tool will assume that point as an origin (0, 0) and important think is when we make the CNC program in absolute mode G90. Origin (0, 0) will take as a reference point for all coordinates in the CNC program.

T0808; This programming line is saying to machine that turret (tool post) will change the cutting tool and take the tool number eight (8) for machining. In this line, First two digits of T**0808** are representing the tool station number, where from tool is coming for machining and second two digits of T08**08** are representing the tool offset, tool offset means location number of tool data, where the tool data (tool geometry) is stored, finally we can say tool offset is the brain of cutting tool.

G00 X70 Z50; In Figure 9.151, the G00 command is used for rapid motion (non-cutting motion). Here, the cutting tool T0808 will take position rapidly in X-axis at diameter 70 mm and Z-axis at distance 50 mm from the origin (0, 0). The tool will use these coordinates to come near the job for cutting.

G97 M03 S1000; After execution of this programming line, in Figure 9.152, the spindle will rotate in clockwise direction at constant surface speed **S1000 rpm**.

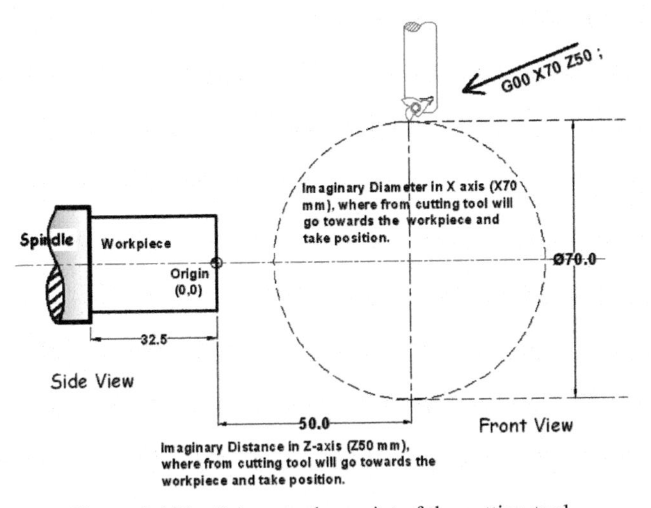

Figure 9.151 It is a starting point of the cutting tool.

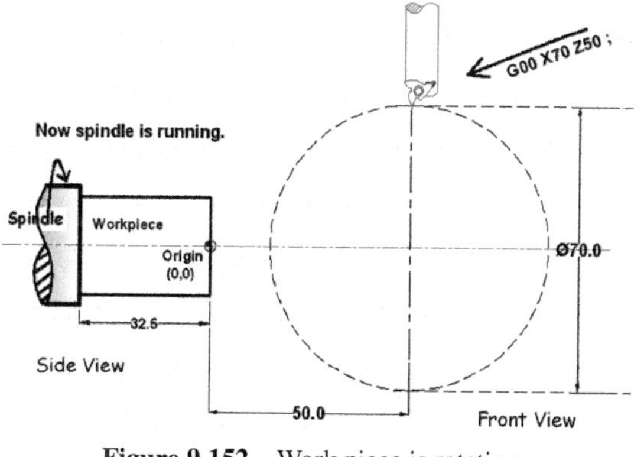

Figure 9.152 Work piece is rotating.

Note:

RPM should be constant during threading operation otherwise thread will completely damage.

G00 X34 Z5.0; In Figure 9.153, the threading operation, the cutting tool will take the final position in rapid motion G00 at the coordinates X34.0 and Z5.0.

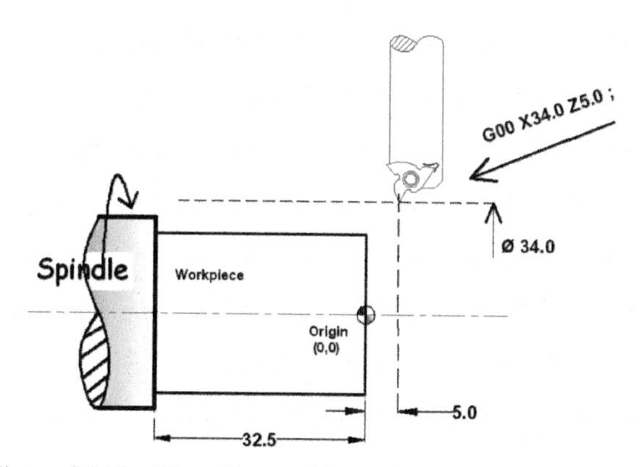

Figure 9.153 Threading tool is coming near the work piece.

Note:

During the threading cycle, when cutting tool retract from the work piece surface (thread) in turning machine, always should keep the safe distance between the threading tool and thread surface (thread). Generally, that safe distance will be diametrically 10 mm more from the crest diameter of the thread.

Example

If thread crest diameter is Ø19.0 mm. after making the thread, retraction diameter of thread tool tip will be Ø29.0 mm. So that thread tool does not damage the thread (work piece thread).

M08; When this programming line executes, it means coolant motor ON.

G92 X23.8 Z-29.0 F2.0; In Figure 9.154, the programming line G92 is a thread cutting code. When **G92** programming line will execute. First, threading tool will take **position in X-axis the direction at diameter 23.8 mm and in Z-axis at distance Z5.0**. Now threading tool will move in Z-axis direction with **feed F2.0** from Z5.0 mm to Z5.0 mm to Z-29.0 mm (thread length) and make the thread. It means tool reached at actual Z-29.0 mm. So finally, we can say that in the first cut, threading tool will make thread at Ø23.8 mm on the length 29.0 mm.

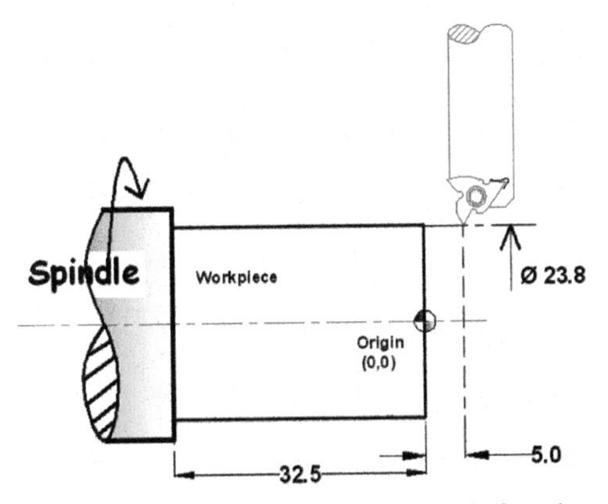

Figure 9.154 Threading tool took the cutting position before threading operation.

Note:

When G92 command will execute, threading tool will start the thread cutting from the diameter 23.8 and Z5.0 and will go at the length Z-29.0 on same diameter.

In Figure 9.154, before threading, threading tool is taking position in X-axis direction at diameter 23.8 mm but in Z-axis, the value of Z for tool tip always 5.0 mm far from the work piece face when tool starts the thread or retracts after threading.

In Figure 9.155, cutting tool tip position with the zoomed figure.

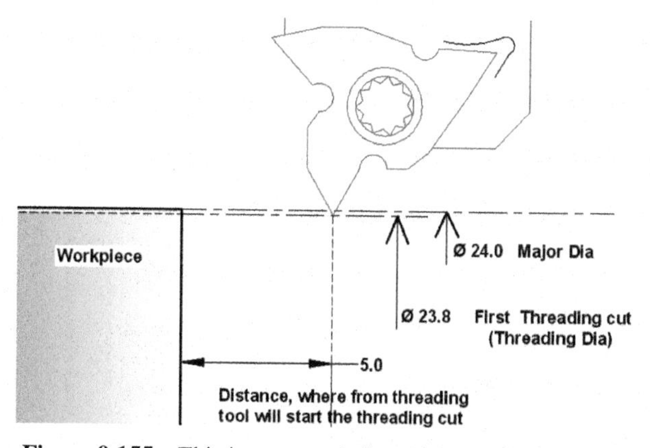

Figure 9.155 This is a zoomed view of the previous figure.

In Figure 9.156, after taking position in X and Z-axis now cutting tool is going in Z-axis direction at the threading length Z-29.0 and removing the material. This time X-axis value will be X23.8. and feed 2.0 mm/revolution.

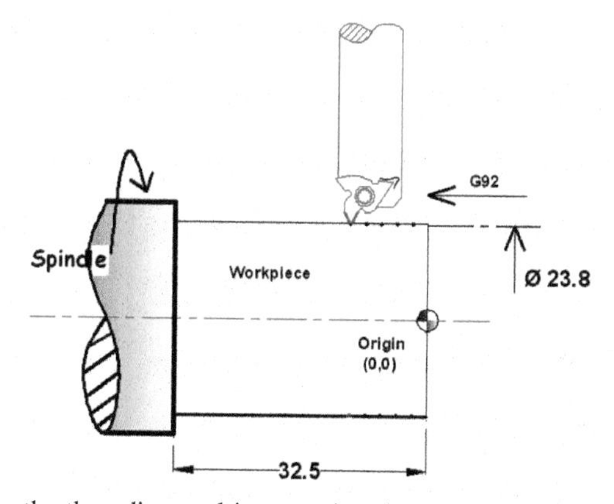

Figure 9.156 Now the threading tool is removing the extra material and making thread on the work piece surface at Ø23.8.

It is happening under G92 command (the whole threading operation under controlled by G92 command).

In Figure 9.157, threading tool removed the material in the first cut at diameter 23.8 mm and made the thread profile at the length Z-29.0 with feed F2.0.

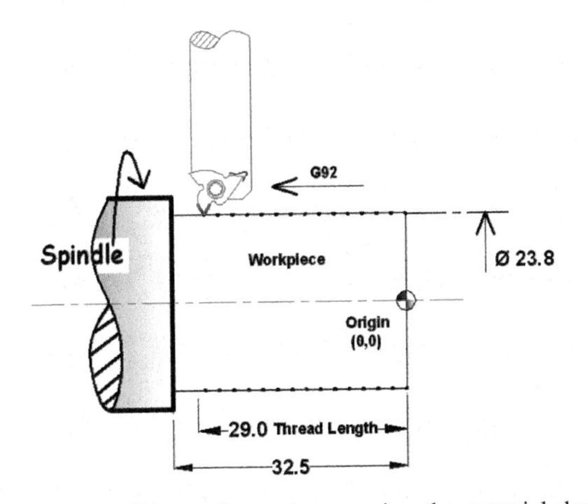

Figure 9.157 Threading tool is continuously removing the material during the threading operation.

During the first threading cut in Figure 9.158, cutting tool tip position with the zoomed figure.

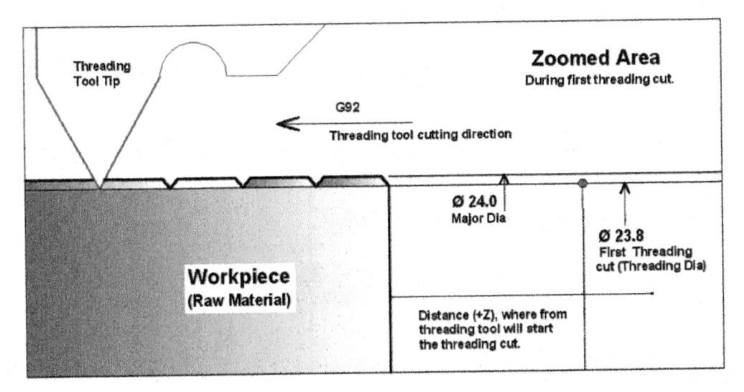

Figure 9.158 This is a zoomed figure of the previous figure for better understanding.

After first threading cut, in Figure 9.159, first of all threading tool is retracting in X-axis at Ø34.0 mm.

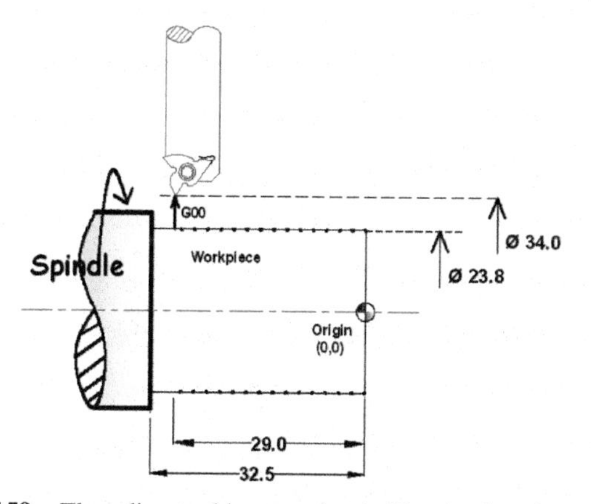

Figure 9.159 Threading tool is retracting in X-axis after thread making.

Note:

After every cut of the thread, programming line (**G00 X34.0 Z5.0;**]) will be executed and when programming line will execute, it means **first of all tool will retract in X-axis at diameter (X34.0)** and after that tool will retract in Z-axis (Z5.0) in length.

After first threading cut, cutting tool **retracted** in X-axis (Ø34.0 mm). Now in Figure 9.160, the **cutting tool is retracting in Z-axis direction at Z5.0**.

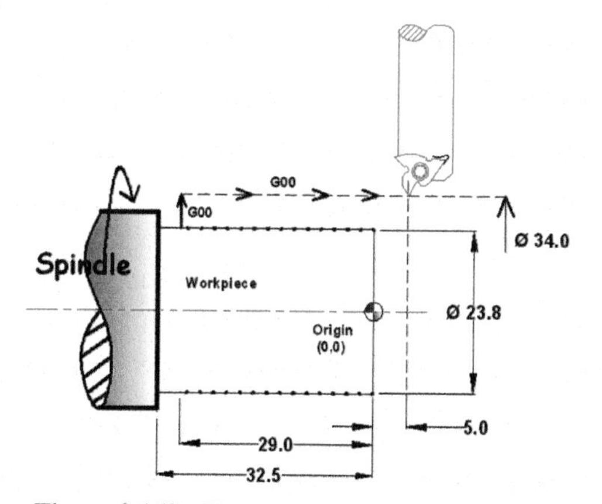

Figure 9.160 Now tool is retracting in Z-axis.

Note:

It is very important that after every threading cut, programming line (G00 X34.0 **Z5.0;**) will be executed and when programming line will execute, it means **first of all X-axis will retract (as per programming line ex: X34) and after then Z-axis will retract (as per programming line ex: Z5.0).** See Figure 9.160.

9.1.8.2 A brief note about threading procedure

Thread cutting procedure are showing in the above figures. This thread cutting procedure starts from the first positioning of the thread tool at Ø34.0 mm with distance Z5.0 (machine took these co-ordinates from the programming line **G00 X34.0 Z5.0;** this line came before **G92 X23.8 Z-29.0 F2;**) and at the end of the first thread cutting procedure, the tool will come again at same dimensions. It will repeat again and again after each threading cut.

We can say that during threading procedure, first of all tool takes the position at Ø34.0 mm and Z5.0 mm now after this step, next step is tool takes the position at threading diameter X23.8 and Z5.0 and start thread cutting at same previous diameter 23.8 in Z direction till length Z-29.0. After thread cutting, first of all tool will retract in X-axis at Ø34.0. When tool reached at Ø34.0 now tool retracts in Z-axis at the Z5.0 (retraction value in X and Z-axis will take from the given programming line G00 X34.0 **Z5.0;** this value can be different in another threading program according to thread size.)

Above figures and statements are about first threading cut of the whole threading operation.

Note:

Now from below programming line (X23.6;), the programmer will give only cutting thread diameter in each programming line and threading operation will perform automatically by threading tool in sequence, in the first step, the tool takes the cutting diameter position. In the second step, the tool starts the thread cutting on required length. In the third step, after thread cutting, tool retracts in the X axis. In the Fourth step, now tool retracts in Z-axis. It is only one complete thread cutting path (one cycle). This procedure will repeat again and again with new thread cutting diameters until tool does not reach at the minor (root) diameter as per drawing.

X23.6; In this programming line. We write only thread diameter X23.6, other cutting parameters like thread length (Z), feed and Pitch will take automatically from the first line (G92 X 23.6 Z -29.0 F2.0;)

of the threading program. Now threading tool makes the thread at diameter X23.6 and length Z -29.0.

Note:

No need to write the G92 X 23.6 Z -29.0 F2.0;

In Figure 9.161, before making the thread, the cutting tool is taking position in X-axis and Z-axis at diameter 23.6 and Z5.0 for thread cutting. Before making the thread, the tool will take this position (X and Z) automatically in every cut.

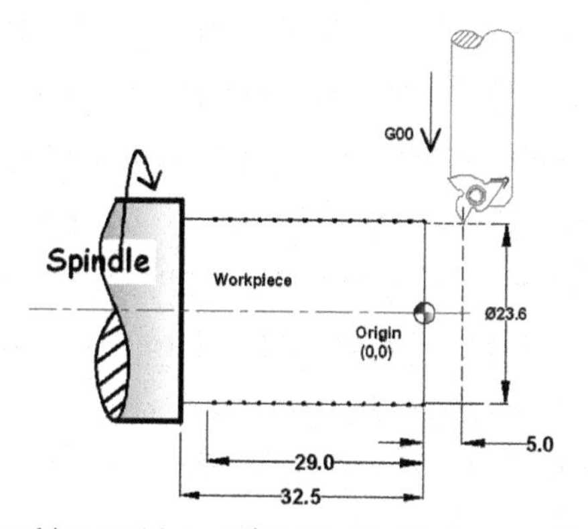

Figure 9.161 Tool is taking position at Ø23.6 in X-axis for next thread cutting operation.

Here, in Figure 9.162 the threading tool is moving continuously towards the Z-axis direction and making the thread.

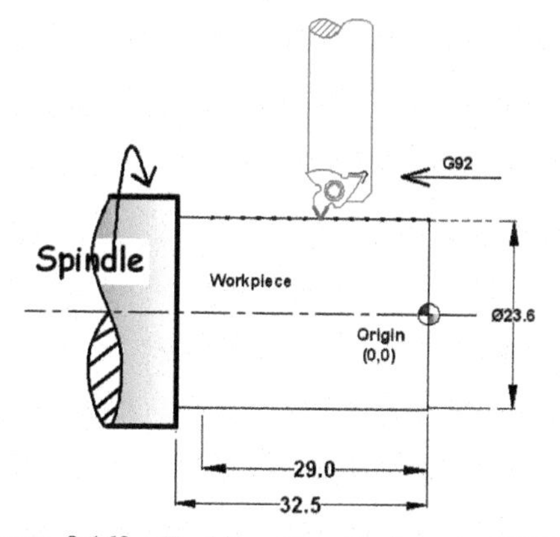

Figure 9.162 Tool is cutting the thread at Ø23.6.

In Figure 9.163, the threading tool reached at the last end of the thread length Z-29.0 and made the thread profile at Ø23.6.

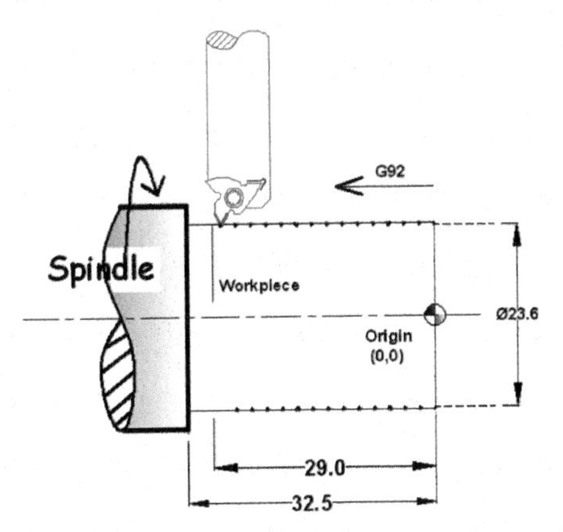

Figure 9.163 Tool is cutting material continuously.

In Figure 9.164, the threading tool is retracting at Ø34.0 mm in X-axis direction.

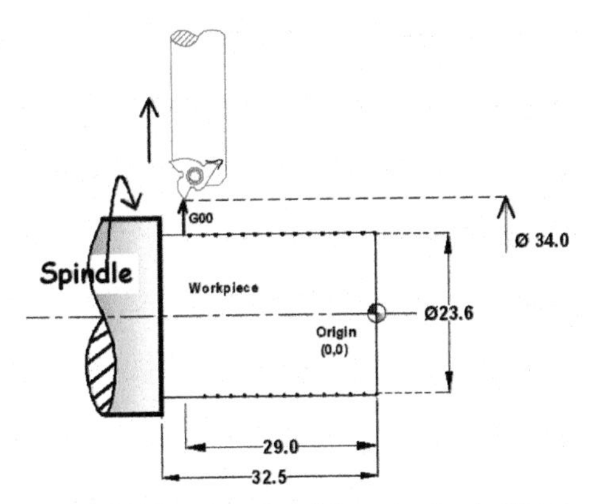

Figure 9.164 Thread tool is retracting in X-axis.

After retracted in X-axis direction now threading tool is retracting in Z-axis and will reach at the positive Z distance Z5.0 in Z-axis (Figure 9.165).

X23.4; In this programming line. Threading tool will cut the thread diameter at X23.4. This line is a complete for threading operation. All previous procedure of tool positioning will apply here during threading operation.

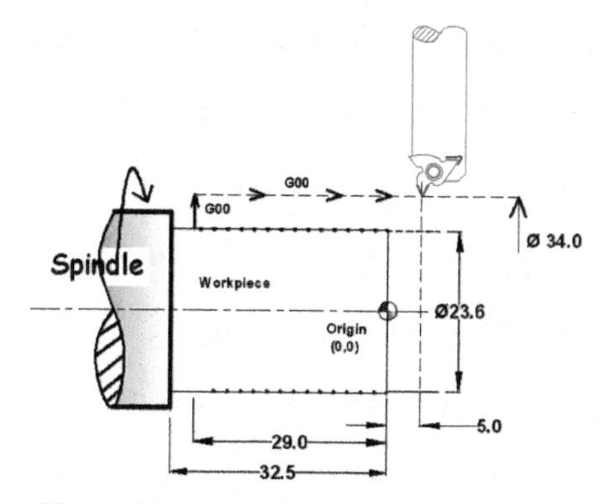

Figure 9.165 Tool is retracting in Z-axis.

In Figure 9.166, the thread cutting tool is taking position rapidly in X-axis and Z-axis at diameter X23.4 and Z5.0.

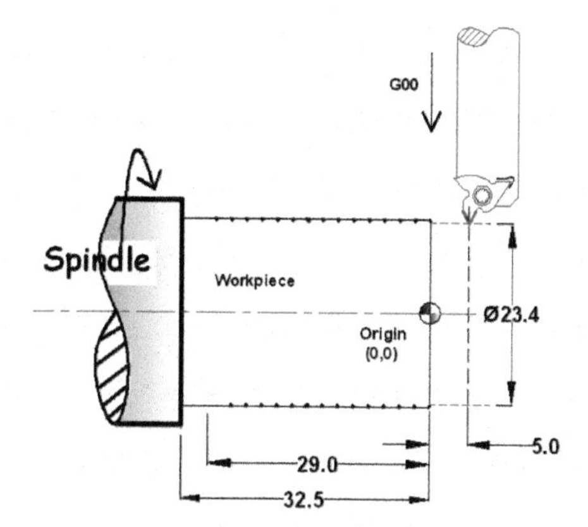

Figure 9.166 Tool is taking position at Ø23.4 in X-axis for next thread cutting operation.

In Figure 9.167, the threading tool is cutting the thread continuously.

In Figure 9.168, the threading tool reached at the last end of the thread length Z-29.0 and made the thread profile at Ø23.4.

In Figure 9.169, the threading tool is retracting at Ø34.0 mm in X-axis direction.

After retracted in X-axis direction now threading tool is retracting in Z-axis and will reach at the positive Z distance Z5.0 (Figure 9.170).

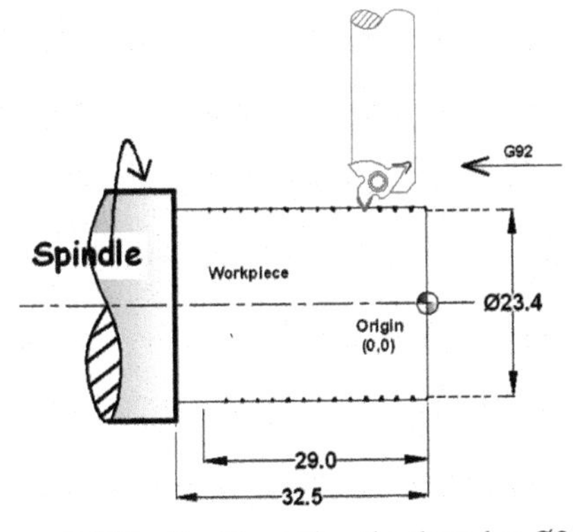

Figure 9.167 Tool is cutting the thread at Ø23.4.

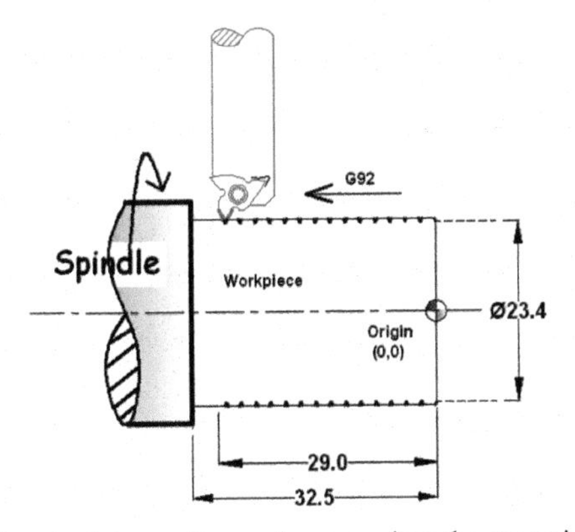

Figure 9.168 Threading tool is continuously removing the material during the threading operation.

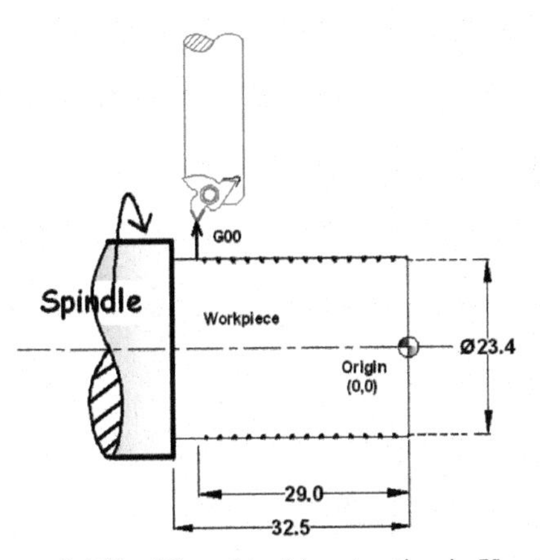

Figure 9.169 Thread tool is retracting in X-axis.

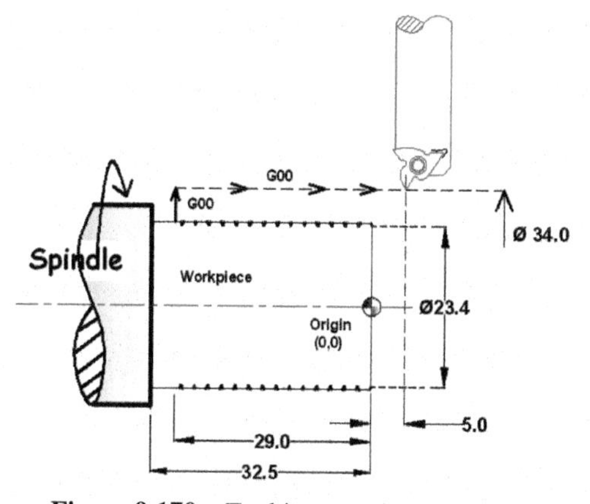

Figure 9.170 Tool is retracting in Z-axis.

X23.2; In this programming line. Threading tool will cut the thread diameter at X23.2. This line is a complete for threading operation. All previous procedure of tool positioning will apply here during threading operation.

In Figure 9.171, the thread cutting tool is taking position rapidly in X-axis and Z-axis at diameter X23.2 and Z5.0.

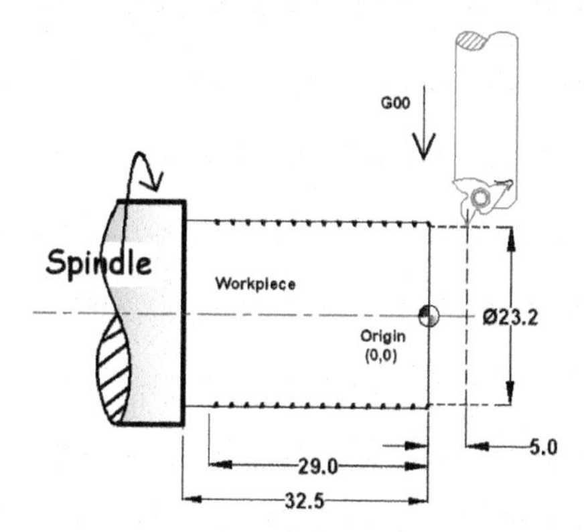

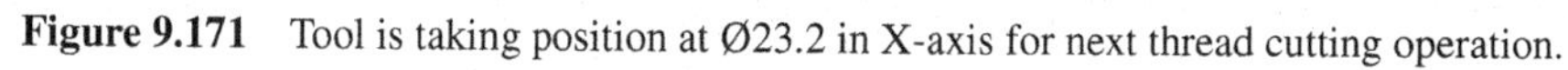
Figure 9.171 Tool is taking position at Ø23.2 in X-axis for next thread cutting operation.

In Figure 9.172, threading tool is cutting the thread continuously.

In Figure 9.173, the threading tool reached at the last end of the thread length Z-29.0 and made the thread profile at Ø23.2.

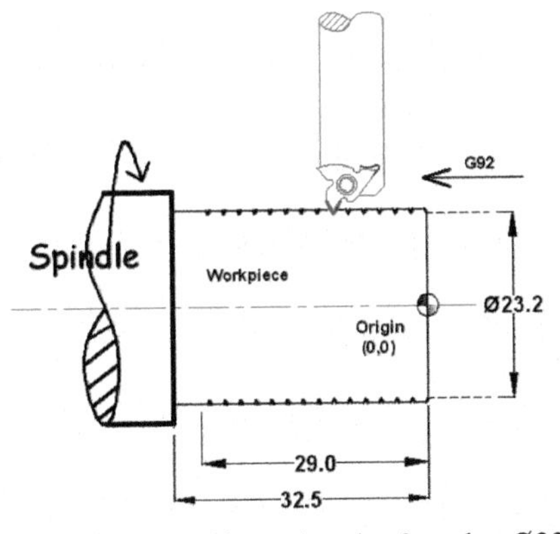

Figure 9.172 Tool is cutting the thread at Ø23.2.

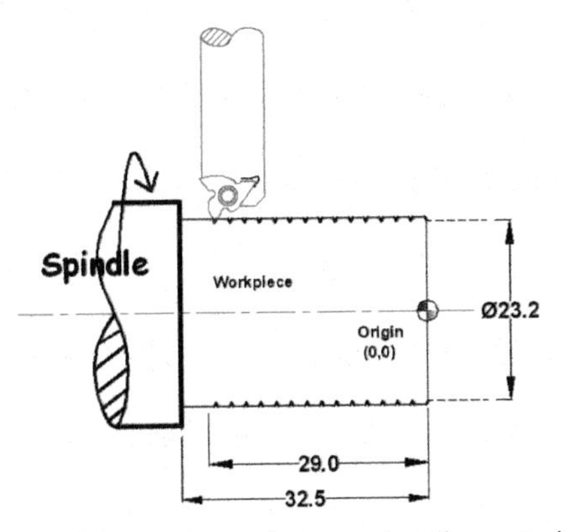

Figure 9.173 Threading tool is continuously removing the material during the threading operation.

In Figure 9.174, the threading tool is retracting at Ø34.0 mm in X-axis direction.

After retracted in X-axis direction now threading tool is retracting in Z-axis and will reach at the positive Z distance Z5.0. See Figure 9.175.

X23.0; In this programming line (Figure 9.176). Threading tool will cut the thread diameter at X23.0. This line is a complete for threading operation. All previous procedure of tool positioning will apply here during the threading operation.

In Figure 9.176, the thread cutting tool is taking position rapidly in X-axis and Z-axis at diameter X23.0 and Z5.0.

In Figure 9.177, the threading tool is cutting the thread continuously.

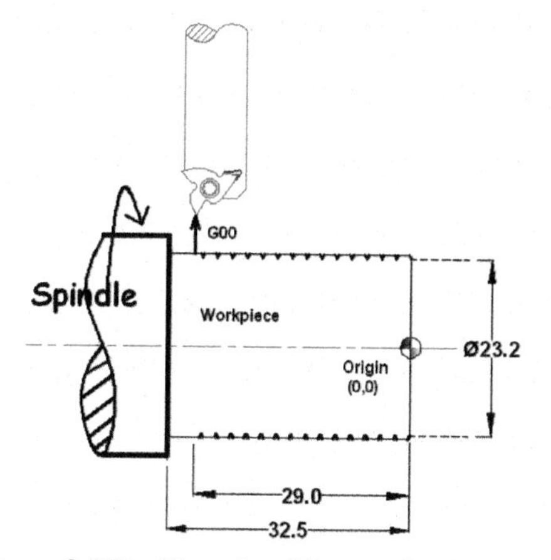

Figure 9.174 Thread tool is retracting in X-axis.

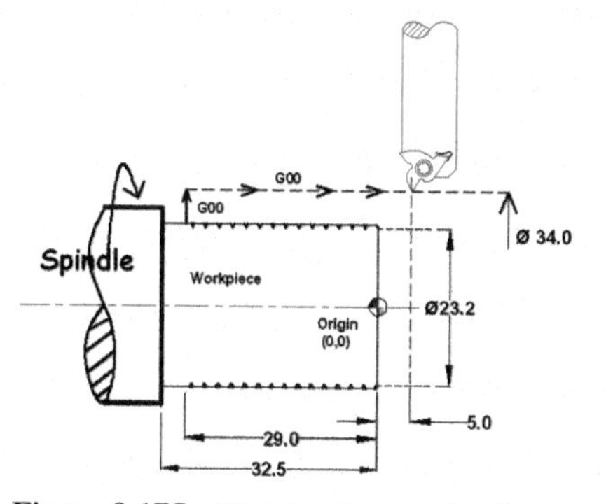

Figure 9.175 Tool is retracting in Z-axis.

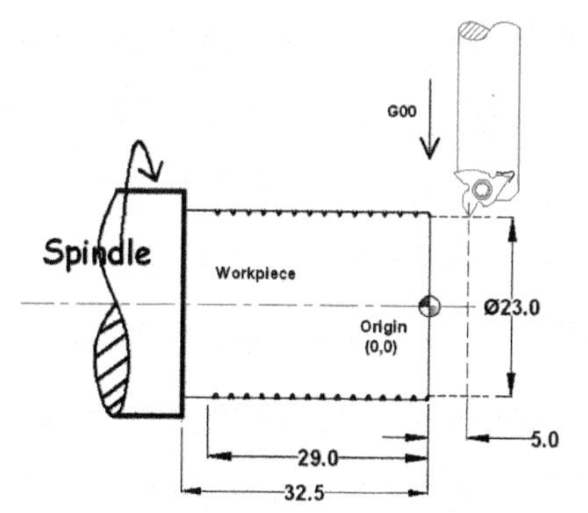

Figure 9.176 Tool is taking position at Ø23.0 in X-axis for next thread cutting operation.

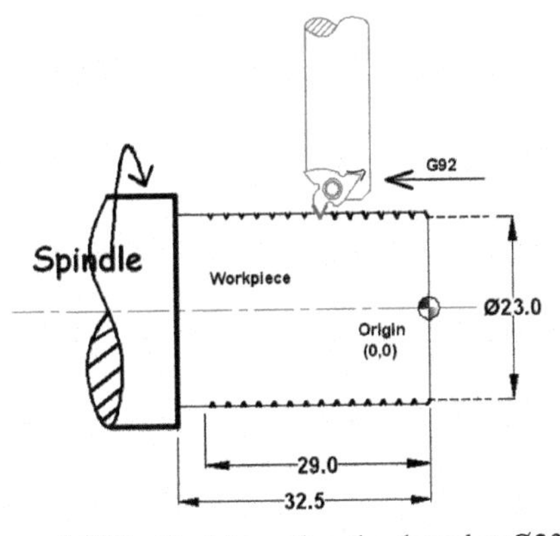

Figure 9.177 Tool is cutting the thread at Ø23.0.

In Figure 9.178, the threading tool reached at the last end of the thread length Z-29.0 and made the thread profile at Ø23.0.

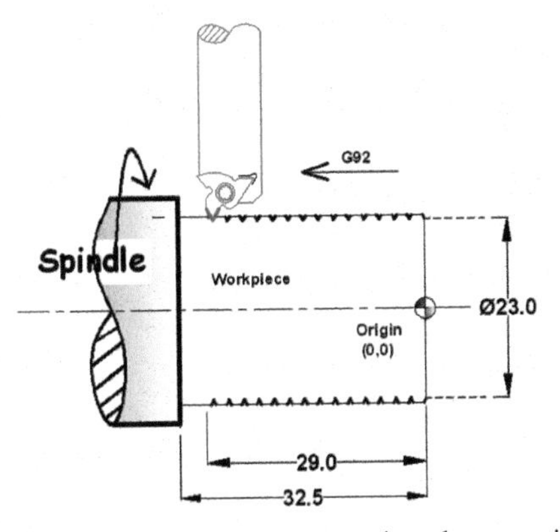

Figure 9.178 Threading tool is continuously removing the material during the threading operation.

In Figure 9.179, the threading tool is retracting at Ø34.0 mm in X-axis direction.

After retracted in X-axis direction now threading tool is retracting in Z-axis and will reach at the positive Z distance Z5.0 (Figure 9.180).

X22.8; In this programming line. Threading tool will cut the thread diameter at X22.8. This line is a complete for threading operation. All previous procedure of tool positioning will apply here during threading operation.

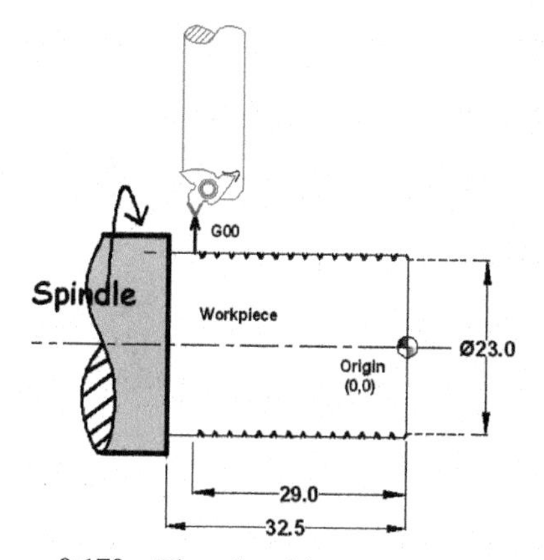

Figure 9.179 Thread tool is retracting in X-axis.

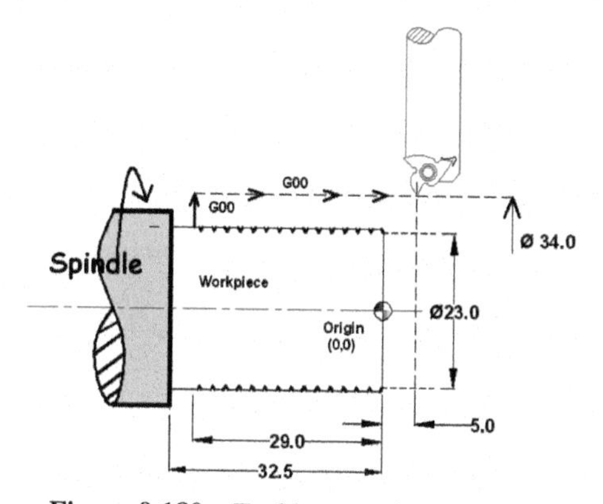

Figure 9.180 Tool is retracting in Z-axis.

In Figure 9.181, the thread cutting tool is taking position rapidly in X-axis and Z-axis at diameter X22.8 and Z5.0.

In Figure 9.182, the threading tool is cutting the thread continuously.

In Figure 9.183, the threading tool reached at the last end of the thread length Z-29.0 and made the thread profile at Ø22.8.

In Figure 9.184, the threading tool is retracting at Ø34.0 mm in X-axis direction.

After retracted in X-axis direction now threading tool is retracting in Z-axis and will reach at the positive distance Z5.0 (Figure 9.185).

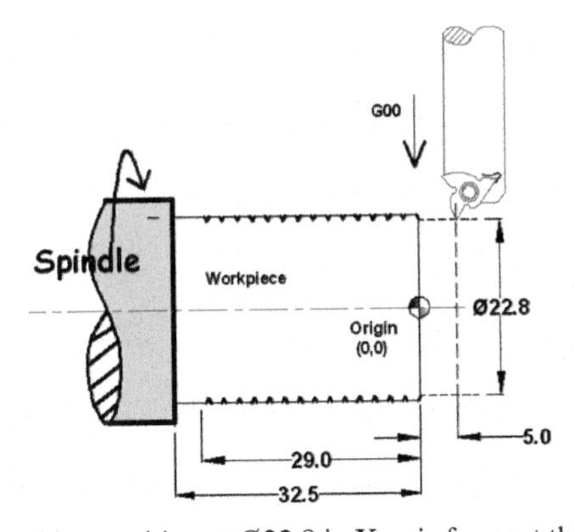

Figure 9.181 Tool is taking position at Ø22.8 in X-axis for next thread cutting operation.

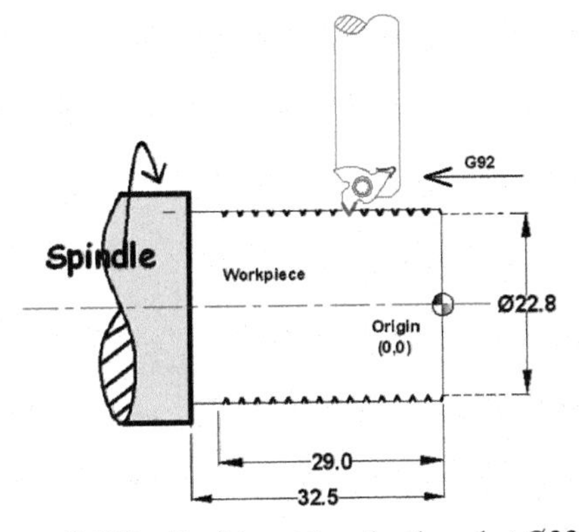

Figure 9.182 Tool is cutting the thread at Ø22.8.

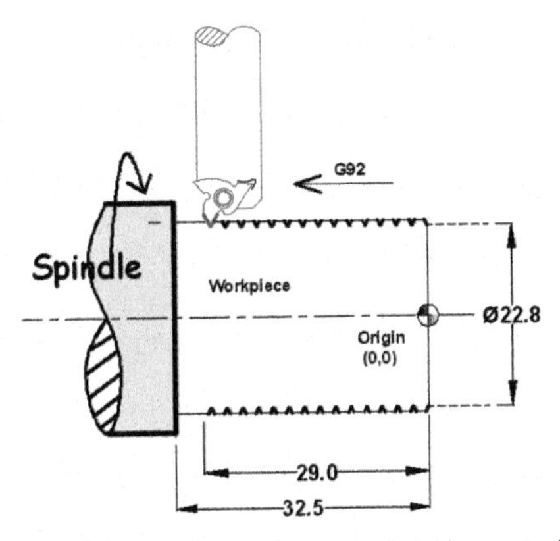

Figure 9.183 Threading tool is continuously removing the material during the threading operation.

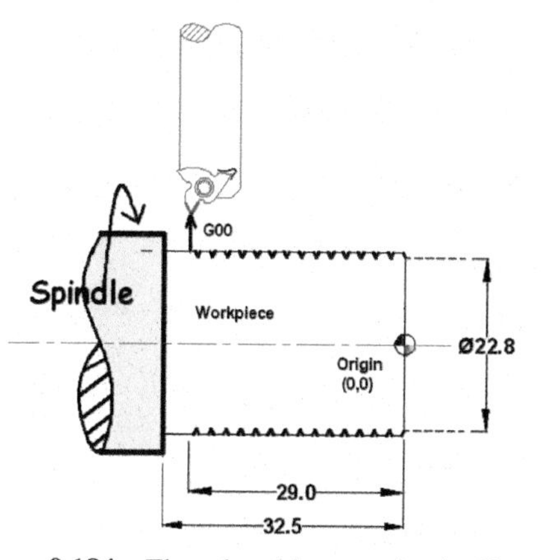

Figure 9.184 Thread tool is retracting in X-axis.

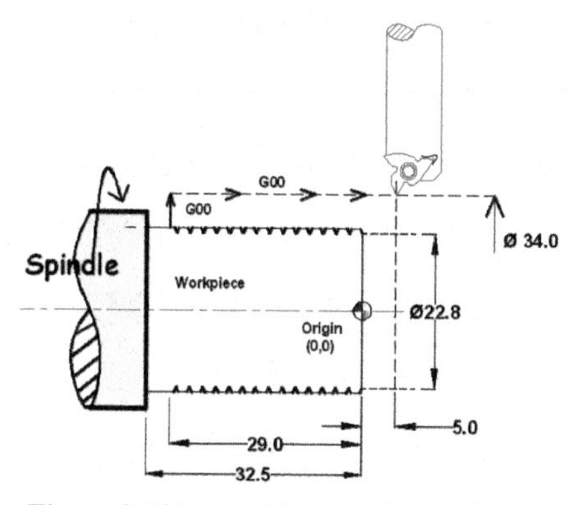

Figure 9.185 Tool is retracting in Z-axis.

X22.6; In this programming line. Threading tool will cut the thread diameter at X22.6. This line is a complete for threading operation. All previous procedures of tool positioning will apply here during threading operation.

In Figure 9.186, the thread cutting tool is taking rapidly position in X-axis and Z-axis at diameter X22.6 and Z5.0.

In Figure 9.187, the threading tool is cutting the thread continuously.

In Figure 9.188, the threading tool reached at the last end of the thread length Z-29.0 and made the thread profile at Ø22.6.

In Figure 9.189, the threading tool is retracting at Ø34.0 mm in X-axis direction.

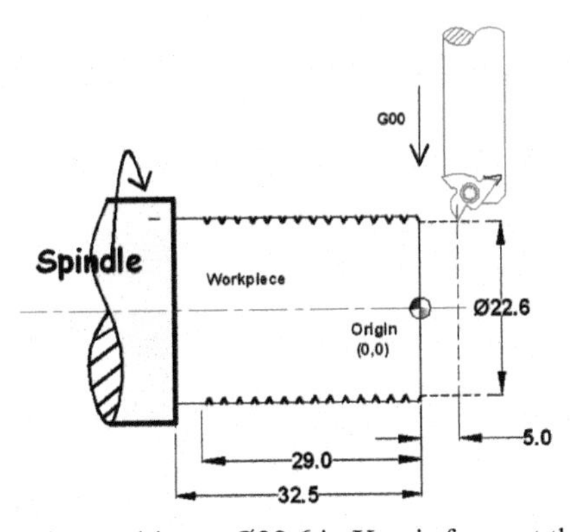

Figure 9.186 Tool is taking position at Ø22.6 in X-axis for next thread cutting operation.

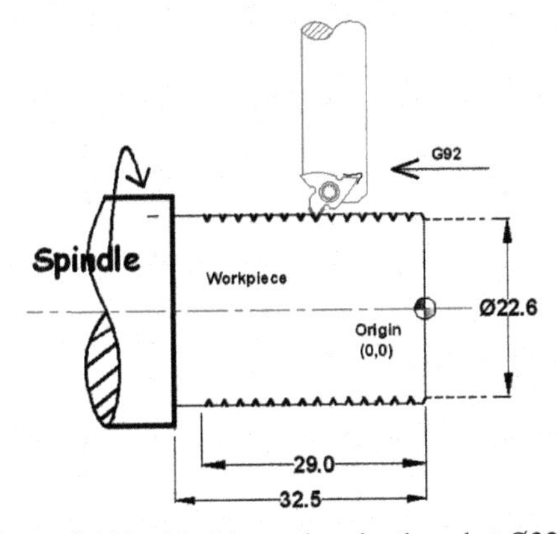

Figure 9.187 Tool is cutting the thread at Ø22.6.

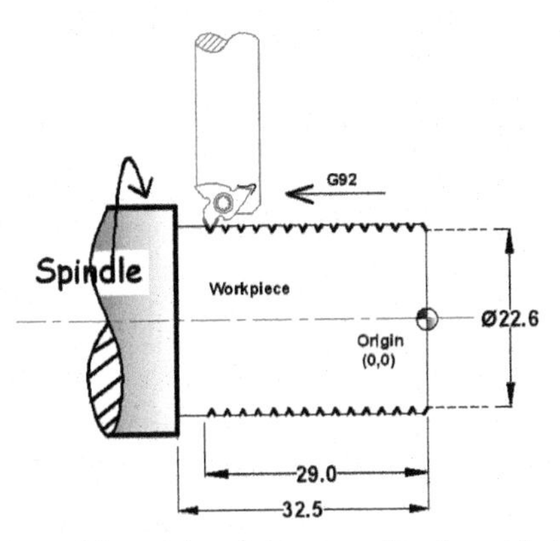

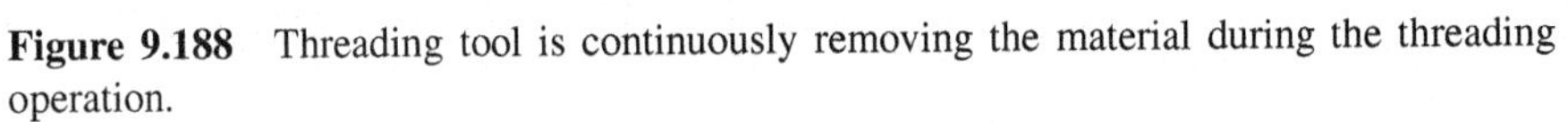

Figure 9.188 Threading tool is continuously removing the material during the threading operation.

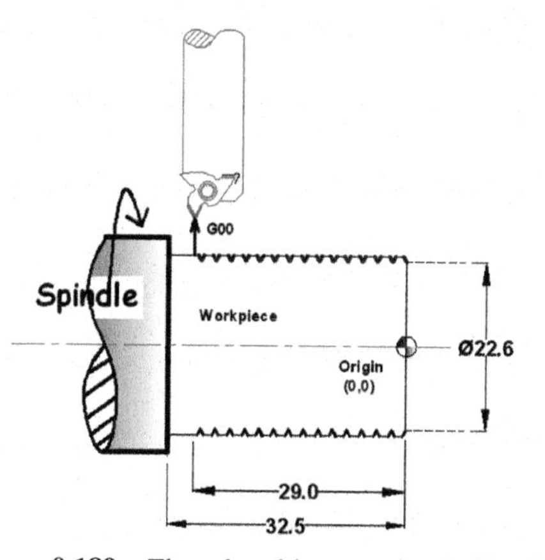

Figure 9.189 Thread tool is retracting in X-axis.

After retraction in X-axis direction now threading tool is retracting in Z-axis and will reach at the positive distance Z5.0 (Figure 9.190).

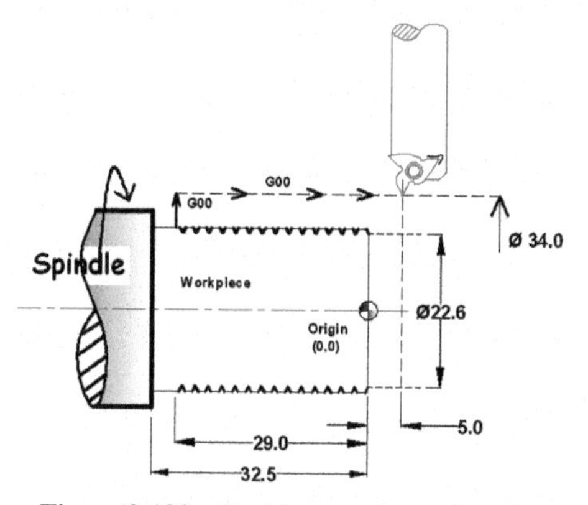

Figure 9.190 Tool is retracting in Z-axis.

X22.4; In this programming line, the threading tool will cut the thread diameter at X22.4. This line is a complete for threading operation. All previous procedure of tool positioning will apply here during threading operation.

In Figure 9.191, the thread cutting tool is taking position rapidly in X-axis and Z-axis at diameters X22.4 and Z5.0.

In Figure 9.192, the threading tool is cutting the thread continuously.

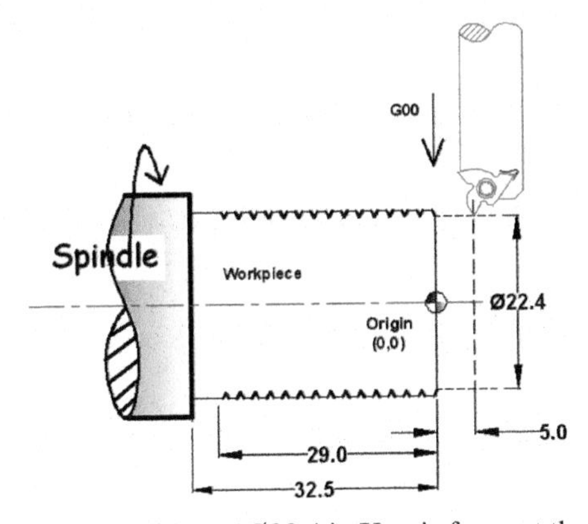

Figure 9.191 Tool is taking position at Ø22.4 in X-axis for next thread cutting operation.

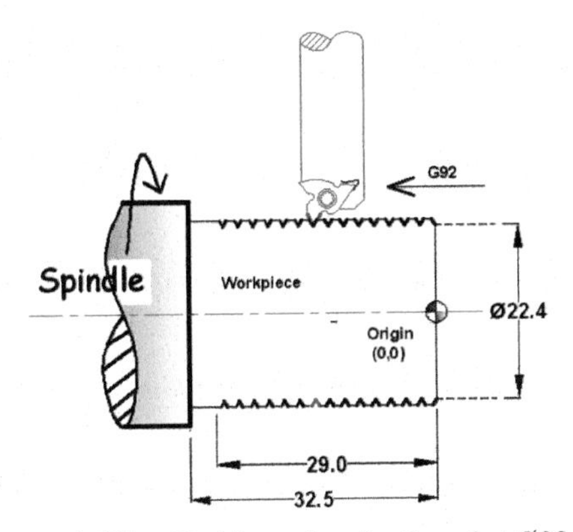

Figure 9.192 Tool is cutting the thread at Ø22.4.

In Figure 9.193, the threading tool reached at the last end of the thread length Z-29.0 and made the thread profile at Ø22.4.

In Figure 9.194, the threading tool is retracting at Ø34.0 mm in X-axis direction.

After retracted in X-axis direction now threading tool is retracting in Z-axis and will reach at the positive distance Z5.0 (Figure 9.195).

X22.2; In this programming line, the threading tool will cut the thread diameter at X22.2. This line is a complete for threading operation. All previous procedure of tool positioning will apply here during threading operation.

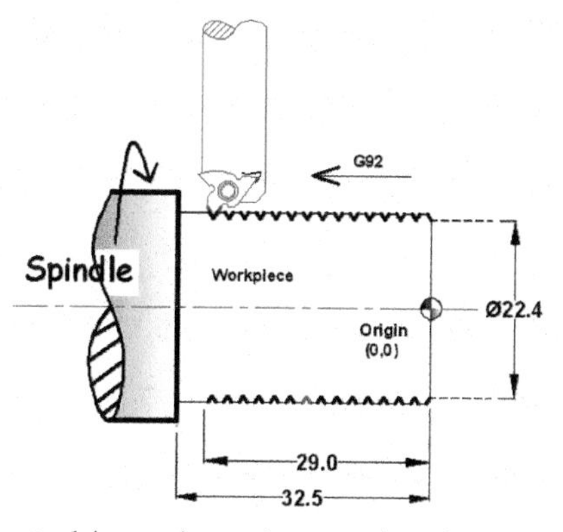

Figure 9.193 Threading tool is continuously removing the material during the threading operation.

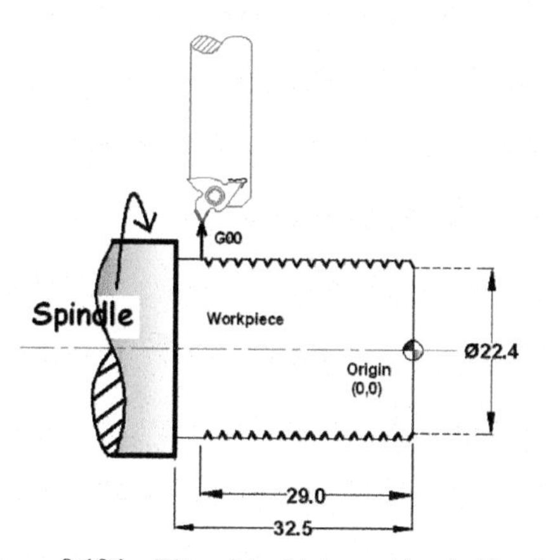

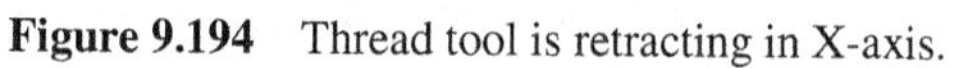

Figure 9.194 Thread tool is retracting in X-axis.

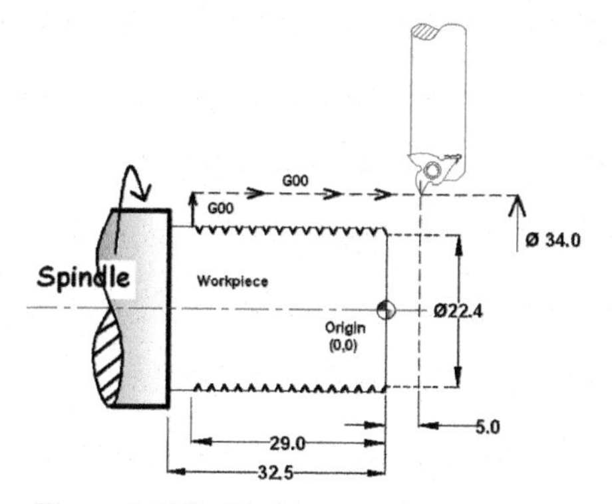

Figure 9.195 Tool is retracting in Z-axis.

In Figure 9.196, the thread cutting tool is taking position rapidly in X-axis and Z-axis at diameter X22.2 and Z5.0.

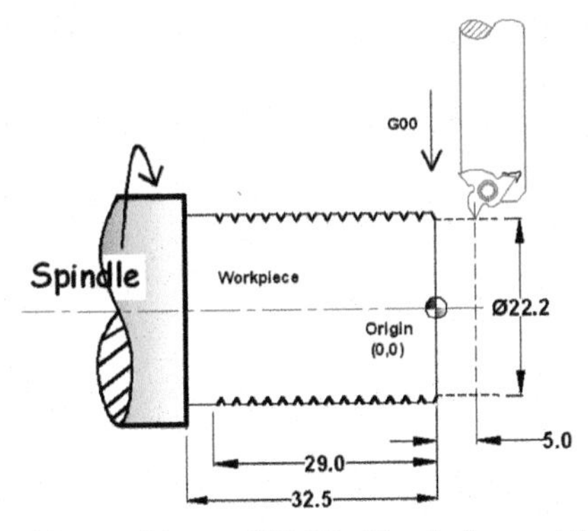

Figure 9.196 Tool is taking position at Ø22.2 in X-axis for next thread cutting operation.

In Figure 9.197, the threading tool is cutting the thread continuously.

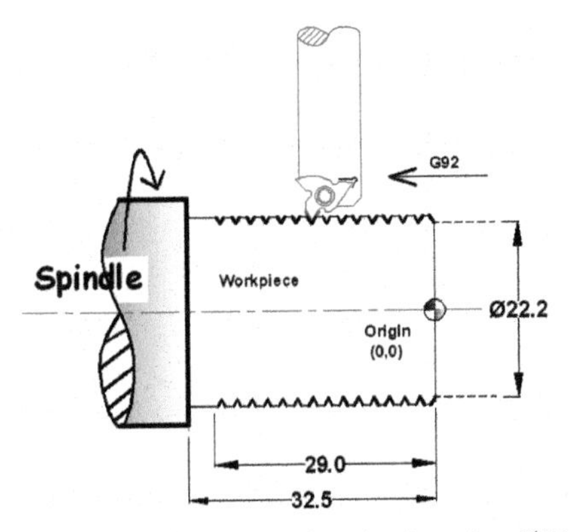

Figure 9.197 Tool is cutting the thread at Ø22.2.

In Figure 9.198, the threading tool reached at the last end of the thread length Z-29.0 and made the thread profile at Ø22.2.

In Figure 9.199, the threading tool is retracting at Ø34.0 mm in X-axis direction.

After retraction, in X-axis direction now threading tool is retracting in Z-axis and will reach at the positive distance Z5.0 (Figure 9.200).

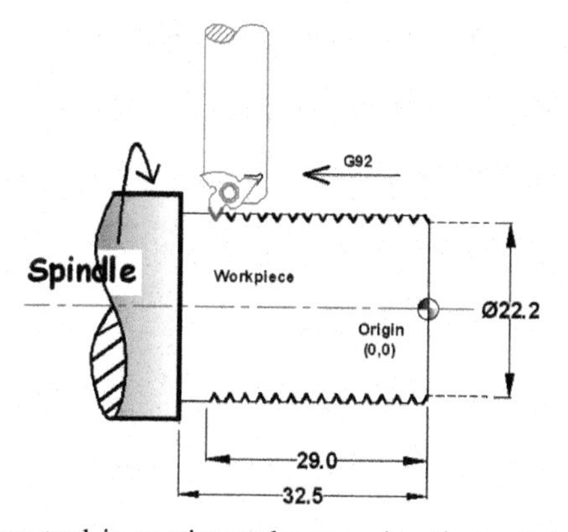

Figure 9.198 Threading tool is continuously removing the material during the threading operation.

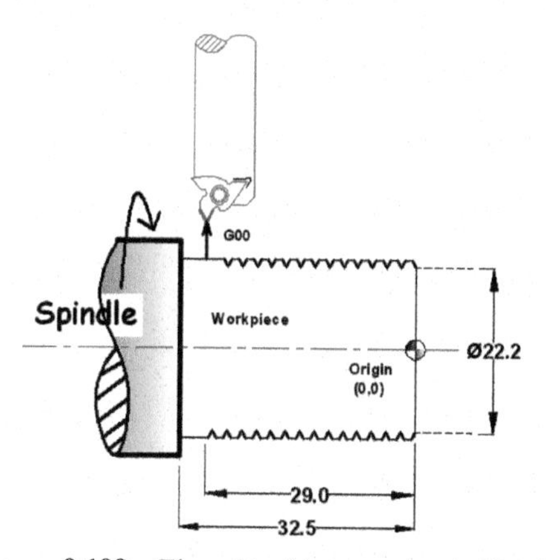

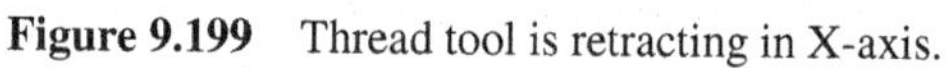

Figure 9.199 Thread tool is retracting in X-axis.

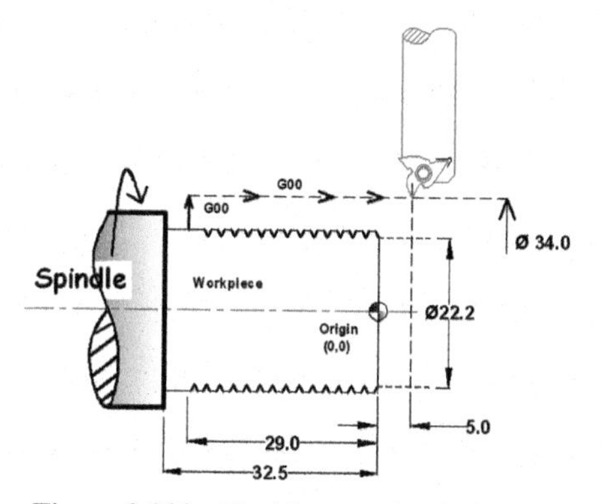

Figure 9.200 Tool is retracting in Z-axis.

X22.0; In this programming line, the threading tool will cut the thread diameter at X22.0. This line is a complete for threading operation. All previous procedure of tool positioning will apply here during the threading operation.

In Figure 9.201, the thread cutting tool is taking position rapidly in X-axis and Z-axis at diameters X22.0 and Z5.0.

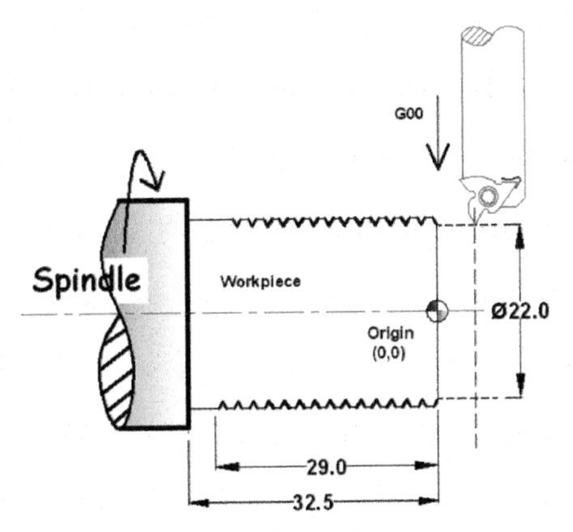

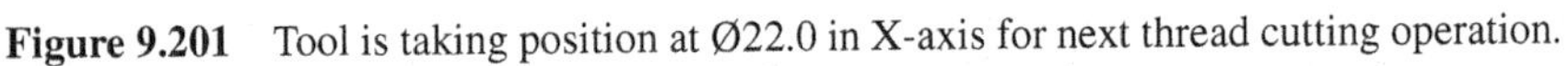
Figure 9.201 Tool is taking position at Ø22.0 in X-axis for next thread cutting operation.

In Figure 9.202, the threading tool is cutting the thread continuously.

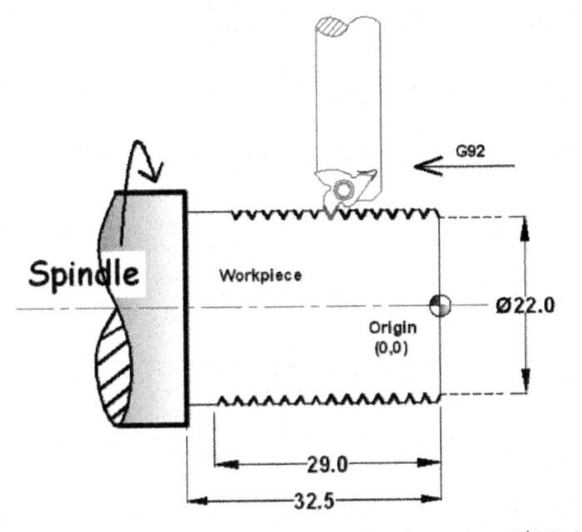

Figure 9.202 Tool is cutting the thread at Ø22.0.

In Figure 9.203, the threading tool reached at the last end of the thread length Z-29.0 and made the thread profile at Ø22.0.

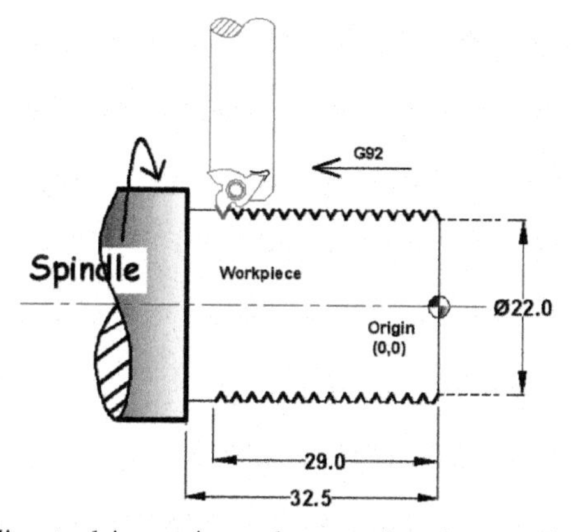

Figure 9.203 Threading tool is continuously removing the material during the threading operation.

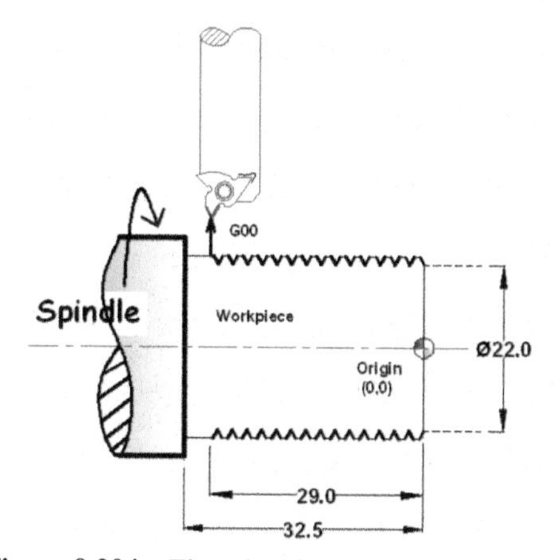

Figure 9.204 Thread tool is retracting in X-axis.

In Figure 9.204, the threading tool is retracting at Ø34.0 mm in X-axis direction.

In Figure 9.205, after retracted in X-axis direction now threading tool is retracting in Z-axis and will reach at the positive distance Z5.0.

X21.8; In this programming line, the threading tool will cut the thread diameter at X21.8. This line is a complete for threading operation. All previous procedure of tool positioning will apply here during threading operation.

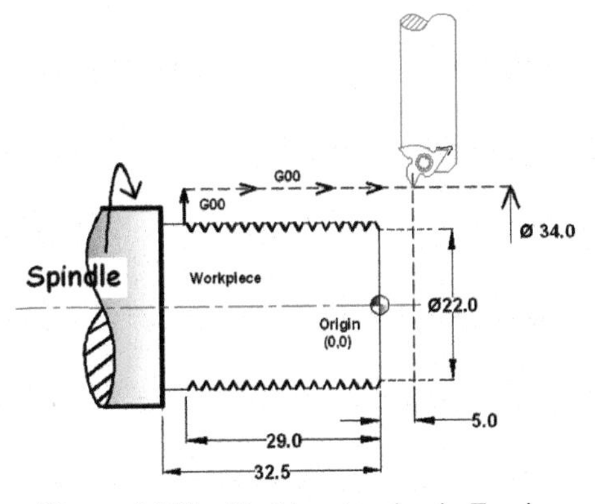

Figure 9.205 Tool is retracting in Z-axis.

In Figure 9.206, the thread cutting tool is taking position rapidly in X-axis and Z-axis at diameters X21.8 and Z5.0.

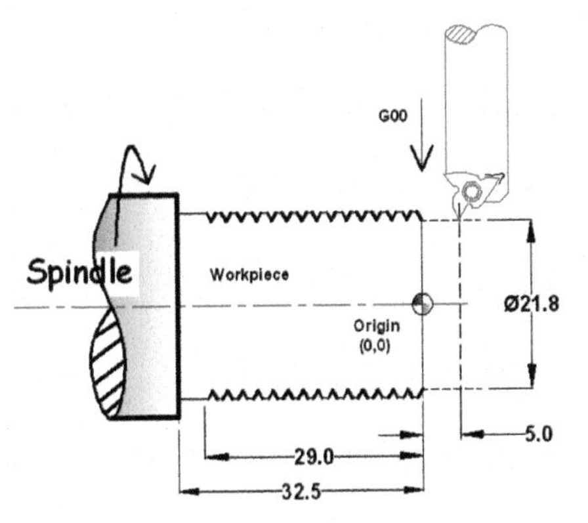

Figure 9.206 Tool is taking position at Ø21.8 in X-axis for next thread cutting operation.

In Figure 9.207, the threading tool is cutting the thread continuously.

In Figure 9.208, the threading tool reached at the last end of the thread length Z-29.0 and made the thread profile at Ø21.8.

In Figure 9.209, the threading tool is retracting at Ø34.0 mm in X-axis direction.

After retracted in X-axis direction now threading tool is retracting in Z-axis and will reach at the positive distance Z5.0 (Figure 9.210).

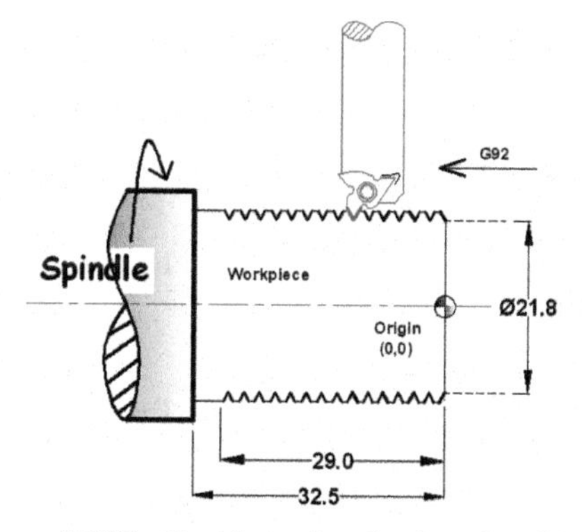

Figure 9.207 Tool is cutting the thread at Ø21.8.

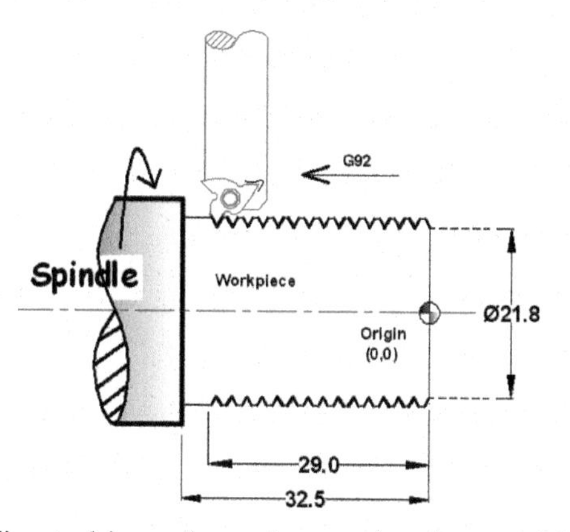

Figure 9.208 Threading tool is continuously removing the material during the threading operation.

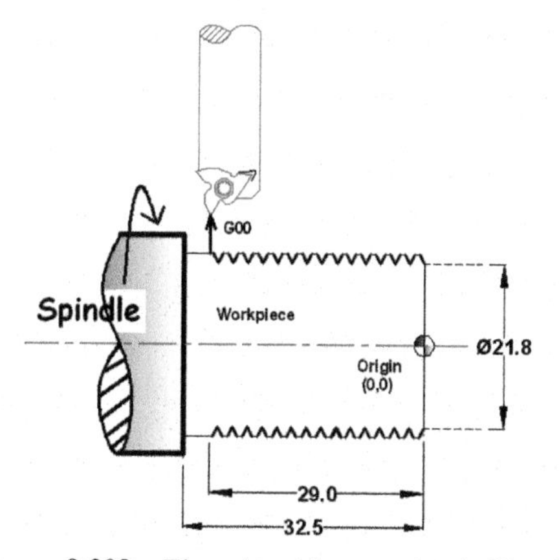

Figure 9.209 Thread tool is retracting in X-axis.

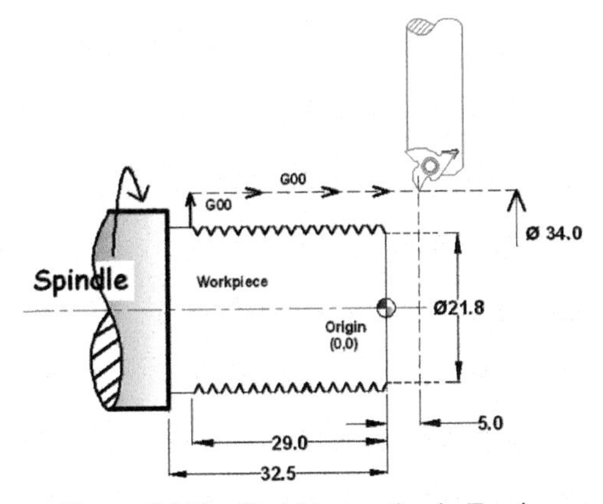

Figure 9.210 Tool is retracting in Z-axis.

X21.6; In this programming line. Threading tool will cut the thread diameter at X21.6. This line is a complete for threading operation. All previous procedure of tool positioning will apply here during threading operation.

In Figure 9.211, the thread cutting tool is taking position rapidly in X-axis and Z-axis at diameters X21.6 and Z5.0.

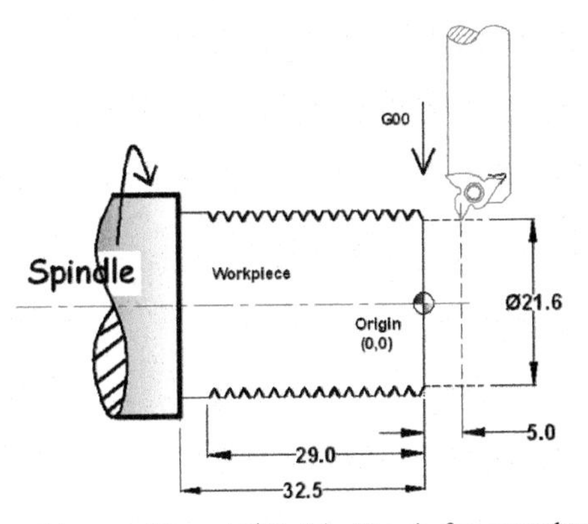

Figure 9.211 Tool is taking position at Ø21.6 in X-axis for next thread cutting operation.

In Figure 9.212, the threading tool is cutting the thread continuously.

In Figure 9.213, the threading tool reached at the last end of the thread length Z-29.0 and made the thread profile at Ø21.6.

In Figure 9.214, the threading tool is retracting at Ø34.0 mm in X-axis direction.

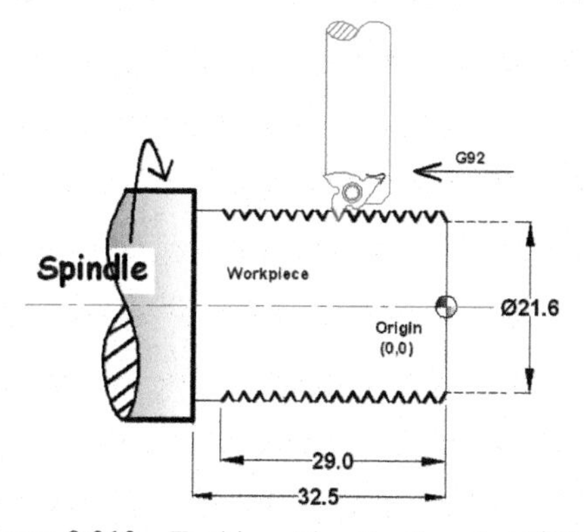

Figure 9.212 Tool is cutting the thread at Ø21.6.

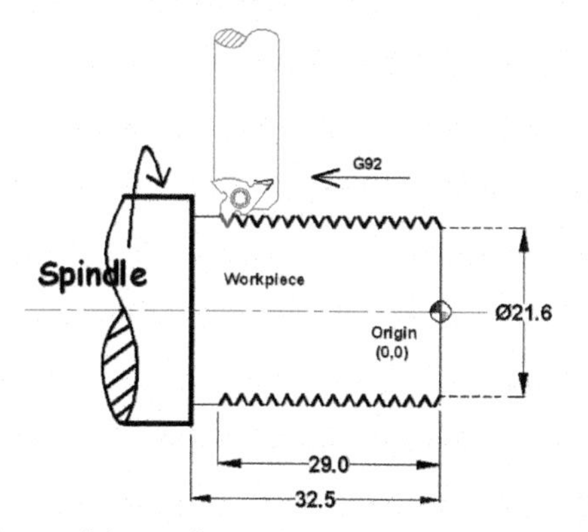

Figure 9.213 Threading tool is continuously removing the material during the threading operation.

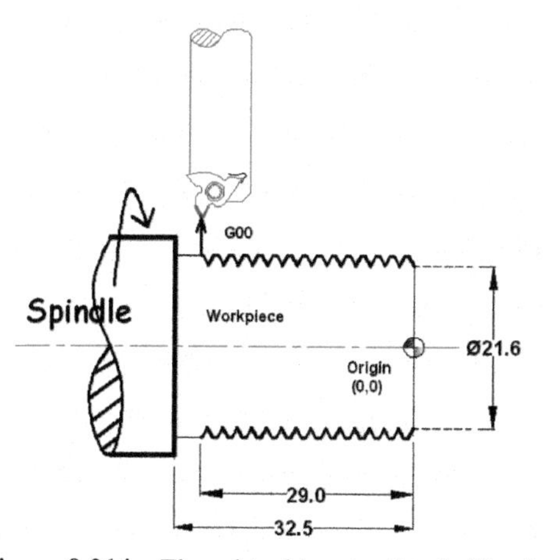

Figure 9.214 Thread tool is retracting in X-axis.

After retraction in X-axis direction now threading tool is retracting in Z-axis and will reach at the positive distance Z5.0 (Figure 9.215).

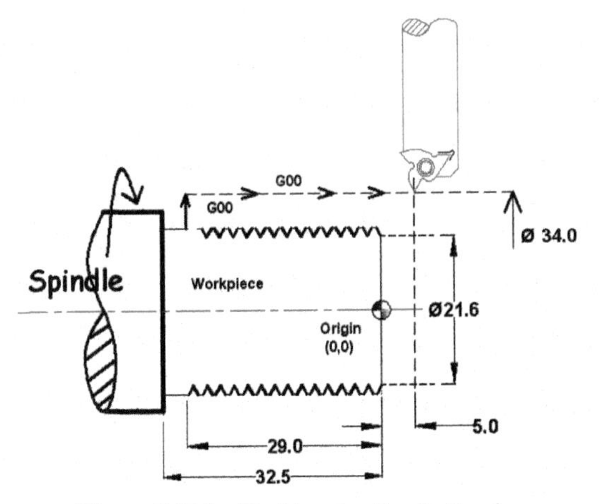

Figure 9.215 Tool is retracting in Z-axis.

X21.548; In this programming line. Threading tool will cut the thread diameter at X21.548. This line is a complete for threading operation. All previous procedures of tool positioning will apply here during the threading operation.

In Figure 9.216, the thread cutting tool is taking position rapidly in X-axis and Z-axis at diameters X21.548 and Z5.0.

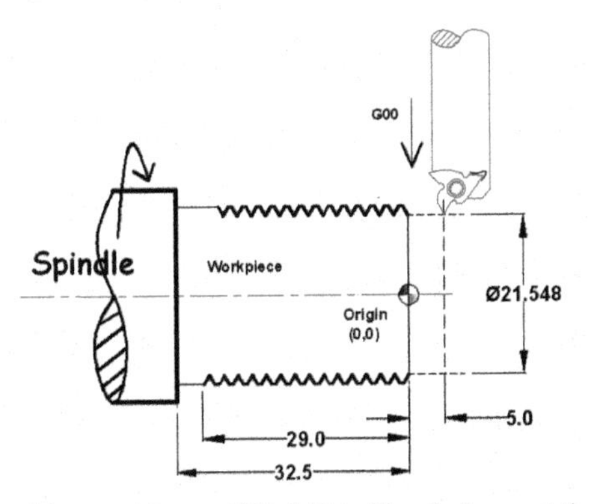

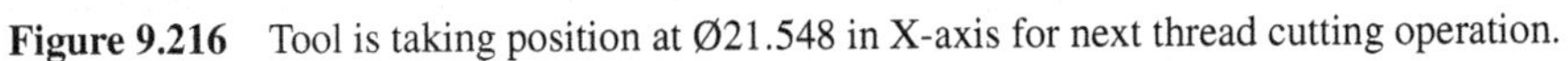
Figure 9.216 Tool is taking position at Ø21.548 in X-axis for next thread cutting operation.

In Figure 9.217, the threading tool is cutting the thread continuously.

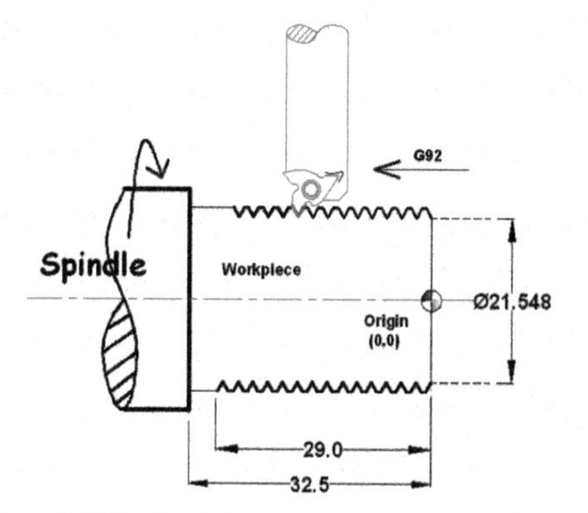

Figure 9.217 Tool is cutting the thread at Ø21.548.

In Figure 9.218, the threading tool reached at the last end of the thread length Z-29.0 and made the thread profile at Ø21.548.

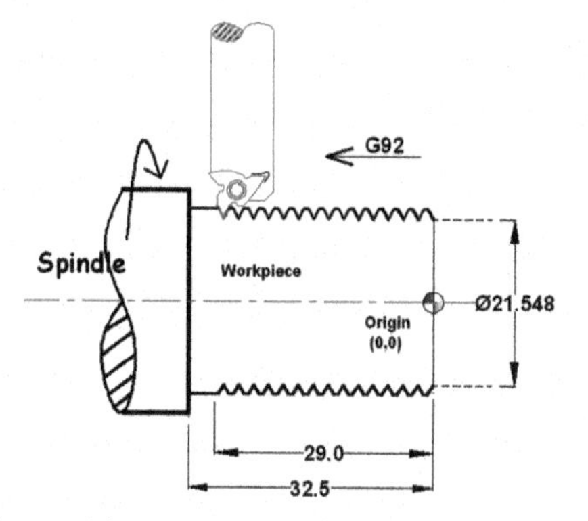

Figure 9.218 Threading tool is continuously removing the material during the threading operation.

In Figure 9.219, the threading tool is retracting at Ø34.0 mm in X-axis direction.

After retraction in X-axis direction now threading tool is retracting in Z-axis and will reach at the positive distance Z5.0. (Figure 9.220)

M09; After retraction the tool in Z-axis, now coolant motor will be stopped.

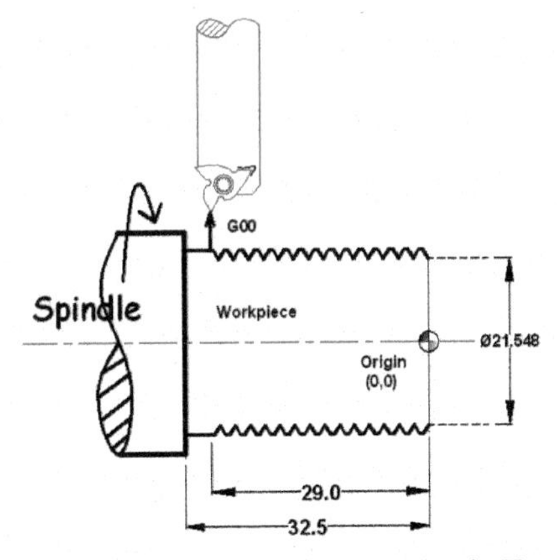

Figure 9.219 Thread tool is retracting in X-axis.

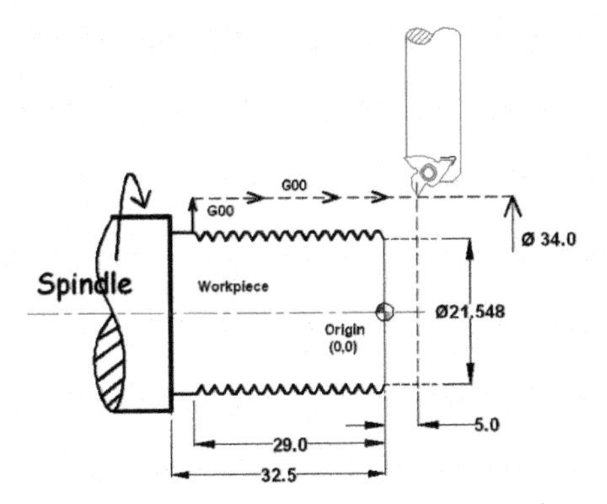

Figure 9.220 Tool is retracting in Z-axis.

M05; After complete machining, rotation of the spindle will be stopped. On the other hand, we can say rotating work piece will stop.

G00 X70 Z50; In the last cutting tool will retract and keep safe distance from the work piece (Figure 9.221)

M30; When this programming line will execute, CNC program will stop and Reset.

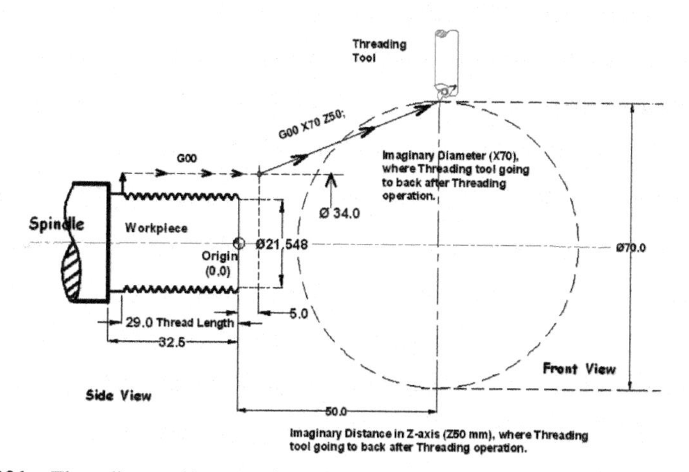

Figure 9.221 Threading tool is retracting towards the starting point of the cutting tool (where from tool start his movement).

Note:

Above CNC program includes complete theoretical procedure for better understanding.

Following the CNC machine program is an actual industrial CNC program. By which, CNC turning machine will be operated and this program will appear on the CNC control screen.

```
O1007;
G54;
T0808;
G00 X70 Z50;
G97 M03 S1000;
G00 X34 Z5.0;
M08;
G92 X23.8 Z-29.0 F2.0;
X23.6;
X23.4;
X23.2;
X23.0;
X22.8;
X22.6;
X22.4;
X22.2;
```

```
X22.0;
X21.8;
X21.6;
X21.548;
M09;
M05;
G00 X70 Z50;
M30;
```

9.2 CNC Programming Examples for Different Machining Operations with Shortest Industrial Form

9.2.1 Program No. 1 (Single Cut Facing Program)

Figure 9.222 shows the raw material drawing.

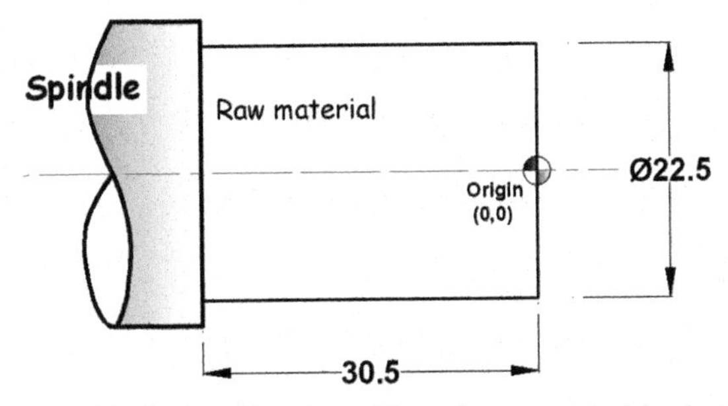

Figure 9.222 Above drawing will use for raw material selection.

Figure 9.223 shows the machining drawing.

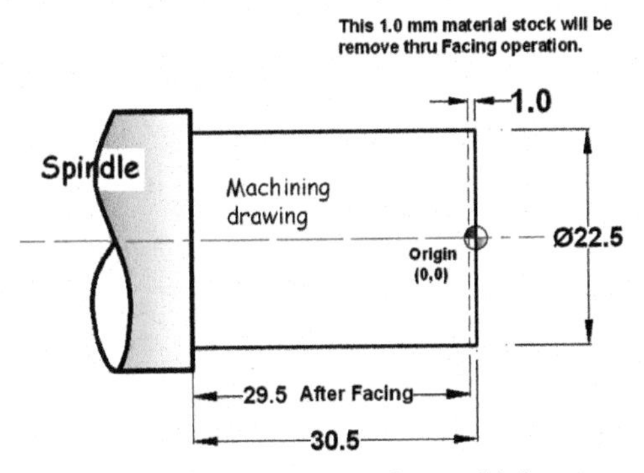

Figure 9.223 Drawing will use for machining purpose.

Machining Operation – Facing (single cut)
Raw Material – Mild steel (MS)
Cutting Tool – Turning tool PCLNR/L with bit (insert) – CNMG
Depth of Cut – 1.0 mm max
Cutting Feed – 0.12 mm/revolution (feed depends on the required surface finish as per drawing. During the facing operation we will give less feed rate due to safety reason)
Spindle Speed – 1500 rpm (RPM is a very important factor. It depends on mainly work piece diameter and holding or gripping of the work piece)

O19547;
G54;
G50 S1500;
T0101 (Turning/Facing tool);
G00 X100 Z150;
G00 X30 Z25;
G00 X26 Z5;
G96 M03 S180;
G00 X26 Z-1.0;
M08;
G01 X0.0 Z-1.0 F0.12; (Here, we can also write program line like **G01 X0.0 F0.12;**)
G00 X0.0 Z1.0; (Here, we can also write program line like **G00 Z1.0 F0.12;**)
M09;
M05;
G00 X100 Z150;
M30;

9.2.2 Program No. 2 (Multi Cut Facing Program)

Figure 9.224 shows the raw material drawing.
Figure 9.225 shows the machining drawing.

Machining Operation – Facing (two cut, 1 mm each cut)
Raw Material – Mild steel (MS)
Cutting Tool – Turning tool PCLNR/L with bit (insert) – CNMG
Depth of Cut – 1.0 mm (max.)
Cutting Feed – 0.12 mm/revolution (feed depends on required surface finish as per drawing.)

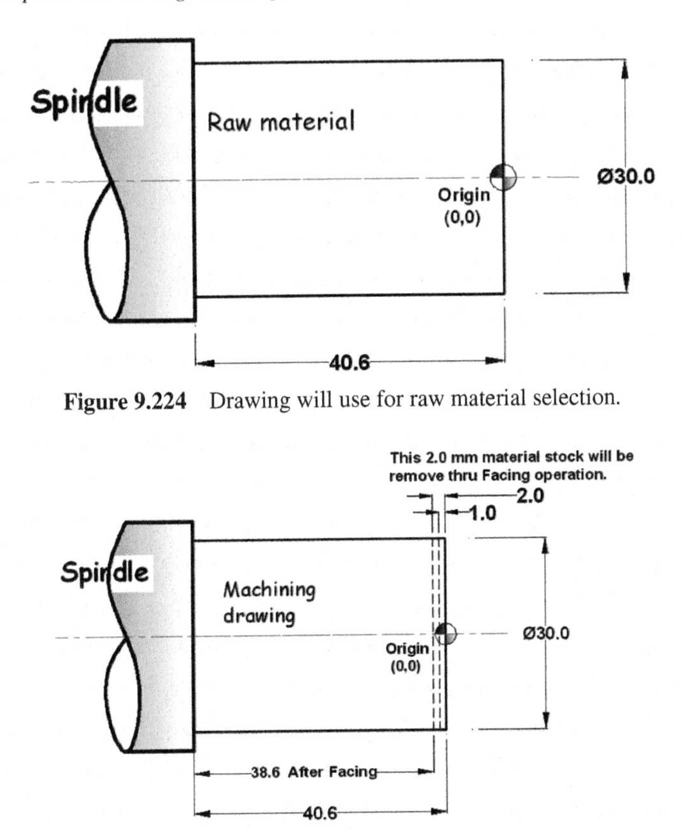

Figure 9.224 Drawing will use for raw material selection.

Figure 9.225 This drawing will use for machining purpose.

Spindle Speed – 1500 rpm (RPM is a very important factor. It depends on mainly work piece diameter and holding or gripping of the work piece) [2].

O02837;
G54;
G50 S1500;
T0101 (Turning/Facing Tool);
G00 X100 Z150;
G00 X35 Z25;
G00 X35 Z5;
G96 M03 S180;

G00 X35 Z-1.0;
M08;
G01 X0.0 F0.12; (Here, we can also write program line like **G01 X0.0 Z-1.0 F0.12;** but we are writing in short form like an industrial program.)
G00 Z1.0; (Here, we can write program line like **G00 X0.0 Z1.0 F0.12;** but we are writing in short form like an industrial program.)
G00 X35;
G00 Z-2.0;
G01 X0.0 F0.12;
G00 Z1.0;
M09;
M05;
G00 X100 Z150;
M30;

9.2.3 Program No. 3 (Multiple Facing Cut Program)

Figure 9.226 shows the raw material drawing.

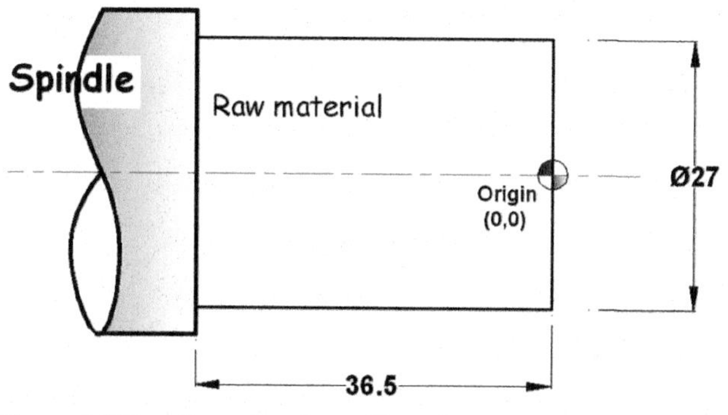

Figure 9.226 Above drawing will use for raw material selection.

Figure 9.227 shows the machining drawing.

Machining Operation – Facing (three cut, 1 mm each cut)
Raw Material – Mild steel (MS)
Cutting Tool – Turning tool PCLNR/L with bit (insert) – CNMG
Depth of Cut – 1.0 mm (maximum)

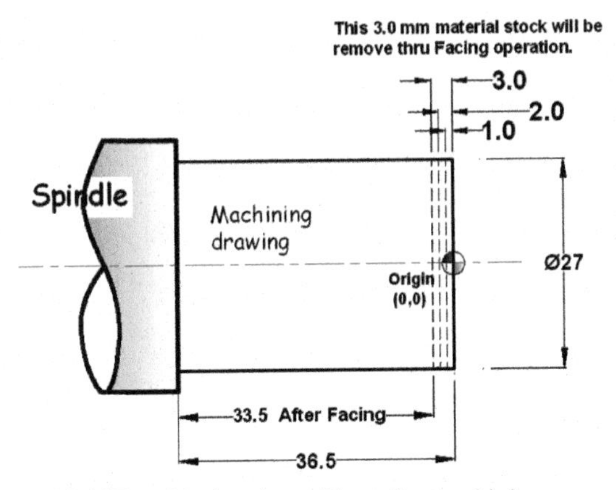

Figure 9.227 This drawing will use for machining purpose.

Cutting Feed – 0.12 mm/revolution (feed depends on the required surface finish as per drawing.)

Spindle Speed – 1500 rpm (RPM is a very important factor. It depends on mainly work piece diameter and holding or gripping of the work piece)

Note:

We can write following program in more short form but it is for beginners.

```
O59848;
G54;
G50 S1500;
T0101 (Turning/Facing Tool);
G00 X100 Z150;
G00 X32 Z25;
G00 X32 Z5
G96 M03 S180;
G00 X32 Z-1.0;
M08;
G01 X0.0 F0.12;
G00 Z1.0;
G00 X32;
G00 Z-2.0;
G01 X0.0 F0.12;
G00 Z1.0;
```

```
G00 X32;
G00 Z-3.0;
G01 X0.0 F0.12;
G00 Z1.0;
M09;
M05;
G00 X100 Z150;
M30;
```

9.2.4 Program No. 4 (Straight Turning Program)

Figure 9.228 shows the raw material drawing.

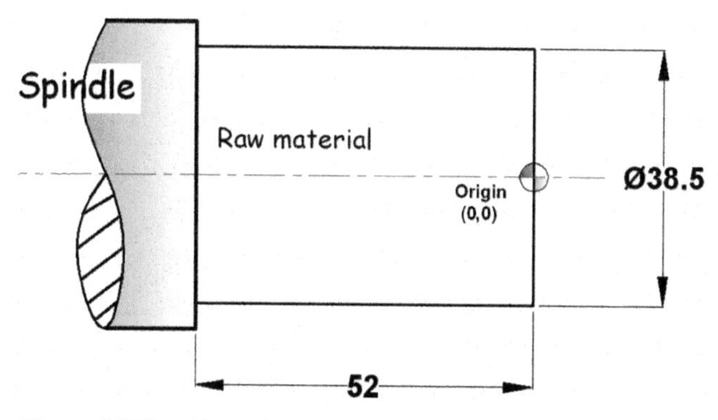

Figure 9.228 Above drawing will use for raw material selection.

Figure 9.229 shows the machining drawing.

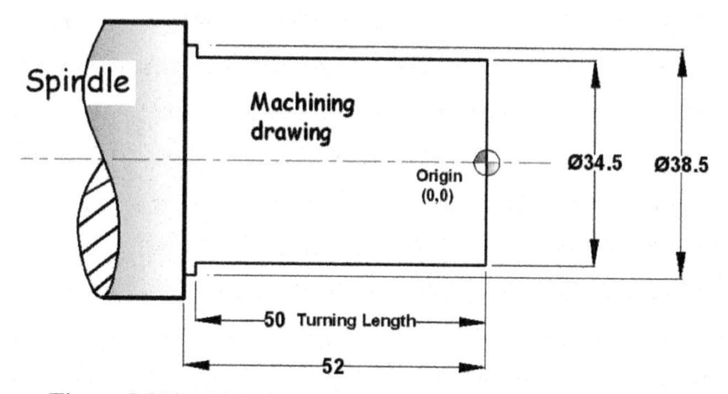

Figure 9.229 This drawing will use for machining purpose.

Machining Operation – Straight turning
Raw Material – Mild steel (MS)
Cutting Tool – Turning tool PCLNR/L with bit (insert) – CNMG/DNMG/WNMG etc. (depends on the depth of cut and tool)
Depth of Cut – 2.0 mm one side/4.0 mm diametrical (both side)
Cutting Feed – 0.15 mm/revolution (feed depends on the required surface finish as per drawing.)
Spindle Speed – 2000 rpm (RPM is a very important factor. It depends on mainly work piece diameter and holding or gripping of the work piece)

```
O11621;
G54;
G50 S2000;
T0101 (Turning Tool);
G00 X100 Z150;
G00 X45 Z10;
G96 M03 S210;
G00 X34.5 Z1.0;
M08;
G01 X34.5 Z-50.0 F0.15;
G01 X40 F0.12;
G00 Z1.0;
M09;
M05;
G00 X100 Z150;
M30;
```

9.2.5 Program No. 5 (Straight Turning Program with Rough Cutting)

Figure 9.230 shows the raw material drawing.
Figure 9.231 shows the machining drawing.

Machining Operation – Straight turning (in three cut)
Raw Material – Mild steel (MS)
Cutting Tool – Turning tool PCLNR/L with bit (insert) – CNMG/DNMG/WNMG etc. (depends on the depth of cut and tool)
Depth of Cut –1.5 mm one side/3.0 mm diametrical (both side)

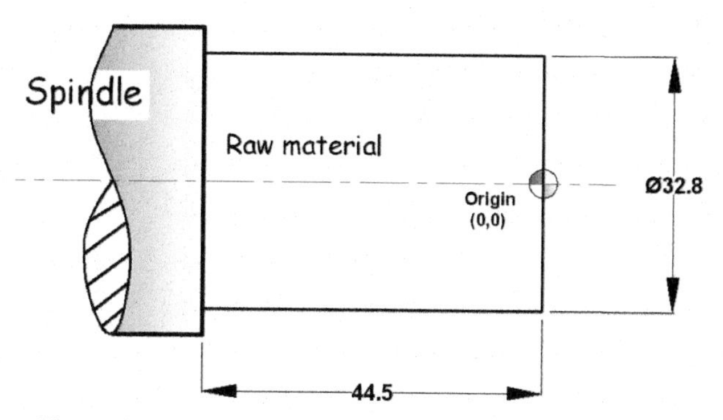

Figure 9.230 Drawing will use for raw material selection.

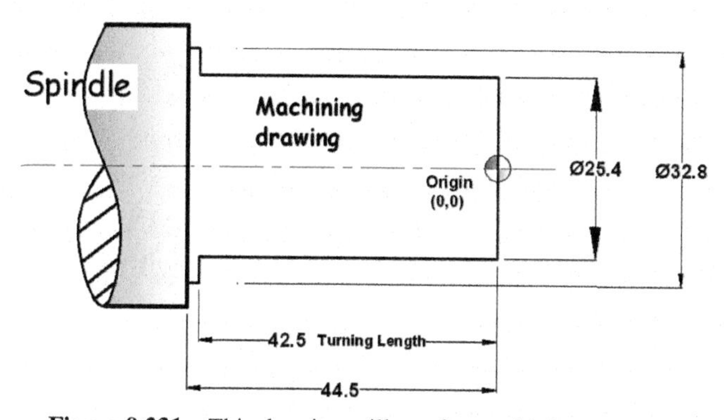

Figure 9.231 This drawing will use for machining purpose.

Cutting Feed – 0.15 mm/revolution (feed depends on the required surface finish as per drawing)
Spindle Speed – 2200 rpm (RPM is a very important factor. It depends on mainly work piece diameter and holding or gripping of the work piece)[2].
Cutting oil (coolant) –

O11621;
G54;
G50 S2200;
T0101 (Turning Tool);
G00 X100 Z150;
G00 X50 Z10;
G96 M03 S230;

```
G00 X29.8 Z1.0;
M08;
G01 Z-42.5 F0.15;
G01 X34 F0.12;
G00 Z1.0;
G00 X26.8;
G01 Z-42.5 F0.15;
G01 X34 F0.12;
G00 Z1.0;
G00 X25.4;
G01 Z-42.5 F0.15;
G01 X34 F0.12;
G00 Z1.0;
M09;
M05;
G00 X100 Z150;
M30;
```

9.2.6 Program No. 6 (Straight Turning Program with Multiple Rough Cutting)

Figure 9.232 shows the raw material drawing.

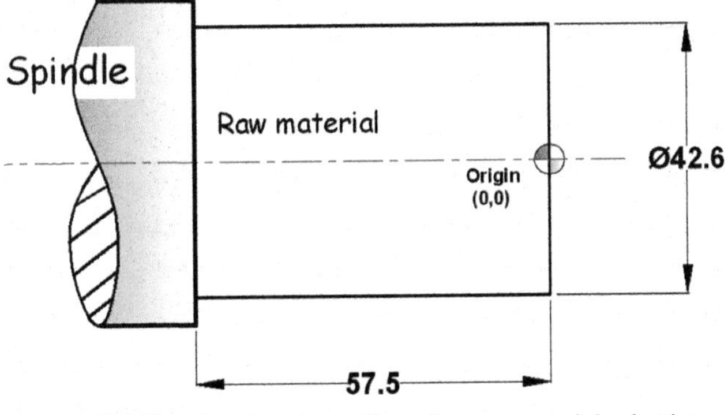

Figure 9.232 This drawing will use for raw material selection.

Figure 9.233 shows the machining drawing.

Machining Operation – Straight turning (in four cut)
Raw Material – Mild steel (MS)

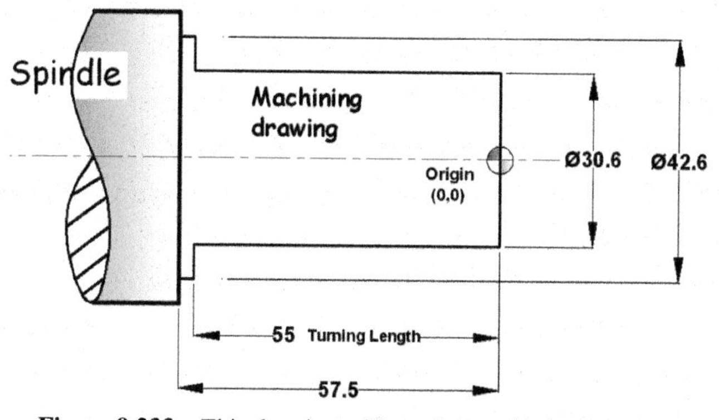

Figure 9.233 This drawing will use for machining purpose.

Cutting Tool – Turning tool PCLNR/L with bit (insert) – CNMG/DNMG/WNMG etc. (depends on the depth of cut and tool)
Depth of Cut – 1.5 mm one side/3.0 mm diametrical (both side)
Cutting Feed – 0.15 mm/revolution (feed depends on the required surface finish as per drawing.)
Spindle Speed – 2000 rpm (RPM is a very important factor. It depends on mainly work piece diameter and holding or gripping of the work piece)

```
O88255;
G54;
G50 S2000;
T0101 (Turning Tool);
G00 X100 Z150;
G00 X70 Z10;
G96 M03 S210;
G00 X39.6 Z1.0;
M08;
G01 Z-55 F0.15;
G01 X44 F0.12;
G00 Z1.0;
G00 X36.6;
G01 Z-55 F0.15;
G01 X44 F0.12;
G00 Z1.0;
G00 X33.6;
```

```
G01 Z-55 F0.15;
G01 X44 F0.12;
G00 Z1.0;
G00 X30.6;
G01 Z-55 F0.15;
G01 X44 F0.12;
G00 Z1.0;
M09;
M05;
G00 X100 Z150;
M30;
```

9.2.7 Program No. 7 (Step Turning Program)

Figure 9.234 shows the raw material drawing.

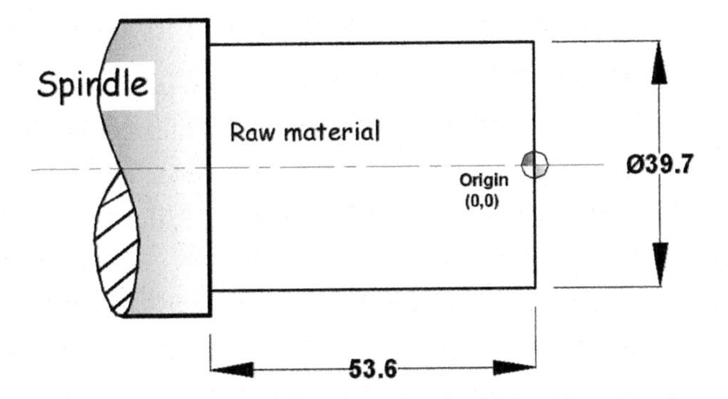

Figure 9.234 Above drawing will use for raw material selection.

Figure 9.235 shows the machining drawing.

Machining Operation – Step turning
Raw Material – Mild steel (MS)
Cutting Tool – Turning tool PCLNR/L with bit (insert) – CNMG/DNMG/WNMG etc. (depends on the depth of cut and tool)
Depth of Cut – 1.5 mm one side/3.0 mm diametrical (both side)
Cutting Feed – 0.15 mm/revolution (feed depends on the required surface finish as per drawing)

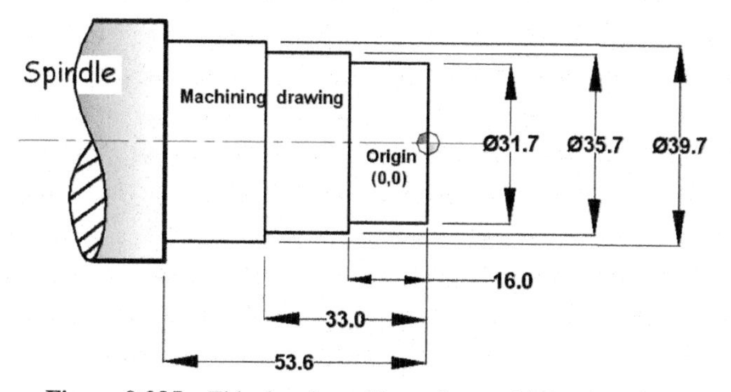

Figure 9.235 This drawing will use for machining purpose.

Spindle Speed – 1800 rpm (RPM is a very important factor. It depends on mainly work piece diameter and holding or gripping of the work piece)

```
O49771;
G54;
G50 S1800;
T0101 (Turning Tool);
G00 X100 Z150;
G00 X70 Z10;
G96 M03 S210;
G00 X36.7 Z1.0;
M08;
G01 Z-33 F0.15;
G01 X42 F0.12;
G00 Z1.0;
G00 X35.7;
G01 Z-33 F0.15;
G01 X42 F0.12;
G00 Z1.0;
G00 X32.7;
G01 Z-16 F0.15;
G01 X38 F0.12;
G00 Z1.0;
G00 X31.7;
G01 Z-16 F0.15;
G01 X38 F0.12;
```

```
G00 Z1.0;
M09;
M05;
G00 X100 Z150;
M30;
```

9.2.8 Program No. 8 (Multi Step Turning Program)

Figure 9.236 shows the raw material drawing.

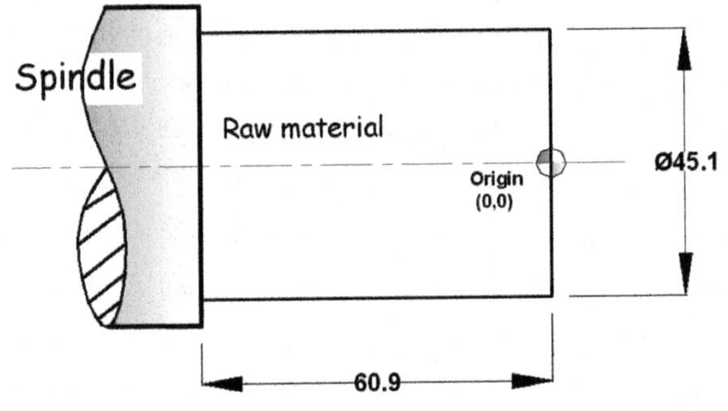

Figure 9.236 This drawing will use for raw material selection.

Figure 9.237 shows the machining drawing.

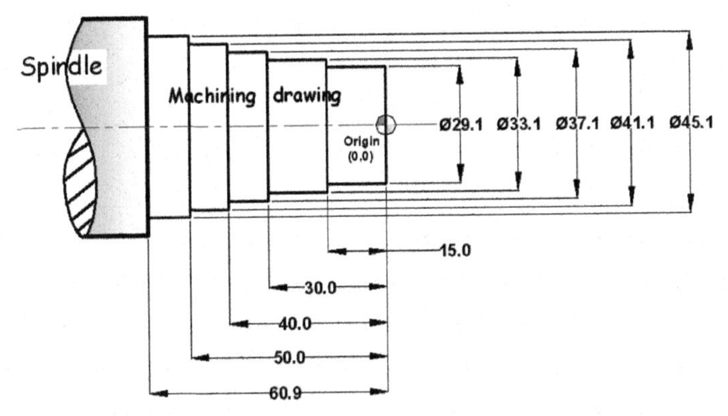

Figure 9.237 This drawing will use for machining purpose.

Machining Operation – Step turning (step turning with rough cut)
Raw Material – Mild steel (MS)

Cutting Tool – Turning tool PCLNR/L with bit (insert) – CNMG/DNMG/WNMG etc. (depends on the depth of cut and tool)
Depth of Cut – 1.5 mm one side/3.0 mm diametrical (both side)
Cutting Feed – 0.15 mm/revolution (feed depends on the required surface finish as per drawing.)
Spindle Speed – 1800 rpm (RPM is a very important factor. It depends on mainly work piece diameter and holding or gripping of the work piece)

```
O32326;
G54;
G50 S1800;
T0101 (Turning Tool);
G00 X100 Z150;
G00 X70 Z10;
G96 M03 S210;
G00 X42.1 Z1.0;
M08;
G01 Z-50 F0.15;
G01 X47 F0.12;
G00 Z1.0;
G00 X41.1;
G01 Z-50 F0.15;
G01 X47 F0.12;
G00 Z1.0;
G00 X38.1;
G01 Z-40 F0.15;
G01 X43 F0.12;
G00 Z1.0;
G00 X37.1;
G01 Z-40 F0.15;
G01 X43 F0.12;
G00 Z1.0;
G00 X34.1;
G01 Z-30 F0.15;
G01 X39 F0.12;
G00 Z1.0;
G00 X33.1;
G01 Z-30 F0.15;
```

```
G01 X39 F0.12;
G00 Z1.0;
G00 X30.1;
G01 Z-15 F0.15;
G01 X35 F0.12;
G00 Z1.0;
G00 X29.1;
G01 Z-15 F0.15;
G01 X35 F0.12;
G00 Z1.0;
M09;
M05;
G00 X100 Z150;
M30;
```

9.2.9 Program No. 9 (Multiple Steps Turning Program)

Figure 9.238 shows the raw material drawing.

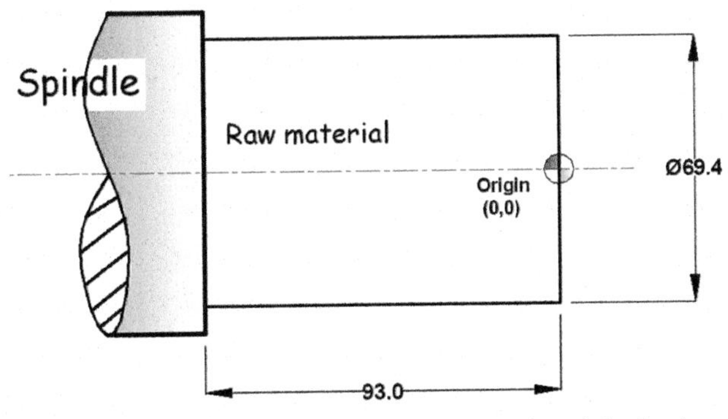

Figure 9.238 This drawing will use for raw material selection.

Figure 9.239 shows the machining drawing.

Machining Operation – Step turning (step turning with rough cut)
Raw Material – Mild steel (MS)
Cutting Tool – Turning tool PCLNR/L with bit (insert) – CNMG/DNMG/WNMG etc. (depends on the depth of cut and tool)
Depth of Cut – 1.5 mm one side/3.0 mm diametrical (both side)
Cutting Feed – 0.15 mm/revolution (feed depends on the required surface finish as per drawing.)

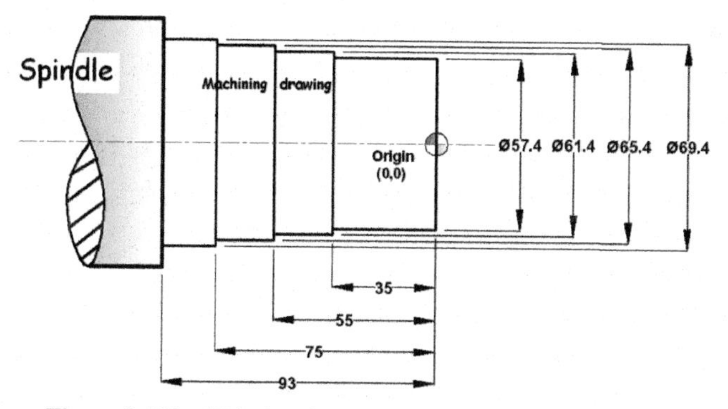

Figure 9.239 This drawing will use for machining purpose.

Spindle Speed – 1800 rpm (RPM is a very important factor. It depends on mainly work piece diameter and holding or gripping of the work piece)

O70042;
G54;
G50 S1800;
T0101 (Turning Tool);
G00 X150 Z150;
G00 X80 Z10;
G96 M03 S200;
G00 X66.4 Z1.0;
M08;
G01 Z-75 F0.15;
G01 X71.5 F0.12;
G00 Z1.0;
G00 X65.4;
G01 Z-75 F0.15;
G01 X71.5 F0.12;
G00 Z1.0;
G00 X62.4;
G01 Z-55 F0.15;
G01 X67 F0.12;
G00 Z1.0;
G00 X61.4;
G01 Z-55 F0.15;

```
G01 X67 F0.12;
G00 Z1.0;
G00 X58.4;
G01 Z-35 F0.15;
G01 X63 F0.12;
G00 Z1.0;
G00 X57.4;
G01 Z-35 F0.15;
G01 X63 F0.12;
G00 Z1.0;
M09;
M05;
G00 X150 Z150;
M30;
```

9.2.10 Program No. 10 (Continuous Drilling Program)

Figure 9.240 shows the raw material drawing.

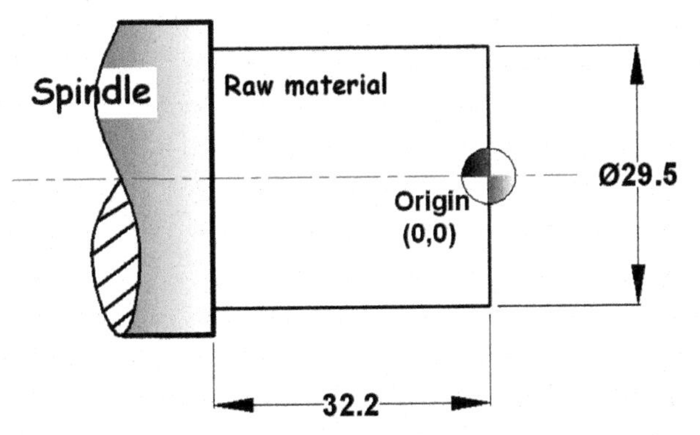

Figure 9.240 This drawing will use for raw material selection.

Figure 9.241 shows the machining drawing.

Machining Operation – Continuous drilling
Raw Material – Mild steel (MS)
Cutting Tool – Drilling tool
Drill Diameter – 10.0 mm
Cutting Feed – 0.05 mm/revolution (drilling operation should be at the low feed.)
Spindle Speed – 450 rpm (spindle rpm should be constant during the drilling operation)

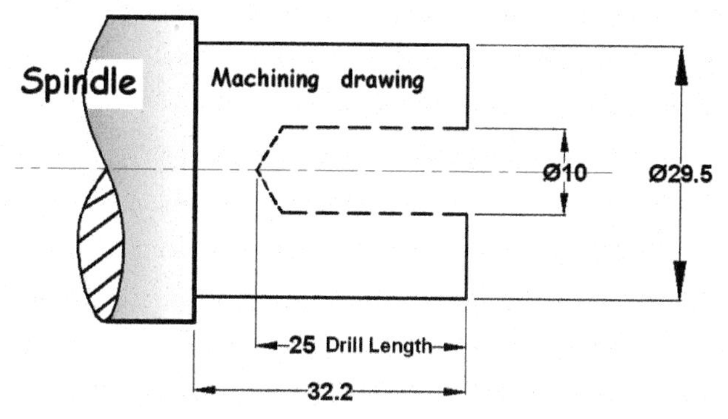

Figure 9.241 This drawing will use for machining purpose.

O56541;
G54;
T0707 (Drilling Tool, Ø10 mm);
G00 X150 Z250;
G00 X60 Z25;
G97 M03 S450; (Must be constant **rpm** during Drilling operation)
G00 X0.0 Z5.0;
M08;
G00 X0.0 Z1.0;
G01 Z-25 F0.05;
G00 Z5.0;
M09;
M05;
G00 X150 Z250;
M30;

9.2.11 Program No. 11 (Peck Drilling Program)

Figure 9.242 shows the raw material drawing.
Figure 9.243 shows the machining drawing.

Machining Operation – Peck drilling
Raw Material – Mild steel (MS)
Cutting Tool – Drilling tool
Drill Diameter – 12.0 mm
Cutting Feed – 0.05 mm/revolution (drilling operation should be at the low feed.)
Spindle Speed – 450 rpm (spindle rpm should be constant during the drilling operation.)

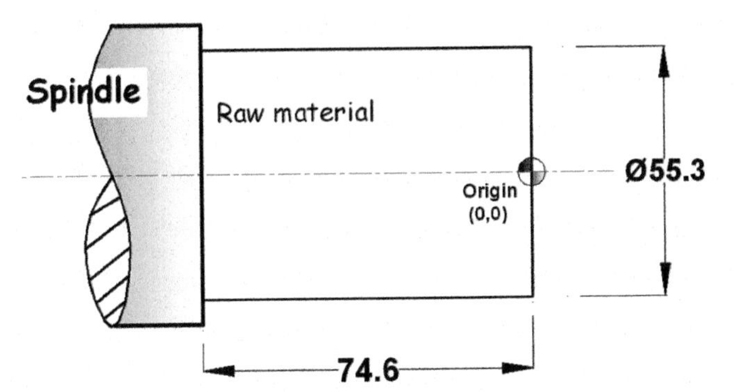

Figure 9.242 This drawing will use for raw material selection.

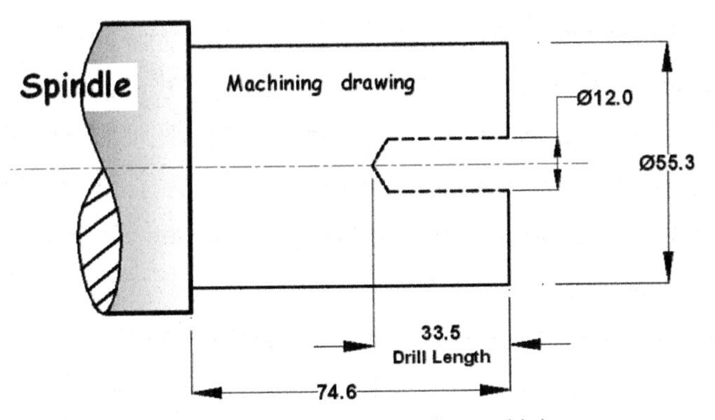

Figure 9.243 This drawing will use for machining purpose.

O22877;
G54;
G21 G90;
T0707 (Drill Tool, Ø12 mm);
G00 X150 Z250;
G00 X60 Z25;
G97 M03 S450; (Must be constant **rpm** during drilling operation)
G00 X0.0 Z5.0;
M08;
G00 X0.0 Z1.0; (Drill will take position in X-axis at the center of the work piece X0.0)
G01 Z-5.0 F0.05;
G00 Z5.0;
G00 Z-4.0;

```
G01 Z-10 F0.05;
G00 Z5.0;
G00 Z-9.0;
G01 Z-15 F0.05;
G00 Z5.0;
G00 Z-14;
G01 Z-20 F0.05;
G00 Z5.0;
G00 Z-19;
G01 Z-25 F0.05;
G00 Z5.0;
G00 Z-24;
G01 Z-30 F0.05;
G00 Z5.0;
G00 Z-29;
G01 Z-33.5 F0.05;
G00 Z5;
M09;
M05;
G00 X150 Z250;
M30;
```

9.2.12 Program No. 12 (Peck Drilling Program with Deep Hole)

Figure 9.244 shows the raw material drawing.

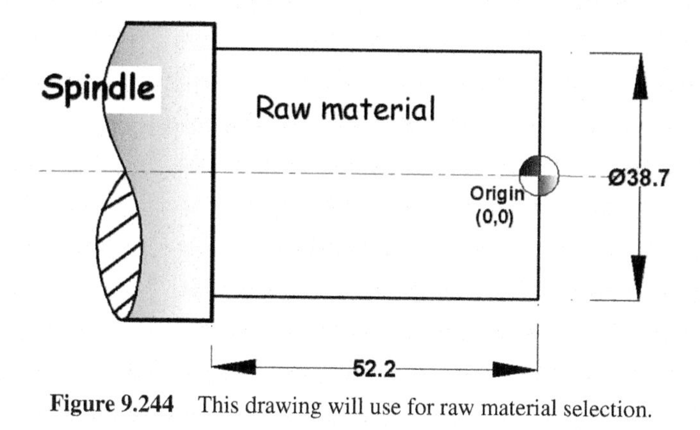

Figure 9.244 This drawing will use for raw material selection.

Figure 9.245 shows the machining drawing.

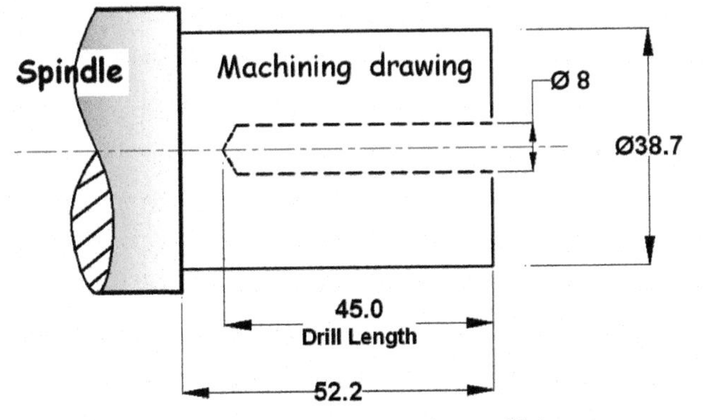

Figure 9.245 This drawing will use for machining purpose.

Machining Operation – Peck drilling
Raw Material – Mild steel (MS)
Cutting Tool – Drilling tool
Drill Diameter – 8.0 mm
Cutting Feed – 0.05 mm/revolution (drilling operation should be at the low feed.)
Spindle Speed – 650 rpm (spindle rpm should not vary during the drilling operation.)

O34649;
G54;
T0707 (Drill Tool, Ø8 mm);
G00 X150 Z250;
G00 X60 Z25;
G97 M03 S650; (Must be constant **rpm** during drilling operation)
G00 X0.0 Z5.0;
M08;
G00 X0.0 Z1.0;
G01 Z-5.0 F0.05;
G00 Z5.0;
G00 Z-4.0;
G01 Z-10 F0.05;
G00 Z5.0;
G00 Z-9.0;
G01 Z-15 F0.05;

```
G00 Z5.0;
G00 Z-14;
G01 Z-20 F0.05;
G00 Z5.0;
G00 Z-19;
G01 Z-25 F0.05;
G00 Z5.0;
G00 Z-24;
G01 Z-30 F0.05;
G00 Z5.0;
G00 Z-29;
G01 Z-35 F0.05;
G00 Z5.0;
G00 Z-34;
G01 Z-40 F0.05;
G00 Z5.0;
G00 Z-39;
G01 Z-45 F0.05;
G00 Z5.0;
M09;
M05;
G00 X150 Z250;
M30;
```

9.2.13 Program No. 13 (Chamfering Program)

Figure 9.246 shows the raw material drawing.

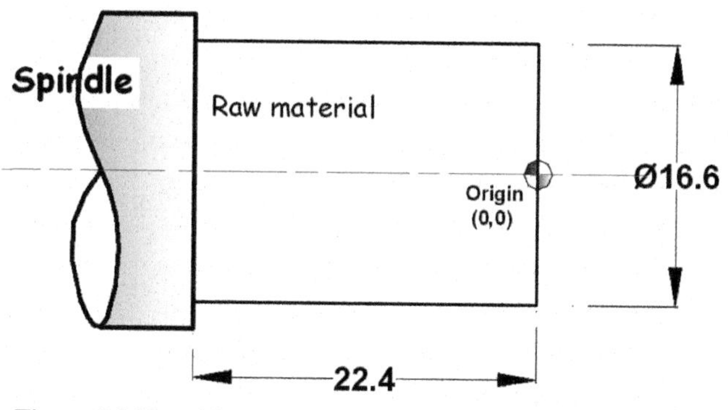

Figure 9.246 This drawing will use for raw material selection.

Figure 9.247 shows the machining drawing.

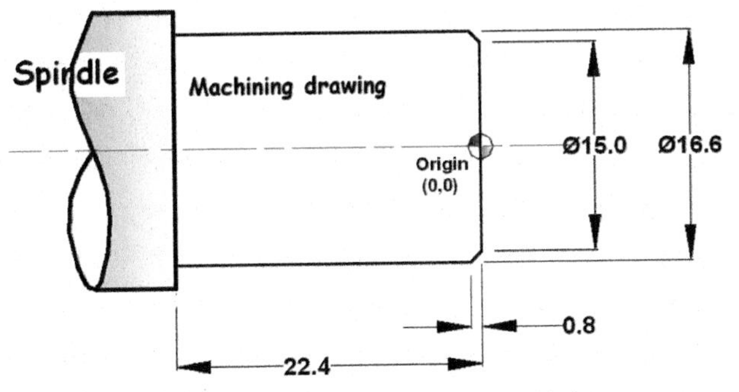

Figure 9.247 This drawing will use for machining purpose.

Machining Operation – Chamfering at 45°
Raw Material – Mild steel (MS)
Cutting Tool – Turning tool PCLNR/L with bit (insert) – CNMG/DNMG/WNMG etc. (depends on the depth of cut and tool)
Depth of Cut – 1.5 mm one side/3.0 mm diametrical (both side)
Cutting Feed – 0.12 mm/revolution (feed depends on the required surface finish as per drawing.)
Spindle Speed – 2600 rpm (RPM is a very important factor. It depends on mainly work piece diameter and holding or gripping of the work piece)

```
O34649;
G54;
G50 S2600;
T0101 (Turning Tool);
G00 X100 Z100;
G00 X25 Z5.0;
G96 M03 S230;
G00 X14 Z0.5; (Cutting tool positioning before chamfering)
M08;
G01 X15 Z0 F0.12; (Starting point of chamfer)
G01 X16.6 Z-0.8 F0.12; (Ending point of chamfer)
G01 X17.6 Z-1.3 F0.12; (After chamfering, retraction of the tool from the
                        work piece surface will be at same 45° angle and
                        direction)
```

G00 Z1.0;
M09;
M05;
G00 X100 Z100;
M30;

OR

Following method is used generally in industry.

O34649;
G54;
G50 S2600;
T0101 (Turning Tool);
G00 X100 Z100;
G00 X25 Z5.0;
G96 M03 S230;
G00 X14 Z0.5; (Cutting tool positioning before chamfering)
M08;
G01 X15 Z0 F0.12; (Starting point of chamfer)
G01 X17.6 Z-1.3 F0.12; [Ending point (After chamfering, retraction of the tool from the work piece surface will be same at 45° angle and direction)]
G00 Z1.0;
M09;
M05;
G00 X100 Z100;
M30;

9.2.14 Program No. 14 (Chamfering Program)

Figure 9.248 shows the raw material drawing.
Figure 9.249 shows the machining drawing.

Machining Operation – Chamfering at 45°
Raw Material – Mild steel (MS)
Cutting Tool – Turning tool PCLNR/L with bit (insert) – CNMG/DNMG/WNMG etc. (depends on the depth of cut and tool)
Depth of Cut – 1.5 mm one side/3.0 mm diametrical (both side)
Cutting Feed – 0.12 mm/revolution (feed depends on the required surface finish as per drawing)

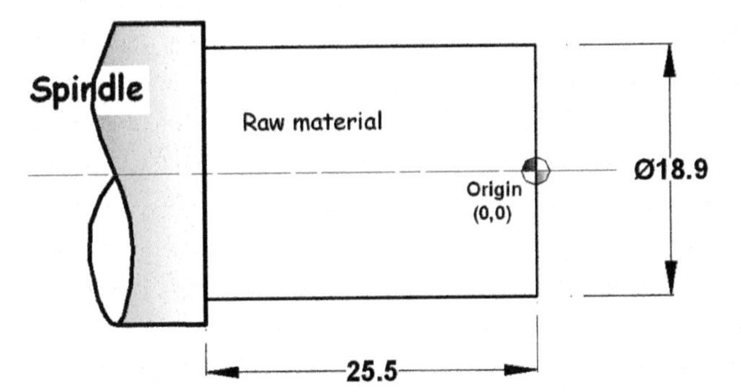

Figure 9.248 This drawing will use for raw material selection.

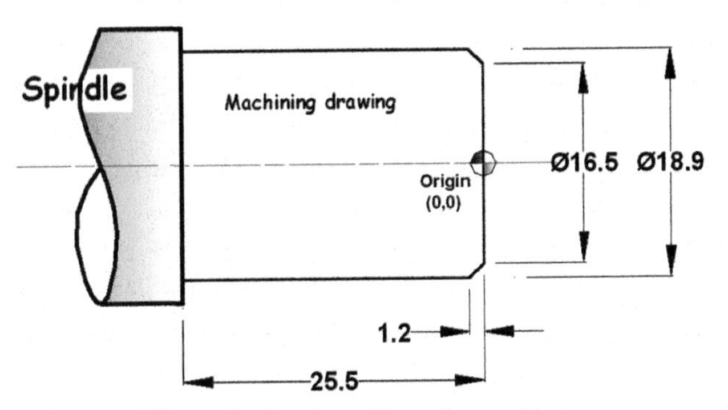

Figure 9.249 This drawing will use for machining purpose.

Spindle Speed – 2600 rpm (RPM is a very important factor. It depends on mainly work piece diameter and holding or gripping of the work piece.)

O11181;
G54;
G50 S2600;
T0101 (Turning Tool);
G00 X100 Z100;
G00 X25 Z5.0;
G96 M03 S230;
G00 X15.5 Z0.5; (Cutting tool positioning before chamfering)
M08;
G01 X16.5 Z0 F0.12; (Starting point of chamfer)
G01 X19.9 Z-1.7 F0.12; (After chamfering, retraction of the tool from the surface will be same at 45° angle and direction)

```
G00 Z1.0;
M09;
M05;
G00 X100 Z100;
M30;
```

9.2.15 Program No. 15 (Chamfering Program with Rough Cut)

Figure 9.250 shows the raw material drawing.

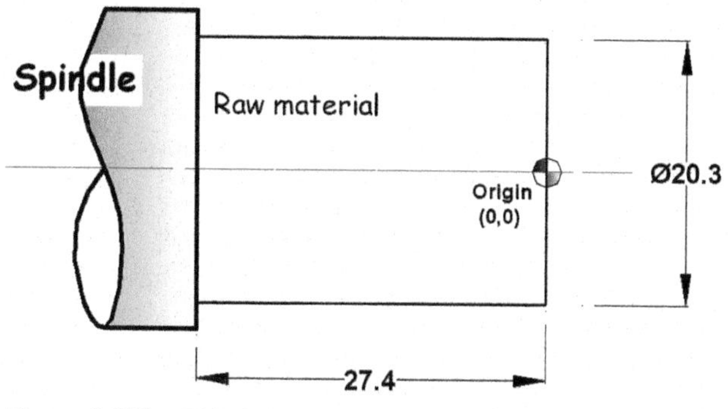

Figure 9.250 This drawing will use for raw material selection.

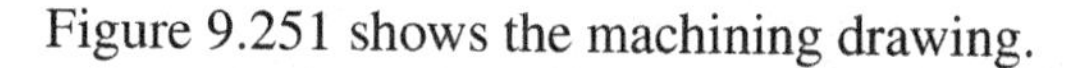
Figure 9.251 shows the machining drawing.

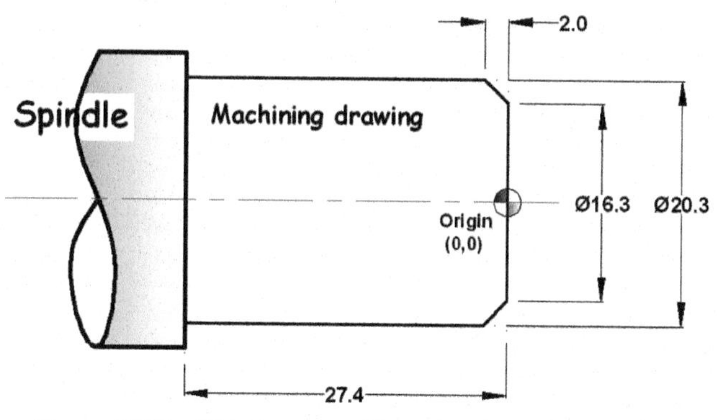

Figure 9.251 This drawing will use for machining purpose.

Machining Operation – Chamfering at 45°
Raw Material – Mild Steel (MS)
Cutting Tool – Turning tool PCLNR/L with bit (insert) – CNMG/DNMG/WNMG etc. (depends on the depth of cut and tool)
Depth of Cut – 1.5 mm one side/3.0 mm diametrical (both side)

Cutting Feed – 0.12 mm/revolution (feed depends on the required surface finish as per drawing.)
Spindle Speed – 2600 rpm (RPM is a very important factor. It depends on mainly work piece diameter and holding or gripping of the
Spindle Speed – work piece)

```
O56661;
G54;
G50 S2600;
T0101 (Turning Tool);
G00 X100 Z100;
G00 X25 Z5.0;
G96 M03 S230;
G00 X17.3 Z0.5; Cutting tool positioning before chamfering
M08;
G01 X18.3 Z0 F0.12; Starting point of the rough chamfer
G01 X20.3 Z-1.0 F0.12; Ending point of the ruough chamfer
G00 Z1.0;
G00 X15.3 Z0.5; Cutting tool is taking position before final chamfering
G01 X16.3 Z0 F0.12; Starting point of the final chamfer
G01 X21.3 Z-2.5 F0.12; Ending point of final chamfer (After chamfering,
                       retraction of the tool from the work piece surface
                       will be same at angle 45° and direction)
M09;
M05;
G00 X100 Z100;
M30;
```

9.2.16 Program No. 16 (Corner Radius Program with Rough Cut and Counter Clock Wise Direction)

Figure 9.252 shows the raw material drawing.
Figure 9.253 shows the machining drawing.

Machining Operation – Corner radius with rough cut (fillet)
Raw Material – Mild steel (MS)
Cutting Tool – Turning tool PCLNR/L with bit (insert) – CNMG/DNMG/WNMG etc. (depends on the depth of cut and tool)
Depth of Cut – 1.5 mm one side/3.0 mm diametrical (both side)
Cutting Feed – 0.12 mm/revolution (feed depends on the required surface finish as per drawing.)

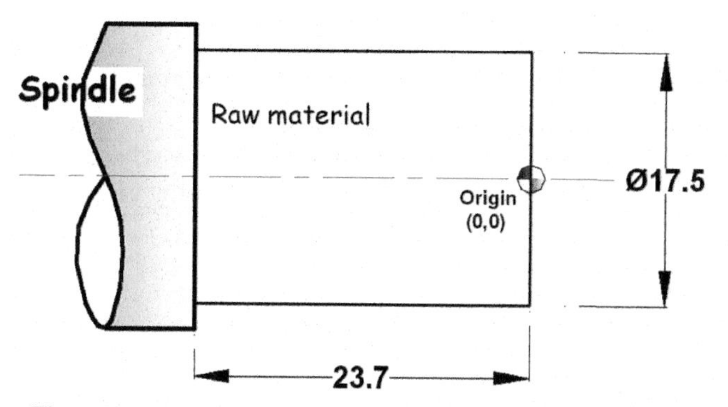

Figure 9.252 This drawing will use for raw material selection.

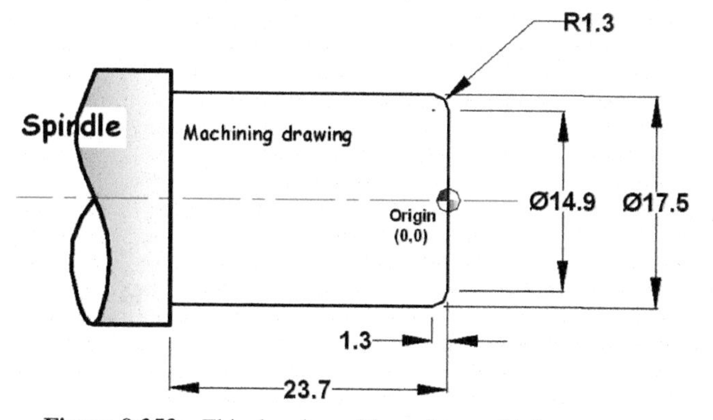

Figure 9.253 This drawing will use for machining purpose.

Spindle Speed – 2600 rpm (RPM is a very important factor. It depends on mainly work piece diameter and holding or gripping of the work piece)

O50060;
G54;
G50 S2600;
T0101 (Turning Tool);
G00 X100 Z100;
G00 X20 Z5.0;
G96 M03 S230;
G00 X16.1 Z1.0; (Cutting tool position before making ***Rough*** *Radius profile*)
M08;
G01 X16.1 Z0.0 F0.12; (**Starting** point of ***Rough*** *Radius*)
G03 X17.5 Z-0.7 **R0.7** F0.12; (**Ending** point of Radius – where cutting tool will go during rough cutting)

G00 Z1.0;
G00 X13.9 Z1.0; (tool is taking position for **final radius cutting**.)
G01 X14.9 Z0.0 F0.12; (**Starting** point of Radius)
G03 X17.5 Z-1.3 **R1.3** F0.12; (**Ending** point of Radius where cutting tool will go during **Final radius cutting**)

G00 Z1.0;
M09;
M05;
G00 X100 Z100;
M30;

9.2.17 Program No. 17 (Corner Radius Program with Counter Clock Wise Direction)

Figure 9.254 shows the raw material drawing.

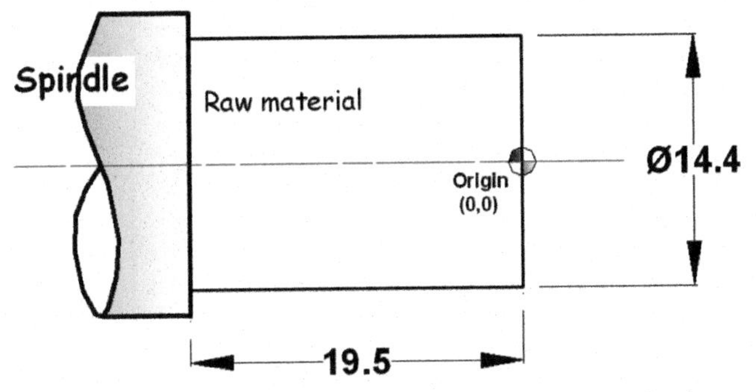

Figure 9.254 This drawing will use for raw material selection.

Figure 9.255 shows the machining drawing.

Machining Operation – Corner radius (fillet)
Raw Material – Mild steel (MS)
Cutting Tool – Turning tool PCLNR/L with bit (insert) – CNMG/DNMG/WNMG etc. (depends on the depth of cut and tool)
Depth of Cut – 1.5 mm one side/3.0 mm diametrical (both side)
Cutting Feed – 0.12 mm/revolution (feed depends on the required surface finish as per drawing.)
Spindle Speed – 2200 rpm (RPM is a very important factor. It depends on mainly work piece diameter and holding or gripping of the work piece)

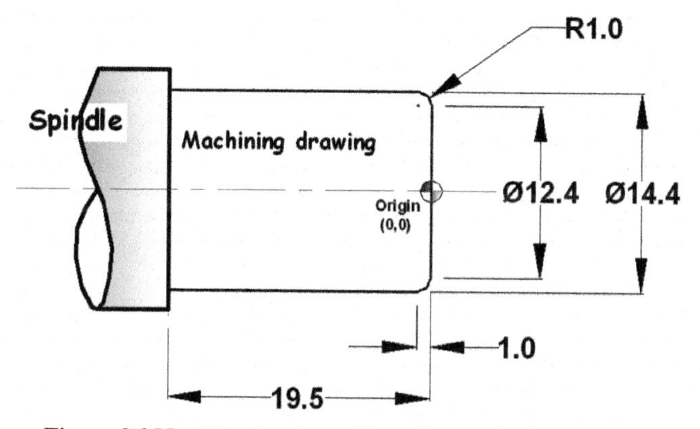

Figure 9.255 This drawing will use for machining purpose.

O50060;
G54;
G50 S2200;
T0101 (Turning Tool);
G00 X100 Z100;
G00 X20 Z5.0;
G96 M03 S230;
M08;
G00 X11.4 Z1.0; (Cutting tool positioning before making Radius)
G01 X12.4 Z0.0 F0.12; (**Starting** point of Radius)
G03 X14.4 Z-1.0 **R1.0** F0.12; (**Ending** point of Radius)
G00 Z1.0;
M09;
M05;
G00 X100 Z100;
M30;

9.2.18 Program No. 18 (Corner Radius Program in Clock Wise Direction)

Figure 9.256 shows the raw material drawing.
Figure 9.257 shows the machining drawing.

Machining Operation – Corner radius (fillet)
Raw Material – Mild steel (MS)

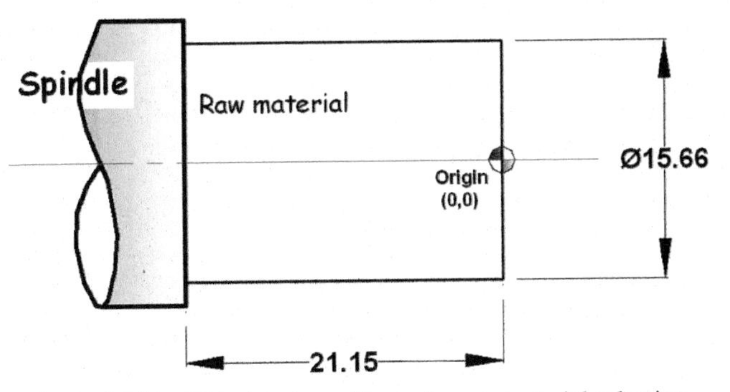

Figure 9.256 This drawing will use for raw material selection.

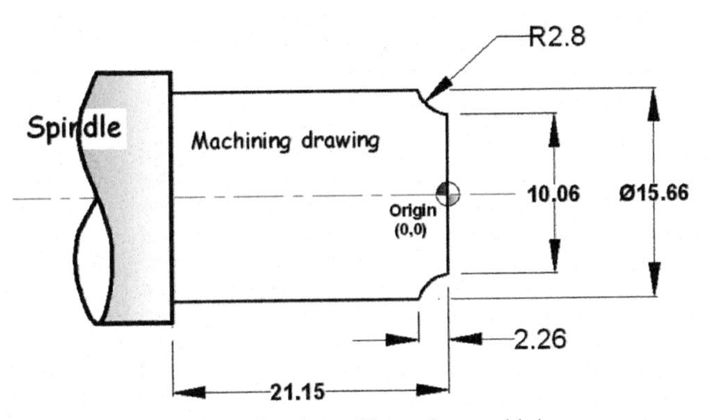

Figure 9.257 This drawing will use for machining purpose.

Cutting Tool – Turning tool PCLNR/L with bit (insert) – CNMG/DNMG/WNMG etc. (depends on the depth of cut)

Depth of Cut – 1.5 mm one side/3.0 mm diametrical (both side)

Cutting Feed – 0.12 mm/revolution (feed depends on the required surface finish as per drawing.)

Spindle Speed – 2200 rpm (RPM is a very important factor. It depends on mainly work piece diameter and holding or gripping of the work piece)

```
O99648;
G54;
G50 S2200;
T0101 (Turning Tool);
G00 X100 Z100;
```

```
G00 X20 Z5.0;
G96 M03 S230;
M08;
G00 X9.06 Z1.0; (Cutting tool positioning before making Radius)
G01 X10.06 Z0.0 F0.12; (Starting point of Radius)
G02 X15.66 Z-2.26 R2.8 F0.12; (Ending point of Radius)
G00 Z1.0;
M09;
M05;
G00 X100 Z100;
M30;
```

9.2.19 Program No. 19 (Taper Turning Program)

Figure 9.258 shows raw material drawing.

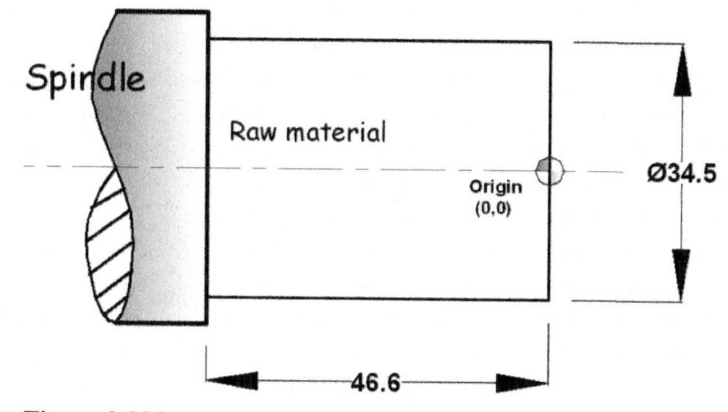

Figure 9.258 This drawing will use for raw material selection.

Figure 9.259 shows the machining drawing.

Machining Operation – Taper Turning
Raw Material – Mild Steel (MS)
Cutting Tool – Turning tool PCLNR/L with bit (insert) – CNMG/DNMG/WNMG etc. (depends on the depth of cut and tool)
Depth of cut – 1.5 mm one side/3.0 mm diametrical (both side)
Cutting Feed – 0.15 mm/revolution (feed depends on the required surface finish as per drawing.)
Spindle speed – 1800 rpm (RPM is a very important factor. It depends on mainly work piece diameter and holding or gripping of the work piece)

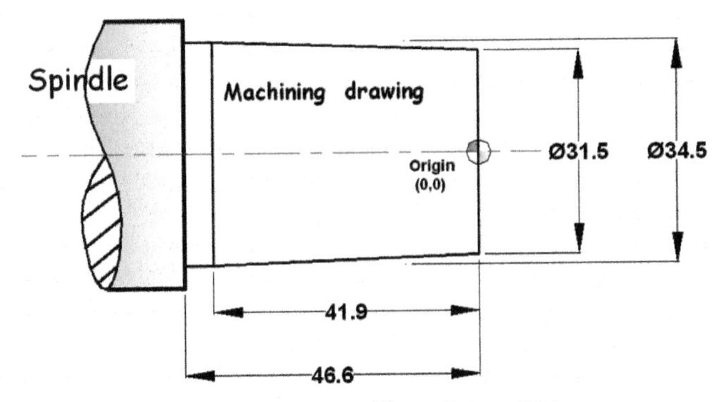

Figure 9.259 This drawing will use for machining purpose.

O92608;
G54;
G50 S1800;
T0101 (Turning Tool);
G00 X100 Z100;
G00 X50 Z10;
G96 M03 S210;
M08;
G00 X31.5 Z1.0;
G01 Z0.0 F0.12; (Starting point of **taper**)
G01 X34.5 Z-41.9 F0.15; (Ending point of **taper**)
G00 Z1.0;
M09;
M05;
G00 X100 Z100;
M30;

9.2.20 Program No. 20 (Taper Turning Program with Rough Cut)

Figure 9.260 is a raw material drawing.
Figure 9.261 shows machining drawing.

Machining Operation – Taper Turning with **rough cut**
Raw Material – Mild Steel (MS)
Cutting Tool – Turning tool PCLNR/L with bit (insert) – CNMG/DNMG/WNMG etc. (depends on the depth of cut and tool)

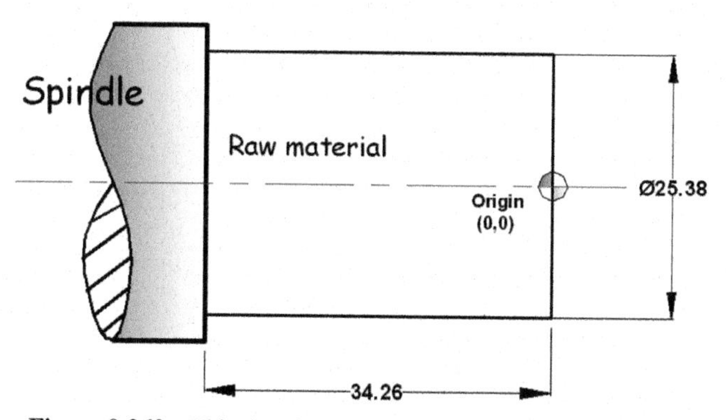

Figure 9.260 This drawing will use for raw material selection.

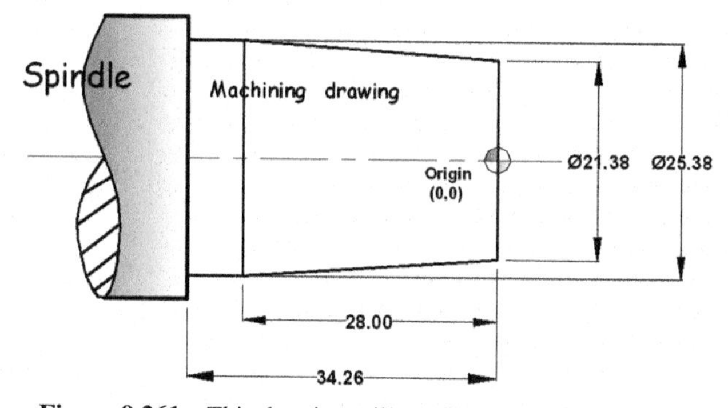

Figure 9.261 This drawing will use for machining purpose.

Depth of Cut – 1.5 mm one side/3.0 mm diametrical (both side)

Cutting Feed – 0.15 mm/revolution (feed depends on the required surface finish as per drawing.)

Spindle Speed – 1800 rpm (RPM is a very important factor. It depends on mainly work piece diameter and holding or gripping of the work piece)

Note:

Remember, before taking rough cut always calculate the machining (turning) length or diameter by trigonometry method.

O22233;
G54;
G50 S1800;
T0101 (Turning Tool);

```
G00 X100 Z100;
G00 X50 Z10;
G96 M03 S210;
M08;
G00 X23.38 Z1.0; (Tool is taking position before making Rough taper)
G01 X23.8 Z0.0 F0.12; [Starting point of Rough taper turning (SPRT)]
G01 X25.38Z-14 F0.15; [Ending point of Rough taper turning (EPRT)]
G00 Z1.0;
G00 X21.38; (Tool is taking position before making Final taper)
G01 X21.38 Z0.0 F0.12; (Starting point of Taper turning)
G01 X25.38 Z-28 F0.15; (Ending point of Taper turning)
G00 Z1.0;
M09;
M05;
G00 X100 Z100;
M30;
```

9.2.21 Program No. 21 (Taper Turning Program)

Figure 9.262 shows the raw material drawing.

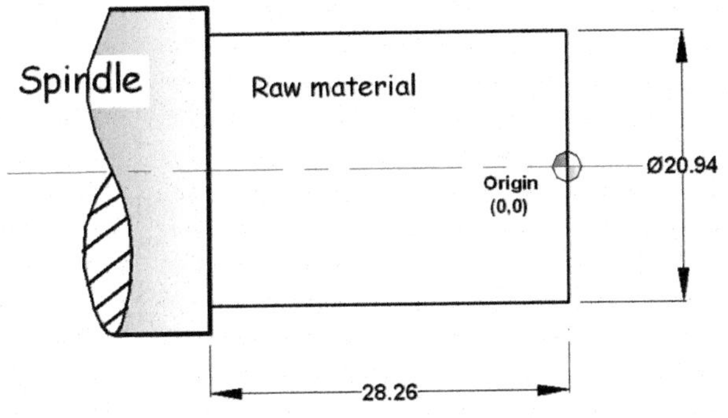

Figure 9.262 This drawing will use for raw material selection.

Figure 9.263 shows the machining drawing.

Machining Operation – Taper Turning
Raw Material – Mild steel (MS)
Cutting Tool – Turning tool PCLNR/L with bit (insert) – CNMG/DNMG/WNMG etc. (depends on the depth of cut and tool)
Depth of Cut – 1.5 mm one side/3.0 mm diametrical (both side)

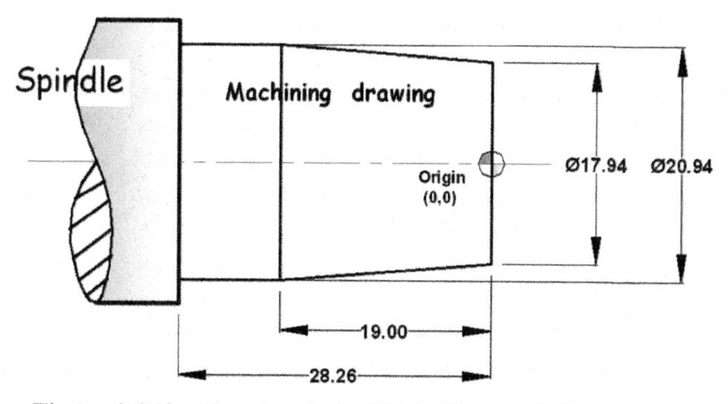

Figure 9.263 This drawing will use for machining purpose.

Cutting Feed – 0.15 mm/revolution (feed depends on the required surface finish as per drawing.)

Spindle Speed – 1800 rpm (RPM is a very important factor. It depends on mainly work piece diameter and holding or gripping of the work piece)

O35421;
G54;
G50 S1800;
T0101 (Turning Tool);
G00 X100 Z100;
G00 X30 Z10;
G96 M03 S210;
M08;
G00 X17.94 Z1.0;
G01 Z0 F0.12;
G01 X20.94 Z-19 F0.15;
G00 Z1.0;
M09;
M05;
G00 X100 Z100;
M30;

9.2.22 Program No. 22 (Threading Program)

Figure 9.264 shows the raw material drawing.
Figure 9.265 shows the machining drawing.

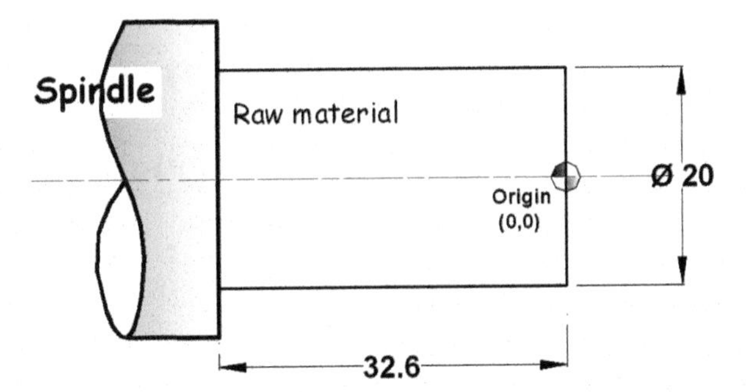

Figure 9.264 This drawing will use for raw material selection.

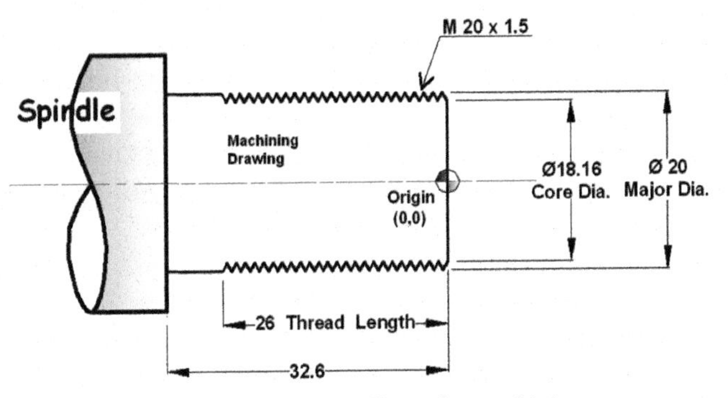

Figure 9.265 This drawing will use for machining purpose.

Machining Operation – Threading
Raw Material – Mild Steel (MS)
Cutting Tool – Threading tool
Thread Profile – Metric thread (60°)
Depth of Cut – 0.010 mm one side/0.020 mm diametrical (both side)
Cutting Feed – Feed = Pitch = 1.5 mm/revolution
Spindle Speed – 1000 rpm (During threading operation, rpm must be constant.)
TPI – 16.9333 Thread per Inch (it means you can see 16.9333 threads in one inch.)
Thread Height – 0.920 mm (one side)
M 20×1.5 – Major Diameter 20.0 mm and Pitch 1.5 mm

Note:

Always use constant rpm during the threading operation.

O72264;
G54;
T0202 (Threading Tool);
G00 X100 Z100;
G00 X50 Z15;
G97 M03 S1000;
M08;
G00 X30 Z5.0;
G92 X20.0 Z-26.0 **F1.5;** (G92 is a thread cutting cycle, We will give only thread cutting diameter in every CNC programming block and CNC machine will take all required parameters automatically like thread length, cutting feed and retraction value in X and Z-axis, because it is already given in the CNC program)
X19.8
X19.6;
X19.4;
X19.2;
X19.0;
X18.8;
X18.6;
X18.4;
X18.2;
X18.16;
G00 Z5.0;
M09;
M05;
G00 X100 Z100;
M30;

9.2.23 Program No. 23 (Threading Program with Pitch 1.25)

Figure 9.266 shows the raw material drawing.
Figure 9.267 shows the machining drawing.

Machining Operation – Threading
Raw Material – Mild Steel (MS)
Cutting Tool – Threading tool
Thread Profile – Metric thread (60°)
Depth of Cut – 0.010 mm one side/0.020 mm diametrical (both side)

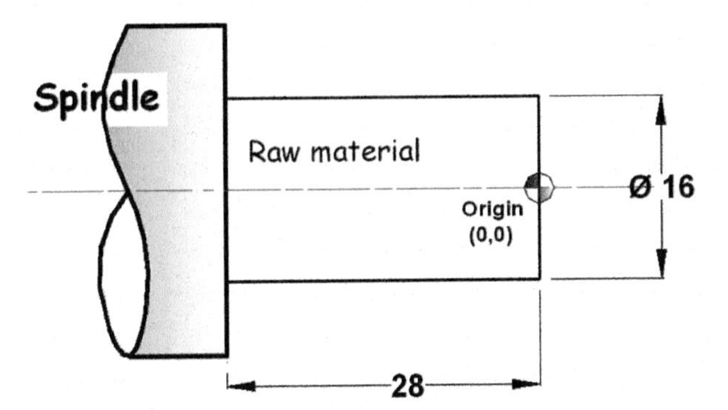

Figure 9.266 This drawing will use for raw material selection.

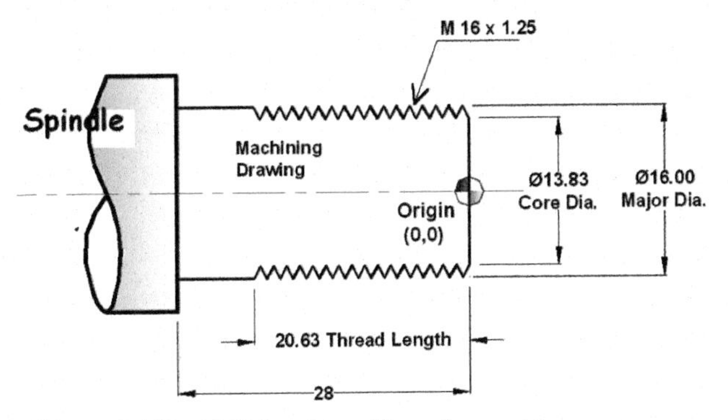

Figure 9.267 This drawing will use for machining purpose.

Cutting Feed – Feed = Pitch = 1.25 mm/revolution

Spindle Speed – 1000 rpm (During threading operation, rpm must be constant.)

TPI – 20.32 Threads per Inch (it means you can see 20.32 threads in one inch.)

Thread Height – 1.085 mm (one side)

M 16 × 1.25 – Major Diameter 16.0 mm and Pitch 1.25 mm

```
O21213;
G54;
T0202 (Threading Tool);
G00 X100 Z100;
G00 X50 Z15;
G97 M03 S1000;
M08;
```

```
G00 X26 Z5.0;
G92 X16.0 Z-20.63 F1.25;
X15.8
X15.6;
X15.4;
X15.2;
X15.0;
X14.8;
X14.6;
X14.4;
X14.2;
X14.0;
X13.83;
G00 Z5.0;
M09;
M05;
G00 X100 Z100;
M30;
```

9.2.24 Program No. 24 (Threading Program with Pitch 2.0)

Figure 9.268 shows the raw material drawing.
Figure 9.269 shows the machining drawing.

Machining Operation – Threading
Raw Material – Mild steel (MS)
Cutting Tool – Threading tool
Thread Profile – Metric thread (60°)
Depth of Cut – 0.010 mm one side/0.020 mm diametrical (both side)

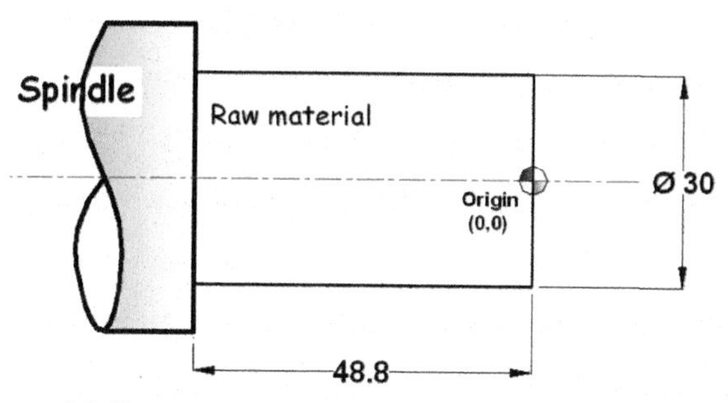

Figure 9.268 This drawing will use for raw material selection.

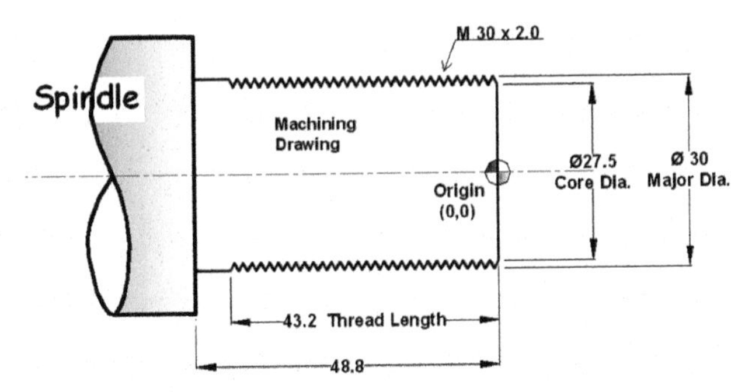

Figure 9.269 This drawing will use for machining purpose.

Cutting Feed – Feed = Pitch = 2.0 mm/revolution
Spindle Speed – 1000 rpm (During threading operation, rpm must be constant.)
TPI – 12.7 Threads per Inch (it means you can see 12.7 threads in one inch.)
Thread Height – 1.25 mm (one side)
M 30 × 2.0 – Major Diameter 30.0 mm and Pitch 2.0 mm

```
O21213;
G54;
T0202 (Threading Tool);
G00 X100 Z100;
G00 X50 Z15;
G97 M03 S1000;
M08;
G00 X40 Z5.0;
G92 X30 Z-43.2 F2.0;
X29.8;
X29.6;
X29.4;
X29.2;
X29.0;
X28.8;
X28.6;
X28.4;
X28.2;
```

```
X28.0;
X27.8;
X27.6;
X27.5;
G00 Z5.0;
M09;
M05;
G00 X100 Z100;
M30; [3]
```

9.3 Profile Turning

Very Important Note

Now from here as per drawing, if programmer will find more than one operation (like facing, drilling, threading and profile turning) in one drawing. Than CNC programmer will use different cutting tools for different machining operations under one CNC program.

9.3.1 Program No. 25 (Profile Turning with Multi Tool Program)

Figure 9.270 shows the raw material drawing.

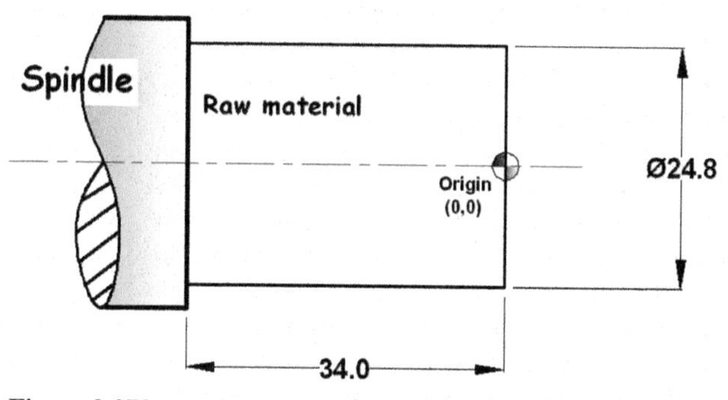

Figure 9.270 This drawing will use for raw material selection.

Figure 9.271 shows the machining drawing.

Operation Planning:

Machining Operation – Facing, Drilling and Profile Turning

Following three operations will be performed on the above job

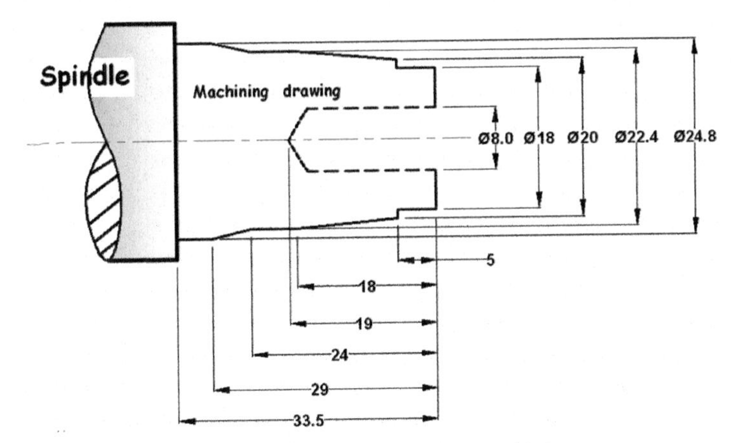

Figure 9.271 This drawing will use for machining purpose.

- **Facing operation** with the depth of cut **0.5 mm** (total material for facing 0.50 mm)
- **Drilling (peck drilling operation)**
- **Profile turning**

Raw Material – Mild steel (MS)
Cutting Tool –

1. Turning Tool (For Facing and Turning profile)
2. Drill (Ø8.0)

Depth of Cut for Roughing – Max.1.5 mm one side/3.0 mm diametrical (both side)
Cutting Feed for Facing – 0.10 mm/revolution
Cutting Feed for Turning – 0.15 mm/revolution
Cutting Feed for Drilling – 0.05 mm/revolution
Drilling Length – 19.0 mm
Spindle Speed for Turning – 1700 rpm
Spindle Speed for Drilling – 450 rpm

During following operations programmer will use two cutting tools.

1. Turning tool (T0101)
2. Drill tool (T0909)

O46715;
G54;
G50 S1700 M03;

T0101 (Turning Tool for **Facing**);
G00 X100 Z200;
G00 X30 Z5;
G96 M03 S180;
M08;
G00 **Z0.1;** [0.100 mm material left for final facing cut]
G01 X0.0 F0.10; (extra material removed during facing operation)
G00 Z1.0;
M09;
M05;
G00 X100 Z200;
M00; (Program stop)
T0909 (**Drilling** Tool);
G00 X100 Z200;
G00 X20 Z15;
G97 M03 S450;
G00 X0.0 Z5.0;
M08;
G00 Z1.0;
G01 Z-5.0 F0.05;
G00 Z5.0;
G00 Z-4.0;
G01 Z-10 F0.05;
G00 Z5.0;
G00 Z-9.0;
G01 Z-15 F0.05;
G00 Z5.0;
G00 Z-14.0;
G01 Z-19 F0.05;
G00 Z5.0;
M09;
M05;
G00 X100 Z200;
M00; (Program stop)
T0101 (Turning Tool for Profile **Turning**);
G00 X100 Z200;
G00 X30 Z5;
G96 M03 S220;
M08;

```
G00 Z0.0;
G01 X0.0 F0.10; (Final Facing operation after drill, tool will cut remaining
                 0.100 mm material from the face.)
G00 Z1.0;
G00 X22.4; Profile Turning start
G01 Z-24.0 F0.15;
G01 X24.8 Z-29.0 F0.15;
G00 X26;
G00 Z1.0;
G00 X20.0;
G01 Z-5.0 F0.15;
G01 X22.4 Z-18 F0.15;
G00 Z1.0;
G00 X18;
G01 Z-5.0 F0.15;
G01 X22 F0.15;
G00 Z1.0;
M09;
M05;
G00 X100 Z100;
M30;
```

9.3.2 Program No. 26 (Radius Profile Turning with Multi Tool Program)

Figure 9.272 shows the raw material drawing.

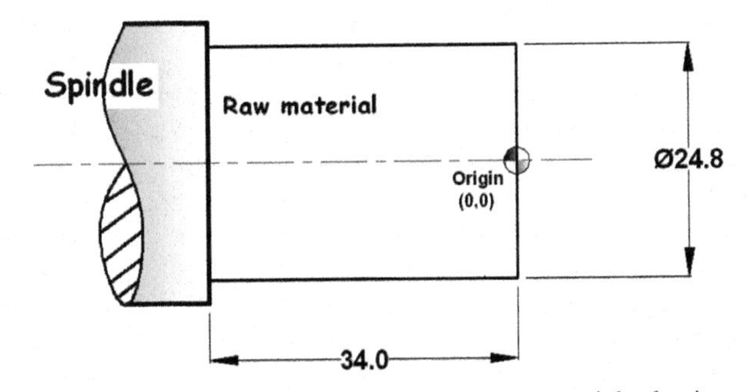

Figure 9.272 This drawing will use for raw material selection.

Figure 9.273 shows the machining drawing.

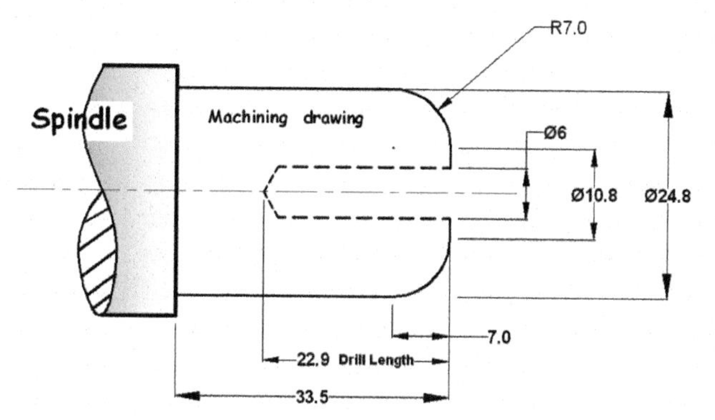

Figure 9.273 This drawing will use for machining purpose.

Operation Planning:

Machining Operation – Facing, Drilling and Radius Profile

Following three operations will be performed on the above job.

- **Facing operation** with the depth of cut **0.5 mm** (total material for facing 0.50 mm)
- **Drilling (peck drilling operation)**
- **Radius Profile (using the G03 command)**

Raw Material – Mild steel (MS)
Cutting Tool –

1. Turning Tool (For Facing and Radius turning profile with rough radius cut)
2. Drill (**Ø6.0**)

Depth of Cut for Roughing – Max.1.5 mm one side/3.0 mm diametrical (both side)
Cutting Feed for facing – 0.10 mm/revolution
Rough Cutting Feed for Radius Turning – 0.15 mm/revolution
Finish Cutting Feed for Radius Turning – 0.12 mm/revolution
Cutting Feed for Drilling – 0.05 mm/revolution
Drilling length – 22.9 mm
Spindle speed for Turning – 1700 rpm
Spindle speed for Drilling – 450 rpm

During the following operations, we will use two cutting tools.

1. Turning tool (T0101)
2. Drill tool (T0909)

O46465;
G54;
G50 S1700 M03;
T0101 (Turning Tool for **Facing**);
G00 X100 Z200;
G00 X30 Z5;
G96 M03 S220;
M08;
G00 **Z0.100;** [0.100 mm material left for final facing cut]
G01 X0.0 F0.12;
G00 Z1.0;
M09;
M05;
G00 X100 Z200;
M00; (Program stop)
T0909; (**Drilling Too**l)
G00 X100 Z200;
G00 X20 Z15;
G97 M03 S450;
G00 X0.0 Z5.0;
M08;
G00 Z1.0;
G01 Z-5.0 F0.05;
G00 Z5.0;
G00 Z-4.0;
G01 Z-10 F0.05;
G00 Z5.0;
G00 Z-9.0;
G01 Z-15 F0.05;
G00 Z5.0;
G00 Z-14.0;
G01 Z-19 F0.05;
G00 Z5.0;
G00 Z-18;

G01 Z-22.9 F0.05;
G00 Z5.0;
M09;
M05;
G00 X100 Z200;
M00; (Program stop)
T0101 (Turning Tool for Radius **Profile Turning**);
G00 X100 Z200;
G00 X30 Z5;
G96 M03 S220;
M08;
G00 **Z0.0;**
G01 X0.0 F0.10; **Final Facing** operation
G00 Z1.0;
G00 X22.8;
G01 X22.8 Z0.0 F0.12; **Radius** Turning Profile start (**Rough**)
G03 X24.8 Z-1.0 **R1.0** F0.15; **Rough Radius** Turning
G00 Z1.0;
G00 X20.8;
G01 X20.8 Z0.0 F0.12;
G03 X24.8 Z-2.0 **R2.0** F0.15; **Rough Radius** Turning
G00 Z1.0;
G00 X18.8;
G01 X18.8 Z0.0 F0.12;
G03 X24.8 Z-3.0 **R3.0** F0.15; **Rough Radius** Turning
G00 Z1.0;
G00 X16.8;
G01 X16.8 Z0.0 F0.12;
G03 X24.8 Z-4.0 **R4.0** F0.15; **Rough Radius** Turning
G00 Z1.0;
G00 X14.8;
G01 X14.8 Z0.0 F0.12;
G03 X24.8 Z-5.0 **R5.0** F0.15; **Rough Radius** Turning
G00 Z1.0;
G00 X12.8;
G01 X12.8 Z0.0 F0.12;
G03 X24.8 Z-6.0 **R6.0** F0.15; **Rough Radius** Turning
G00 Z1.0;
G00 X9.8;

G01 X10.8 Z0.0 F0.10; Starting point of radius
G03 X24.8 Z-7.0 **R7.0** F0.12; **Final Radius** Turning, End point of Radius
G00 Z1.0;
M09;
M05;
G00 X100 Z200;
M30;

9.3.3 Program No. 27 (Turning Profile Program)

Figure 9.274 shows the raw material drawing.

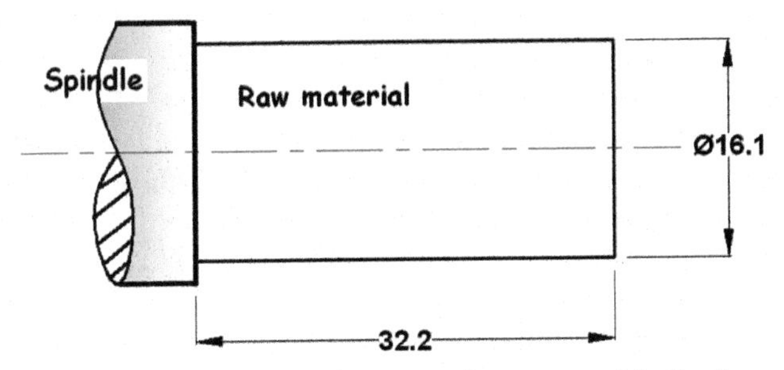

Figure 9.274 This drawing will use for raw material selection.

Figure 9.275 shows the machining drawing.

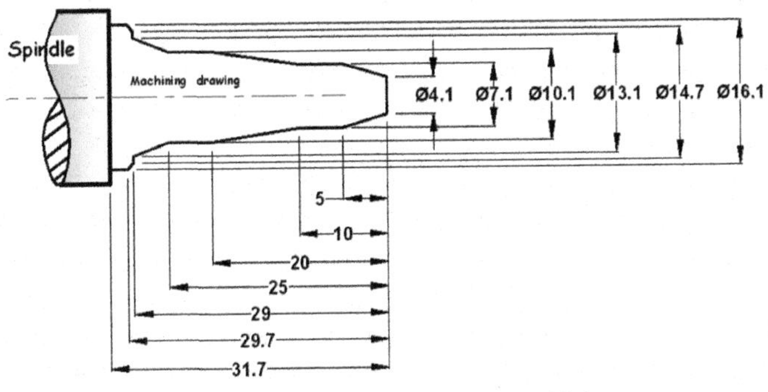

Figure 9.275 This drawing will use for machining purpose.

Operation Planning:

Machining Operation – Facing and Turning Profile

Following two operations will be performed on the above job.

- **Facing operation** with the depth of cut **0.5 mm** (total material for facing 0.50 mm)
- **Turning Profile**

Raw Material – Mild Steel (MS)
Cutting Tool – Turning Tool (Facing and Turning Profile)
Depth of Cut for Roughing – Max.1.5 mm one side/3.0 mm diametrical (both side)
Rough Cutting Feed for Profile Turning – 0.15 mm/revolution
Finish Cutting Feed for Profile Turning – 0.12 mm/revolution
Spindle Speed for Turning – 2300 rpm

During following operations, we will use....

1. Turning tool (T0101)

Note:

During profile turning of above job, first of all tool will cut the extra material from every diameter. As per CNC program (drawing), the tool will leave some extra material on the work piece surface for finishing the operation. Generally this extra material can be 0.200 mm plus (+) in X-axis and 0.100 mm plus (+) in Z-axis. For an example after rough cutting, every diameter should be Ø4.3 mm, Ø7.3 mm, Ø10.3 mm, Ø13.3 mm. and step turning length should 4.9 mm, 9.9 mm, 19.9 mm, 24.9 mm. Now during finishing operation, turning tool will take final cut in X and Z-axis (in X-axis tool will remove 0.200 mm extra material from the diameter and in Z-axis tool will remove 0.100 mm extra material from the work piece face).

O66888;
G54;
G50 S2300 M03;
T0101 (Turning Tool);
G00 X100 Z100;
G00 X20 Z5;
G96 M03 S240;
G00 Z0.0;
M08;
G01 X0.0 F0.12; **Facing** operation
G00 Z1.0;
G00 X13.3 Z1.0; **Rough** turning operation start
G01 Z-28.9 F0.15;
G01 X17.1;

G00 Z1.0;
G00 X10.3;
G01 Z-24.9 F0.15;
G01 X13.3 Z-28.9 F0.15;
G01 X17.1;
G00 Z1.0;
G00 X7.3;
G01 Z-9.9 F0.15;
G01 X10.3 Z-19.9 F0.15;
G00 Z1.0;
G00 X4.3;
G01 Z0.0 F0.15;
G01 X7.3 Z-4.9 F0.15;
G00 Z1.0; **Rough** turning operation complete
G00 X3.1 Z1.0; **Final profile turning operation starts** and tool will cut extra remaining material 0.200 mm from the external diameter (this material was left for finishing operation during rough cutting) and 0.100 mm from the face (length).
G01 X4.1 Z0.0 F0.12;
G01 X7.1 Z-5.0 F0.12;
G01 Z-10.0 F0.12;
G01 X10.1 Z-20 F0.12;
G01 Z-25 F0.12;
G01 X13.1 Z-29 F0.12;
G01 X14.7 F0.12;
G01 X16.1 Z-29.7 F0.12;
G00 Z1.0;
M09;
M05;
G00 X100 Z100;
M30;

9.3.4 Program No. 28 (Multi Tool Program for Profile Turning and Threading Operation)

Figure 9.276 shows the raw material drawing.
Figure 9.277 shows the machining drawing.

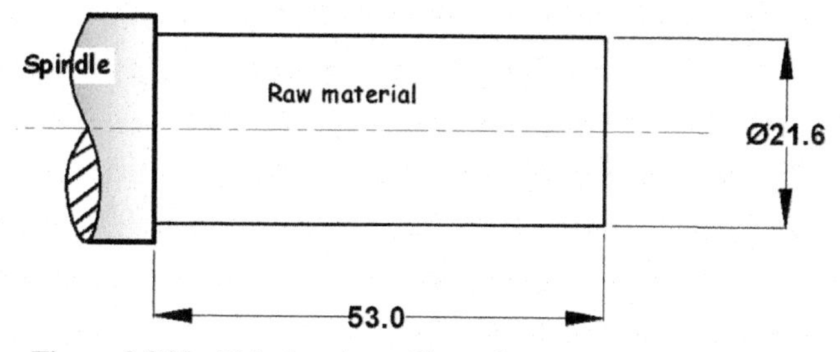

Figure 9.276 This drawing will use for raw material selection.

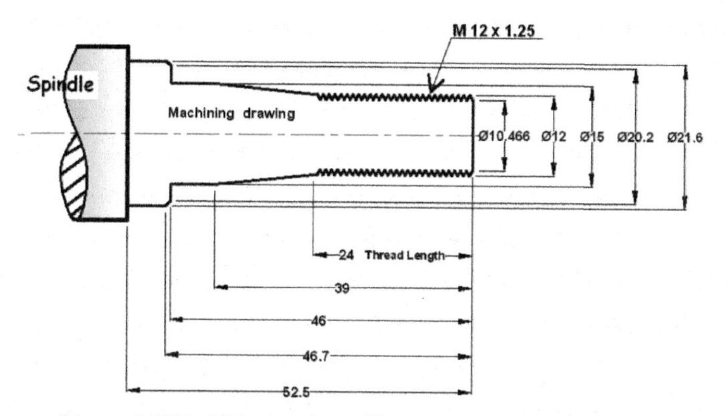

Figure 9.277 This drawing will use for machining purpose.

Operation Planning:

Machining Operation – Facing, Turning Profile and Threading

Following three operations will perform on the above job.

- **Facing operation** with the depth of cut **0.5 mm** (total material for facing 0.50 mm)
- **Turning profile**
- **Threading**

Raw Material – Mild steel (MS)
Cutting Tool –

1. Turning Tool (For Facing and Turning Profile)
2. Threading Tool

Thread Profile –Metric thread (60°)
Depth of Cut for Rough Turning – Max.1.5 mm one side/3.0 mm diametrical (both side)

Depth of Cut for Thread Cutting – 0.100 mm one side/0. 200 mm diametrical (both side)
Cutting Feed for Rough Profile Turning – 0.20 mm/revolution
Cutting Feed for Finish Profile Turning – 0.12 mm/revolution
Cutting Feed for Threading – Feed = Pitch = 1.25 mm/revolution
Spindle Speed for Turning – 2200 rpm
Spindle Speed for Threading – 1000 rpm (During threading operation, **rpm** must be constant.)
Thread Height – 0.767 mm (one side)
TPI – 20.32 Threads per inch (it means you can see 20.32 threads in one inch.)
M 12 × 1.25 – Major Diameter 12.0 mm and Pitch 1.25 mm

During following operations, we will use....

1. Turning tool (T0101)
2. Threading tool (T0909)

Starting diameter of the chamfer (45°) on the starting point of the thread for better application of thread – Ø9.466 mm (*Chamfer will start from 1.0 mm less from the root dia.)*
Ending diameter of thread chamfer for 45° chamfer – Ø12.0 mm
Thread chamfer length at 45° – 1.267mm (*this chamfer is required for proper working of the thread*)

Note:

i. *For better threading result, we should give proper chamfer amount on the starting point of the thread.*
ii. *During rough turning, the programmer will leave 0.200 mm material on the external diameter of the work piece. This extra material (0.200 mm) will remove during final turning.*

O49972;
G54;
G50 S2200 M03;
T0101 (Turning Tool);
G00 X100 Z100;
G00 X26 Z5;
G96 M03 S180;
G00 Z0.0;
M08;

G01 X0.0 F0.12; **Facing** operation
G00 Z1.0;
G00 X18.6; **Rough** turning operation start
G01 Z-45.9 F0.20;
G01 X23.0 F0.2;
G00 Z1.0;
G00 X15.6;
G01 Z-45.9 F0.20;
G01 X23.0;
G00 Z1.0;
G00 X15.2;
G01 Z-45.9 F0.20;
G01 X23.0;
G00 Z1.0; **Rough** turning operation complete
G00 X12.2;
G01 Z-24.0 F0.20;
G01 X15.2 Z-39.0 F0.20;
G00 Z1.0;
G00 X9.466;
G01 X9.466 Z0.0 F0.12; Finish **Turning** Profile start (starting point of chamfer)
G01 X12.0 Z-1.267 F0.12; Ending point of chamfer, Thread Chamfer at 45°
G01 Z-24 F0.12;
G01 X15 Z-39 F0.12;
G01 Z-46 F0.12;
G01 X20.2 F0.12;
G01 X21.6 Z-46.7 F0.12;
G00 Z1.0;
M09;
M05;
G00 X100 Z100;
M00;
T0909 (**Threading Tool**);
G00 X100 Z100;
G00 X50 Z15;
G97 M03 S1000;
M08;
G00 X22.0 Z5.0;
G92 X12.0 Z-24.0 **F1.25;**

```
X11.8
X11.6;
X11.4;
X11.2;
X11.0;
X10.8;
X10.6;
X10.466;
G00 Z5.0;
M09;
M05;
G00 X100 Z100;
M30;
```

9.3.5 Program No. 29 (CNC Programming for Oval Shape)

Figure 9.278 shows the raw material drawing.

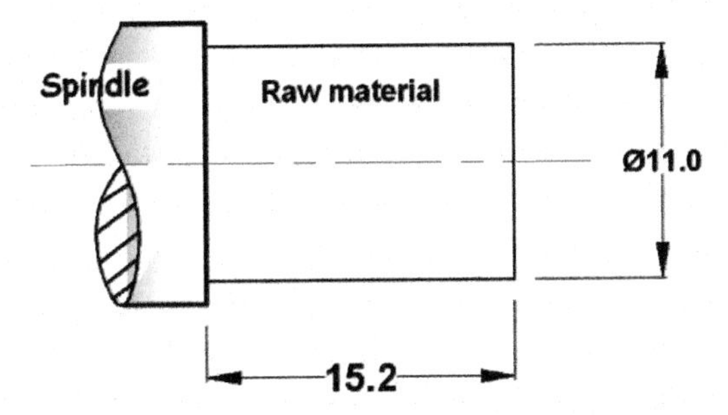

Figure 9.278 This drawing will use for raw material selection.

Figure 9.279 shows the machining drawing.

Operation Planning:

Machining Operation – Facing, Radius Profile (**Counter Clock wise - G03**)

Following two operations will perform on the above job

- **Facing operation** with the depth of cut **0.5 mm** (total material for facing 0.50 mm)
- **Radius profile**

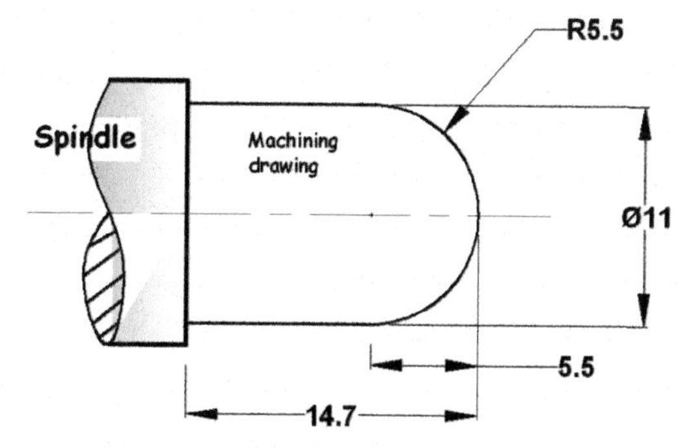

Figure 9.279 This drawing will use for machining purpose.

Raw Material – Mild steel (MS)
Cutting Tool –

- Turning Tool (For Facing and Radius Profile)

Depth of Cut for Roughing – Max.1.5 mm one side/3.0 mm diametrical (both side)
Rough Cutting Feed for Radius Profile – 0.200 mm/revolution
Finish Cutting Feed for Radius Profile – 0.12 mm/revolution
Spindle speed for Turning – 2400 rpm

During following operations, the programmer will use ....

1. Turning tool (T0101)

Note:

First of all programmer will make rough radius profile or we can say programmer will remove rough material as a rough radius profile after this cutting tool will take a final cut of radius profile.

O33118;
G54;
G50 S2400 M03;
T0101 (Turning Tool);
G00 X100 Z100;
G00 X15 Z5;
G96 M03 S240;
M08;
G00 Z0.0;
G01 X0.0 F0.12; Facing operation
G00 Z1.0;

G00 X9.0 Z1.0;
G01X9.0 Z0.0 F0.12; Starting point of Radius (**Rough** material cutting)
G03 X11.0 Z-1.0 **R1.0** F0.20; **Rough** Radius Profile (Ending Point of Radius)
G00 Z1.0;
G00 X7.0;
G01 X7.0 Z0.0 F0.12;
G03 X11.0 Z-2.0 **R2.0** F0.20;
G00 Z1.0;
G00 X5.0;
G01 X5.0 Z0.0 F0.12;
G03 X11.0 Z-3.0 **R3.**0 F0.20;
G00 Z1.0;
G00 X3.0;
G01 X3.0 Z0.0 F0.12;
G03 X11.0 Z-4.0 **R4.**0 F0.20;
G00 Z1.0;
G00 X1.0;
G01 X1.0 Z0.0 F0.12;
G03 X11.0 Z-5.0 **R5.0** F0.20;
G00 Z1.0;
G00 X-2.0; Tool is taking position in X-axis (going down under the **center line in negative direction**) for accurate radius profile.
G01 X0.0 Z0.0 F0.12; **Starting point of Radius** (SPR), tool is taking position for final radius cut.
G03 X11.0 Z-5.5 **R5.5** F0.12; Final Radius Profile [**Ending Point of Radius** (EPR)]
G01 X13.0 F0.12; Tool will take clearance after machining
G00 Z1.0;
M09;
M05;
G00 X100 Z100;
M30;

9.3.6 Program No. 30 (CNC Programming for Circular Clock Wise Direction)

Figure 9.280 shows the raw material drawing.
Figure 9.281 shows the machining drawing.

Operation Planning:

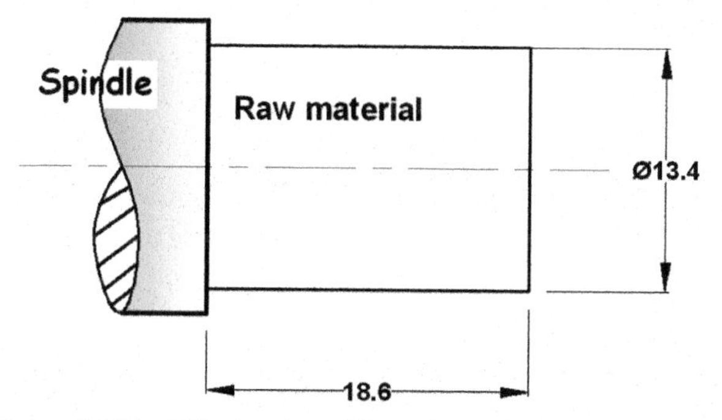

Figure 9.280 This drawing will use for raw material selection.

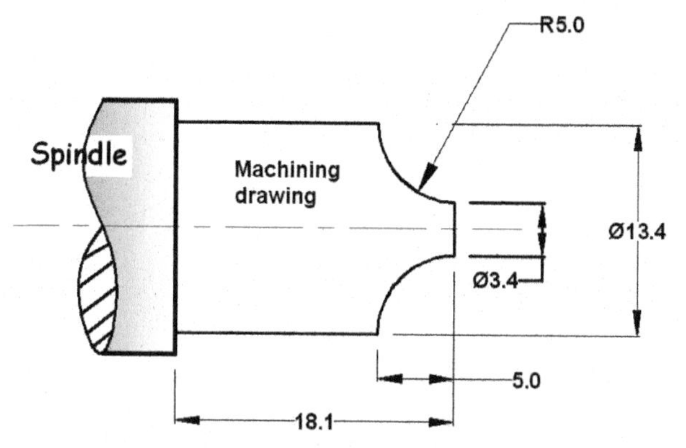

Figure 9.281 This drawing will use for machining purpose.

Machining Operation – Facing, Radius Profile (**Clock wise-G02**)
Following two operations will perform on the above job

- **Facing operation** with the depth of cut **0.5 mm** (total material for facing 0.50 mm)
- **Radius profile**

Raw Material – Mild steel (MS)
Cutting Tool –

- Turning Tool (for Facing and Radius Profile)

Depth of Cut for Roughing – Maximum1.5 mm one side/3.0 mm diametrical (both side)
Rough cutting Feed for Radius Profile – 0.200 mm/revolution
Finish cutting Feed for Radius Profile – 0.12 mm/revolution
Spindle Speed for Turning – 2400 rpm

During following operations, the programmer will use.......

1. Turning tool (T0101)

Note:

First of all programmer will make rough radius profile or we can say programmer will remove rough material as a rough radius profile after this cutting tool will take a final cut of radius profile.

O12125;
G54;
G50 S2400 M03;
T0101 (Turning Tool);
G00 X100 Z100;
G00 X18.0 Z5.0;
G96 M03 S240;
M08;
G00 Z0.0;
G01 X0.0 F0.12; Facing operation
G00 Z1.0;
G00 X11.4;
G01 X11.4 Z0.0 F0.12; Starting point of Radius (**Rough**)
G02 X13.4 Z-1.0 **R1.0** F0.20; **Rough** Radius Profile (Ending Point of Radius)
G00 Z1.0;
G00 X9.4;
G01 X9.4 Z0.0 F0.12;
G02 X13.4 Z-2.0 **R2.0** F0.20;
G00 Z1.0;
G00 X7.4;
G01 X7.4 Z0.0 F0.12;
G02 X13.4 Z-3.0 **R3.0** F0.20;
G00 Z1.0;
G00 X5.4;
G01 X5.4 Z0.0 F0.12;
G02 X13.4 Z-4.0 **R4.0** F0.20;
G00 Z1.0;
G00 X2.4; Tool is taking position before final radius profile cutting
G01 X3.4 Z0.0 F0.12; **Starting** point of Radius (for final cut)
G02 X13.4 Z-5.0 **R5.0** F0.12; **Final** Radius Profile (**Ending** Point of Radius)
G01 X15.0 F0.12; Tool will take clearance after radius profile

```
G00 Z1.0;
M09;
M05;
G00 X100 Z100;
M30;
```

9.3.7 Program No. 31 (CNC Programming for Big Arc Profile)

Figure 9.282 shows the raw material drawing.

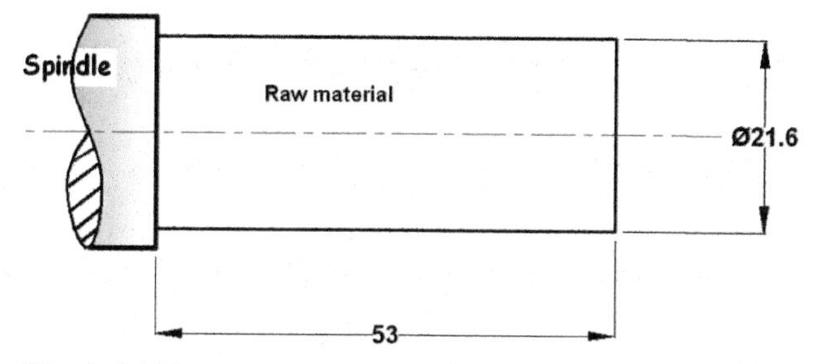

Figure 9.282 This drawing will use for raw material selection.

Figure 9.283 shows the machining drawing.

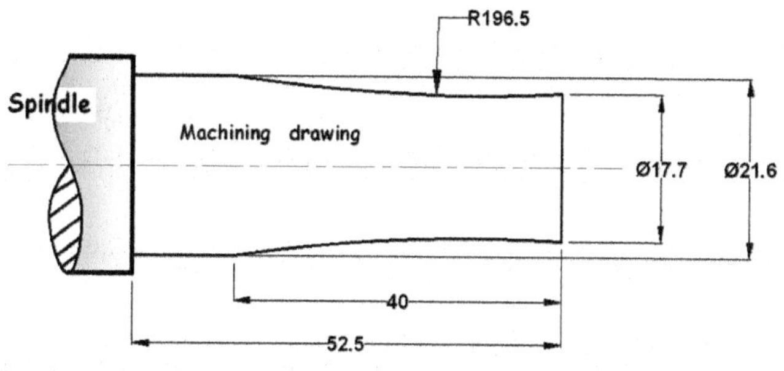

Figure 9.283 This drawing will use for machining purpose.

Operation Planning:

Machining Operation – Facing, Radius Profile
Following two operations will perform on the above job.

- **Facing operation** with the depth of cut **0.5 mm** (total material for facing 0.50 mm)
- **Radius Profile**

Raw Material – Mild steel (MS)

Cutting Tool –

- Turning Tool (For Facing and Radius Profile)

Depth of cut for Roughing – Max.1.5 mm one side/3.0 mm diametrical (both side)
Rough cutting Feed for Radius Profile – 0.200 mm/revolution
Finish cutting Feed for Radius Profile – 0.12 mm/revolution
Spindle Speed for Turning – 2000 rpm

During following operations, the programmer will use ....

1. Turning tool (T0101)

O49137;
G54;
G50 S2000 M03;
T0101 (Turning Tool);
G00 X100 Z100;
G00 X26.0 Z10.0;
G96 M03 S210;
M08;
G00 X26.0 Z0.0;
G01 X0.0 F0.12; Facing operation
G00 Z1.0;
G00 X22.6 Z1.0; If tool comes from upside to down side for taking radius cut. It will give better radius profile at the starting point of the radius.
G01 X21.6 Z0.0 F0.12; Rough cutting is start. Starting Point of **Radius** profile (**Rough** cut)
G02 X21.6 Z-33 **R196.5** F0.20; Ending Point of **Radius** profile
G00 Z1.0;
G00 X21.6 Z1.0;
G01 X20.6 Z0.0 F0.12;
G02 X21.6 Z-36.0 **R196.5** F0.20;
G00 Z1.0;
G00 X20.6 Z1.0;
G01 X19.6 Z0.0 F0.12;
G02 X21.6 Z-37.0 **R196.5** F0.20;
G00 Z1.0;
G00 X19.6 Z1.0;

G01 X18.6 Z0.0 F0.12;
G02 X21.6 Z-38.0 **R196.5** F0.20;
G00 Z1.0
G00 X18.7 Z1.0;
G01 X17.7 Z0.0 F0.12; Starting point of Radius (for **Final cut**)
G02 X21.6 Z-40.0 **R196.5** F0.12; **Finial** Radius Profile (**Ending Point of Radius**)
G01 X23.0 F0.12; Tool will take clearance after machining
G00 Z1.0
M09;
M05;
G00 X100 Z100;
M30;

9.3.8 Program No. 32 (Profile Turning with Multi Cutting Tools)

Important Note

From here we will cut/machine the material of the work piece in two steps. One is called roughing cut procedure and another is called finishing cut procedure. For roughing cut we use the rough cutting tool and for finishing cut we use the final cutting tool. This method gives profile accuracy and better surface finish.

- First of all, we will cut the material as a rough part from the work piece surface. It means the machine will cut the extra material from the work piece surface and makes a rough profile.
- Now after roughing, the machine will take finishing cut. Mostly finishing cut amount gives 0.200 mm Diametrical in (X-axis) and 0.100 mm on length in (Z-axis).

Figure 9.284 is a raw material drawing.

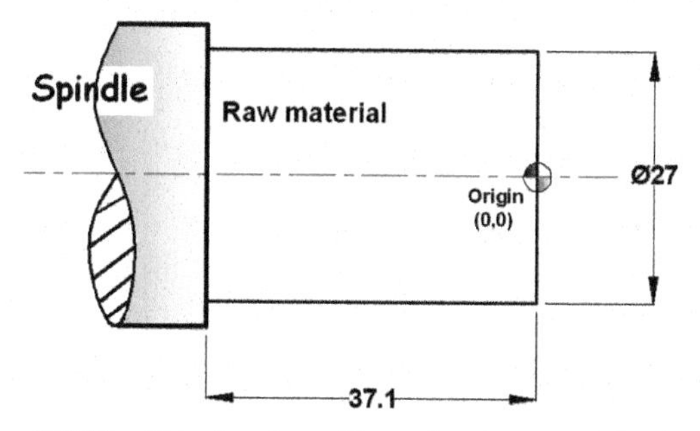

Figure 9.284 This drawing will use for raw material selection.

Figure 9.285 shows machining drawing.

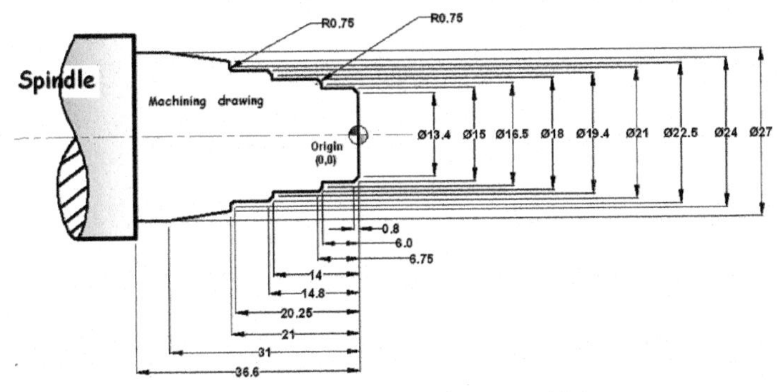

Figure 9.285 This drawing will use for machining purpose.

Following zoomed figure (Figure 9.286) is showing corner radius profile.

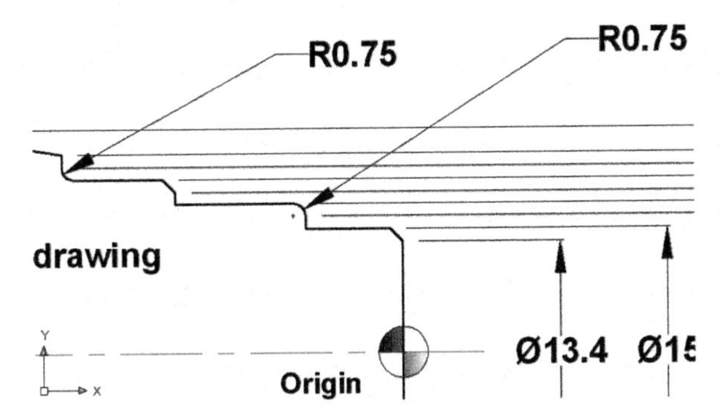

Figure 9.286 It is a zoomed drawing of the previous drawing for better understanding.

Operation Planning:

Machining Operation – Facing and Profile Turning

Following two operations will be performed on the above job.

- **Facing operation** with the depth of cut **0.5 mm** (total material for facing 0.50 mm)
- **Profile turning**

Raw Material – Mild steel (MS)
Cutting Tool – Turning tool (rough turning tool and finishing turning tool)

Note:

Turning tool can use for facing operation.

Depth of cut for Roughing – Max.1.5 mm one side/3.0 mm diametrical (both side)
Finishing allowance in X-axis – 0.200 mm diametrical (both side)
Cutting Feed for roughing – 0.25 mm/revolution
Cutting Feed for finishing – 0.12 mm/revolution
Spindle speed – 1800 rpm

Note:

Finishing allowance means, during roughing operation CNC machine will leave some material on the work piece surface for final cut.

During the following operations, the two cutting tools used are:

1. Rough turning tool (T0101)
2. Finish turning tool for Finishing cut (T0303)

O66971;
G54;
G50 S1800 M03;
T0101 (Turning Tool);
G00 X100 Z100;
G00 X32 Z5;
G96 M03 S210;
M08;
G00 X32 **Z0.05;** [0.050 mm material left for final facing]
G01 X0.0 F0.12; **Rough Facing** operation
G00 Z1.0;
G00 X24.2; **Rough** Profile Turning start
G01 Z-21 F0.25;
G01 X27.2 Z-31.0 F0.25;
G00 Z1.0;
G00 X21.2;
G01 Z-20.25 F0.25;
G02 X22.7 Z-21.0 **R0.75** F0.15;
G01 X25 F0.15;
G00 Z1.0;
G00 X18.2;
G01 Z-14.0 F0.25;
G01 X22.0 F0.15;
G00 Z1.0;
G00 X15.2;

G01 Z-6.0 F0.25;
G00 X16.7;
G03 X18.2 Z-6.75 R0.75 F0.15;
G00 Z1.0;
G00 X100 Z100;
M09;
M05;
M00; (Program stop)
T0303; (Turning Tool for Final finishing)
G00 X100 Z100;
G00 X30 Z5;
G96 M03 S230;
M08;
G00 X20 Z1.0;
G00 Z0.0;
G01 X0.0 F0.12; **Final facing** operation
G00 Z1.0;
G00 X12.4 Z1.0; **Finishing cut** of Profile Turning (for better chamfering profile, cutting tool comes from down side to up side during cutting operation)
G01 X13.4 Z0.0 F0.12;
G01 X15.0 Z-0.8 F0.12;
G01 Z-6.0 F0.12;
G01 X16.5 F0.12;
G03 X18.0 Z-6.75 **R0.75** F0.12;
G01 Z-14.0 F0.12;
G01 X19.4 F0.12;
G01 X21.0 Z-14.8 F0.12;
G01 Z-20.25 F0.12;
G02 X22.5 Z-21.0 **R0.75** F0.25;
G01 X24.0 F0.12;
G01 X27.0 Z-31 F0.12;
G00 Z1.0;
M09;
M05;
G00 X100 Z100;
M30;

9.4 How Will You Take Work Zero Offset G54? or Where You Will Take Z Zero (Z = 0) on the Work piece Surface?

Any type of machining operation in CNC turning machine, first of all we take work zero offset (G54) by the master tool (first cutting tool). **Generally, the turning tool or the first tool becomes the master tool.**

Before machining operation every work piece has little extra raw material on length for final facing as per drawing requirements. So during the procedure of work zero offset G54, when the work piece is revolving, we will touch the master tool on the front face of the work piece and tool removes the little material from the face. After this procedure, we open the software page of G54 on the computer screen and selects the Z value on the screen by the cursor movement. After this we press the Z measure key on the control panel. After pressing the Z key, the value of the Z on the screen, will be change. Now this G54 value will become Z zero for current work piece face and current program (current coordinates). You can also see Z zero value on the controls screen where the machine axis is showing.

Now the master tool (generally turning tool/first cutting tool) gets actual Z zero and it is called origin (0, 0).

The table (Figure 9.287) is showing work zero offset example and this **Z** value or Z = 0 on the face (after removing the material) will be constant for all same work piece/same lot/batch.

When a programmer makes the facing program or any machining program, he will consider the Z zero, on the face of the work piece.

But if work piece has extra length as per drawing then actual work offset Z = 0 (Figure 9.288) will be shift inside the work piece for removing the extra material from the length. See Figure 9.289.

Extra material will show on work piece length after work zero shifting inside the work piece (if the length is extra as per drawing) (Figure 9.289)

Work Zero offset (G54)	
X	**0.0**
Z	**129.354**

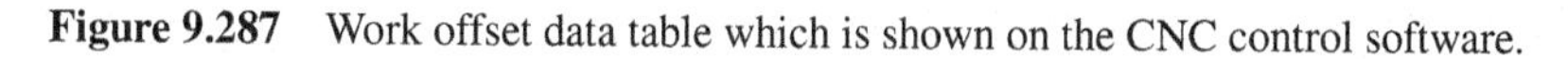

Figure 9.287 Work offset data table which is shown on the CNC control software.

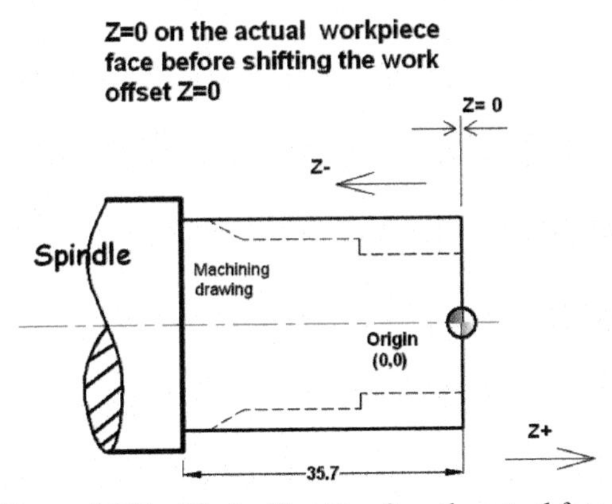

Figure 9.288 Work offset Z = 0 on the actual face.

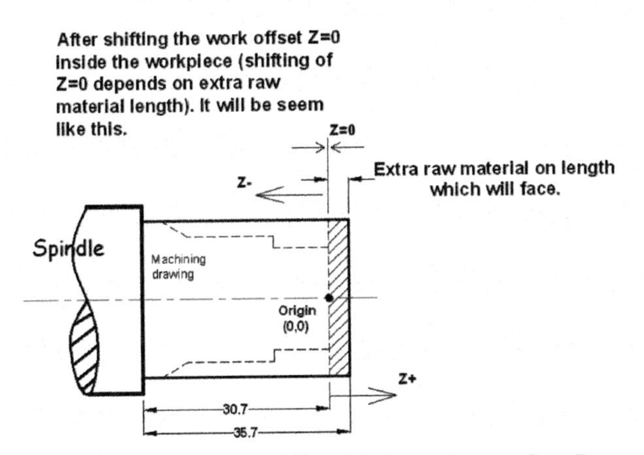

Figure 9.289 Work offset Z = 0 will be shift inside the work piece face. Because work piece has extra length as per drawing).

Figure 9.289 shows the extra material (*always you will take extra length job for machining, this extra material on length will cut during facing operation.*) on the work piece face, but we take the Z = 0 on the face of the work piece and shift the Z value (Z zero) inside the work piece, now our Z value will be inside the work piece. This condition will apply when if we have extra length and we want to cut that extra length from the work piece face.

Important Note

- Whenever the programmer will machine profile facing, always the value of extra material (length) will take in positive direction (Z+) in Z-axis. See Figure 9.289.

Very Important Note

- Only in facing operation you can apply above method or not, it is your choice and what is better in that condition.
- When more than one operation performs in a work piece than you must apply above explained method.

9.5 Practice Drawings with Multiple Cutting Tools

(Fill in the blanks)

Very Important Note

Now from here we will use 2 cutting tools, one is for roughing operation and another is for the finishing operation. Generally it is used in profile turning operation or heavy material removing. During profile turning operation, chances of wear and tear of cutting tool edge are increased. That's why another cutting tool is used as a finishing tool. During finishing operation the cutting tool cuts very less (approximate 0.200 mm) material. Due to very less cutting, cutting tool gets more cutting life than rough tool.

It is very necessary that work piece can achieve the drawing standard as per the given drawing (Figure 9.290).

Extra material is showing on the diameter and length. This extra material leaves during rough cutting and removes during finishing operation. Due to this procedure, we can maintain surface finish and dimensions between tolerance. See Figure 9.290.

- In Figure 9.290, during rough turning operation, rough turning tool will leave some material on the surface of the work piece in X and Z-axis.

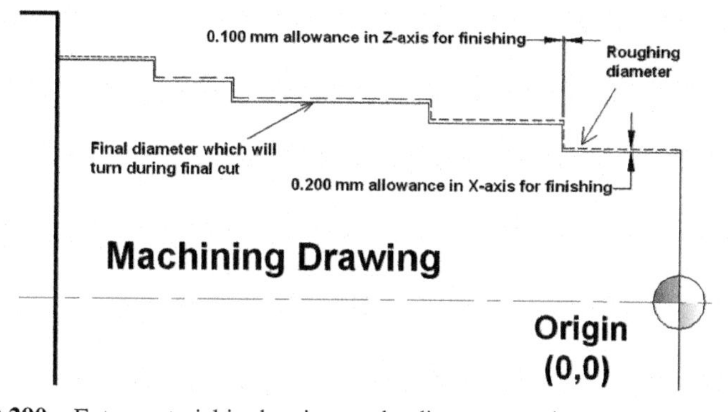

Figure 9.290 Extra material is showing on the diameters and each step turning length.

That leaving material is called finishing allowance and this material will removed during final turning operation.

- Figure 9.290 shows finishing allowance in X and Z-axis.
- Generally we take finishing allowance, 0.200 mm on diameter and 0.100 mm on work piece face.

9.5.1 Choose Correct Answers and Fill in the Blanks where from Program Lines are Missing in following CNC Programs

Note:

Answer keys are available at the last of the CNC programming book.

9.5.1.1 Figure 9.291 shows both the raw material and machining in one drawing

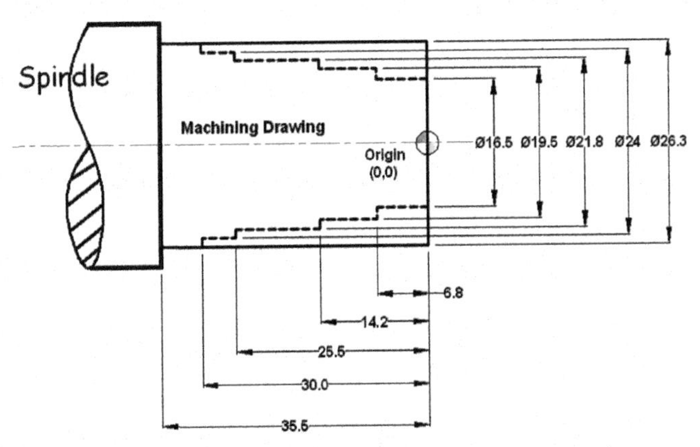

Figure 9.291 Practice drawing for step turning operation.

- Total length of work piece = 36.0 mm

How to set work zero (Z = 0) and face it, if work piece length is bigger according to drawing?

Example

1. First, we set the work zero offset G54 (Z = 0) on the front face of the work piece.
2. Work piece (Figure 9.291) has 0.50 mm extra material on the length. So work offset will be shift 0.500 mm inside the work piece face (it means Z zero will be shift −0.50 mm inside the work piece face).
3. **New work zero offset Z = 0 will be −0.5 mm inside the work piece face.**

4. **Extra material (0.500 mm)** will be face by the turning tool during facing operation in Z positive direction but step profile turning will be machine in Z negative (Z−) direction.

Operation Planning of Facing and Step Turning operation

Machining Operation – Facing and Step Turning
Following two operations will be perform on the above job.

- **Facing operation** with depth of cut **0.5 mm** (total material for facing 0.50 mm)
- **Profile turning**

Raw Material – Mild steel (MS)
Cutting Tool –

- Turning tool (rough turning tool and finish turning tool)

Note:

Turning tool can use for facing operation.

Depth of Cut for Roughing – Max.1.5 mm one side/3.0 mm diametrical (both side)
Finishing Allowance in X-axis – 0.200 mm diametrical (both side)
Finishing Allowance in Z-axis – 0.100 mm
Cutting Feed for Roughing – 0.25 mm/revolution
Cutting Feed for Finishing – 0.12 mm/revolution
Spindle Speed – 1800 rpm

Note:

Finishing allowance means, during roughing operation CNC machine will leave some material on the work piece surface for final cut.

During the following operations, we will use two cutting tools.

i. Rough turning tool (T0101)
ii. Finish turning tool for Finishing cut (T0303)

O66971;
G54;
G50 S1800;
T0101 (Rough Turning Tool);
------- X150 Z150;
G00 X32 Z5;

------- M03 S200;
M08;
G00 X32 **-------;** [0.100 mm material left for final cut]
------- X0.0 F0.12; **Rough Facing** operation
G00 Z1.0;
G00 X24.2;
G01 ------- F0.25;
G01 X28.0 F0.25;
G00 Z1.0;
G00 X22.0;
G01 ------- F0.25;
-------X25.0 F0.25;
G00 Z1.0;
G00 X19.7;
G01 Z-14.1 -------;
------- X23.0 F0.25;
G00 Z1.0;
G00 -------;
G01 ------- F0.25;
G01 X20.5 F0.25
G00 Z1.0;
M09;
-------;
G00 X150 Z150;
M00;
T0303 (Final Turning Tool);
G00 X150 Z150;
G00 X32 Z5;
G96 ------- S230;
M08;
G00 X32.0 **Z0.0; Final Facing** operation
------X0.0 F0.12; Remaining material (0.100 mm) on the work piece face will be remove.
G00 Z1.0;
G00 X16.5;
G01 ------- F0.12;
G01 ------- F0.12;
------- Z-14.2 F0.12;

```
G01 X21.8 -------;
-------Z-25.5 F0.12;
G01 ------- F0.12;
G01 ------- F0.12;
------- X28.0 -------;
G00 Z1.0;
M09;
-------;
G00 X150 Z150;
M30;
```

9.5.1.2 Figure 9.292 shows both the raw material and machining in one drawing

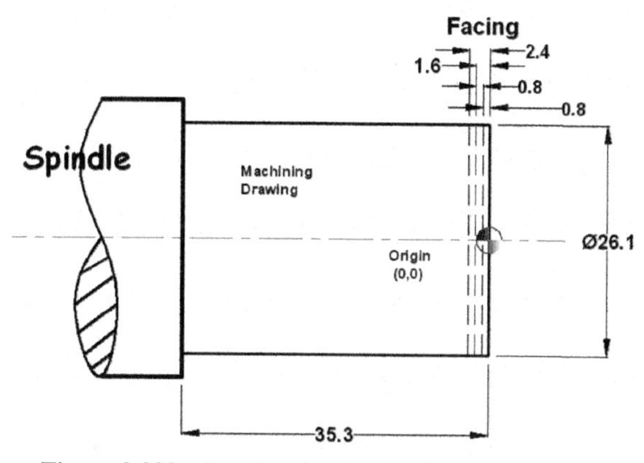

Figure 9.292 Practice drawing for facing operation.

Note:

Here in this case work zero offset G54 (Z = 0) can be take on the work piece face, but during facing operation, cutting value of Z will be negative (Z−) in the CNC program. This condition can possible when only facing operation will perform on the work piece. On the other hand, we can say if we take Z zero on the work piece face then all facing program will make negative in the term of Z.

Operation Planning of Facing Operation

Machining Operation – Facing

Following operation will be perform on the above job.

- **Facing operation** with depth of cut **0.800 mm** (Total material for facing 2.40 mm)

Raw Material – Mild Steel (MS)
Cutting Tool –

- Turning Tool

Cutting Feed – 0.12 mm/revolution
Spindle Speed – 1500 rpm

```
O50014;
G54;
------- S1500;
T0101;
G00 X100 Z100;
G00 X30 Z5;
-------M03 S180;
M08;
G00 X30.0 Z-0.8;
G01 ------- F0.12;
G00 Z1.0;
G00 X30;
G00 -------;
-------X0.0 F0.12;
G00 Z1.0;
G00 X30;
G00 -------;
G01 X0.0 F0.12;
G00 Z1.0;
M09;
-------;
G00 X150 Z150;
M30;
```

9.5.1.3 Figure 9.293 shows both the raw material and machining in one drawing

- Total length of work piece = 35.0 mm

Operation Planning of Chamfering Operation

Machining Operation – Facing and Chamfering
Following two operations will be perform on the above job.

- **Facing operation** with depth of cut **0.7 mm** (Total material for facing 0.70 mm)
- **Chamfering**

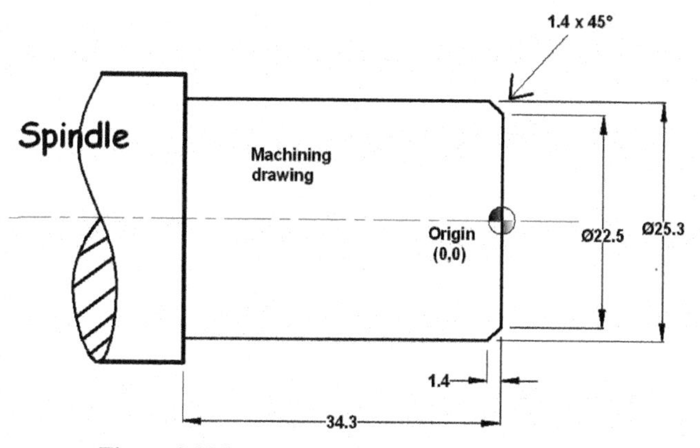

Figure 9.293 Practice drawing for chamfering.

Raw Material – Mild Steel (MS)
Cutting Tool –

- Turning tool

Note:

Turning tool can use for chamfering operation.

Depth of Cut for Roughing – Maximum1.5 mm one side/3.0 mm diametrical (both side)
Cutting Feed – 0.12 mm/revolution
Spindle Speed – 1800 rpm

```
O66971;
-------;
G50 S1800;
T0101;
G00 X150 Z150;
------- X30 Z5;
G96 ------- S200;
M08;
G00 X30.0-------;
-------X0.0 F0.12;
G00 Z1.0;
------- X21.5;
G01 X22.5 Z0.0 F0.12; Starting point of chamfer
```

------- X26.3 ------- F0.12; Ending point of chamfer including clearance with 45° angle yG00 Z1.0;

-------;
M05;
G00 X150 Z150;
M30;

9.5.1.4 Figure 9.294 shows both the raw material and machining in one drawing

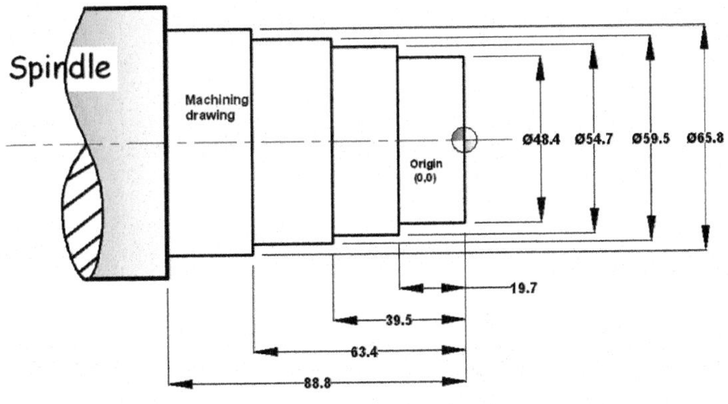

Figure 9.294 Practice drawing.

- Total length of work piece = 90.0 mm

Operation Planning of Step Turning Operation

Machining Operation – Facing and Step Turning
Following two operations will be performed on the above job.

- **Facing operation** with depth of cut **1.0 mm** (Total material for facing 1.20 mm)
- **Profile turning**

Raw Material – Mild steel (MS)
Cutting Tool –

- Turning tool (rough turning tool and finishing turning tool)

Note:

Turning tool can use for facing operation.

Depth of Cut for Roughing – Max.1.5 mm one side/3.0 mm diametrical (both side)

Finishing Allowance in X-axis – 0.200 mm diametrical (both side)
Finishing Allowance in Z-axis – 0.100 mm
Cutting Feed for Roughing – 0.25 mm/revolution
Cutting Feed for Finishing – 0.12 mm/revolution
Spindle Speed – 1500 rpm

Note:

Finishing allowance means, during roughing operation CNC machine will leave some material on the work piece surface for final cut.

During following operations, two cutting tools used are:

- Rough turning tool (T0101)
- Finish turning tool for Finishing cut (T0303)

O51020;
G54;
------- S1500;
T0101 (Rough Turning Tool);
G00 X200 Z250;
-------X70.0 Z25;
G00 X70.0 Z5.0;
G96 ------- S200;
M08;
G00 X70**-------;** [0.200 mm material left for final cut]
G01 X0.0 F0.12; **Rough Facing** operation
G00 Z1.0;
G00 --------;
G01 Z-63.3 F0.25;
G01 ------- F0.25;
G00 Z1.0;
G00 X59.7;
G01 ------- F0.25;
G01 X67.0 F0.25;
-------Z1.0;
------- X56.8;
G01 ------- F0.25;
G01 X60.0 F0.25;
G00 Z1.0;
G00 X54.9;

G01 ------- F0.25;
G01 X60.0 -------;
G00 Z1.0;
G00 X51.9;
G01 ------- F0.25;
------- X56.0 F0.25;
G00 Z1.0;
------- X48.9;
G01 Z-19.6 F0.25;
G01 ------- F0.25;
G00 Z1.0;
G00 X48.6;
G01 ------- F0.25;
G01 X56.0 F0.25;
G00 -------;
M09;
M05;
G00 X200 Z250;
M00;
T0303 (Final Turning Tool);
G00 X150 Z150;
-------X55.0 Z5;
------- M03 S230;
M08;
G00 X55 **Z0.0; Final Facing** operation
------- X0.0 F0.12; Remaining material (0.200 mm) on the work piece face will be remove.
G00 Z1.0;
G00 -------;
G01 ------- F0.12;
------- X54.7 F0.12;
G01 Z-39.5 F0.12;
------- X59.5 F0.12;
G01 ------- F0.12;
G01 X67.0 -------;
G00 Z1.0;
M09;
M05;
G00 X200 Z250;
M30;

9.5.1.5 Figure 9.295 shows both the raw material and machining in one drawing

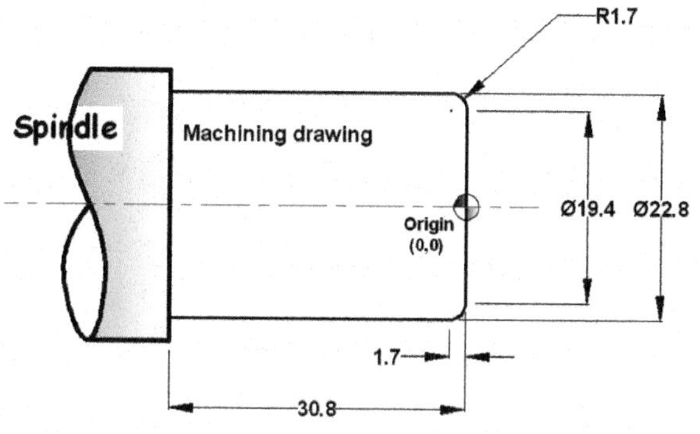

Figure 9.295 Practice drawing.

- Total length of work piece = 31.5 mm

Operation Planning of Radius Profile Operation

Machining Operation – Facing and Corner Radius
Following two operations will be performed on the above job

- **Facing operation** with depth of cut **0.7 mm** (total material for facing 0.7 mm)
- **Corner radius**

Raw Material – Mild steel (MS)
Cutting Tool –

- Turning tool

Note:

Turning tool can use for facing operation.

Depth of Cut for Roughing – Max.1.5 mm one side/3.0 mm diametrical (both side)
Cutting Feed – 0.12 mm/revolution
Spindle speed – 2000 rpm

O58920;

-------;

G50 S2000;

T0101 (Rough Turning Tool);

G00 X100 Z150;

------- M03 S200;
M08;
G00 X27.0 **Z0.0;**
G01 ------- F0.12; **Facing** operation
G00 Z1.0;
G00 X20.8;
-------X20.8------- F0.12; rough cut, starting point of radius
G03 X22.8 -------**R1.0** F0.12; ending point of radius
G00 Z1.0;
G00 -------;
G01 X19.4 Z0.0 -------; Finishing cut, starting point of radius
-------X22.8 Z-1.7 **-------** F0.12; ending point of radius
G00 Z1.0;
M09;
-------;
G00 X100 Z150;
M30;

9.5.1.6 Figure 9.296 shows both the raw material and machining in one drawing

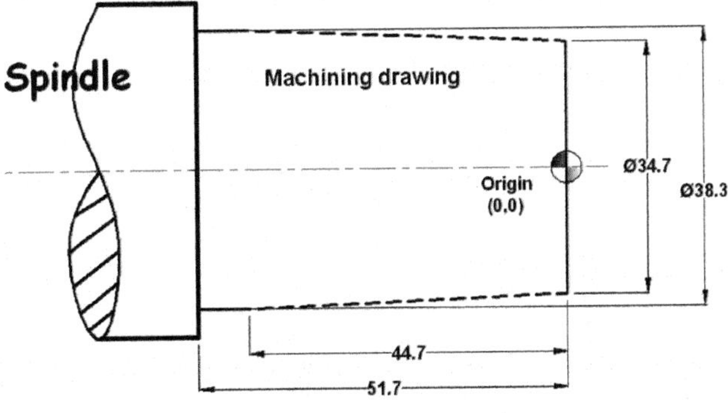

Figure 9.296 Practice drawing.

- Total length of work piece = 52.5 mm

Operation Planning of Taper Turning Operation

Machining Operation – Facing and Taper Turning
Following two operations will be performed on the above job.

- **Facing operation** with depth of cut **0.8 mm** (total material for facing 0.8 mm)
- **Taper turning**

Raw Material – Mild steel (MS)
Cutting Tool –

- Turning tool

Note:

Turning tool can use for facing operation.

Depth of Cut – Maximum1.5 mm one side/3.0 mm diametrical (both side)
Cutting Feed – 0.15 mm/revolution
Spindle Speed – 1500 rpm

O51020;
G54;
G50 -------;
------- (Turning Tool);
G00 X200 Z200;
------- X45.0 Z25;
-------M03 S200;
M08;
-------X45.0 **Z0.0;**
G01 X0.0 F0.12; **Facing** operation
G00 Z1.0;
G00 X35.3;
G01 ------- Z0.0 F0.12; Starting point of Rough taper turning
-------X38.3 Z-43.0 F0.15; Rough taper turning operation, ending point of taper turning
G00 Z1.0;
G00 X33.7;
------- X34.7 Z0.0 -------; Starting point of final taper turning
G01 X38.3 ------- F0.15; Ending point of final taper turning
G00 Z1.0;
M09;
M05;
------- X200 Z250;
-------;

9.5.1.7 Figure 9.297 shows both the raw material and machining in one drawing

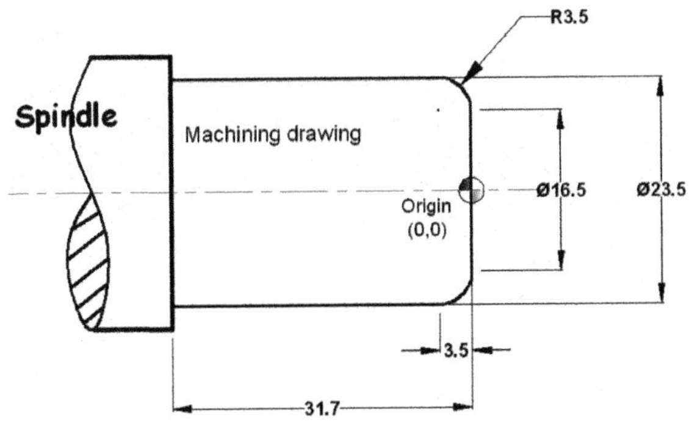

Figure 9.297 Practice drawing.

- Total length of work piece = 32.0 mm

Operation Planning of Radius Profile Operation

Machining Operation – Facing and Corner Radius
Following operation will be performed on the above job.

- **Facing Operation** with depth of cut **0.3 mm** (total material for facing 0.3 mm)

Raw Material – Mild Steel (MS)
Cutting Tool –

- Turning tool

Note:

Turning tool can use for facing operation.

Depth of cut for Roughing – Maximum1.5 mm one side/3.0 mm diametrical (both side)
Cutting Feed – 0.12 mm/revolution
Spindle Speed – 2000 rpm

O02876;
-------;
G50 S2000;
T0101 (Rough Turning Tool);
------- X100 Z250;
G00 X27.0 Z5.0;

G96 ------- S200;
M08;
G00 X27.0**-------;**
------- X0.0 F0.12; **Facing** operation
G00 Z1.0;
------- X21.5;
G01 ------- Z0.0 F0.12; SPR
G03 X23.5 -------**R1.0** F0.12; Rough radius turning operation, EPR
G00 Z1.0;
------- X19.5;
G01 ------- Z0.0 F0.12; SPR
G03 X23.5 Z-2.0 **R2.0**-------; Rough radius turning operation, EPR
-------Z1.0;
G00 X17.0;
G01 -------Z0.0 F012; SPR
-------X23.5 -------**R3.25** F0.12; Rough radius turning operation, EPR
G00 Z1.0;
G00 X15.5;
G01 X16.5 Z1.0 F0.12; **Finish radius** turning operation, SPR
G03 X23.5 Z-3.5 **-------** F0.12; EPR
G00 Z1.0;
M09;
M05;
G00 X100 Z250;
M30;

Note:

SPR – starting point of radius
EPR – ending point of radius

9.5.1.8 Figure 9.298 shows both the raw material and machining in one drawing

- Total length of work piece = 73.5 mm

Operation Planning of Drilling Operation

Machining Operation – Facing and Drilling Operation
Following two operations will be performed on the above job

- **Facing operation** with depth of cut **0.7 mm** (total material for facing 0.7 mm)

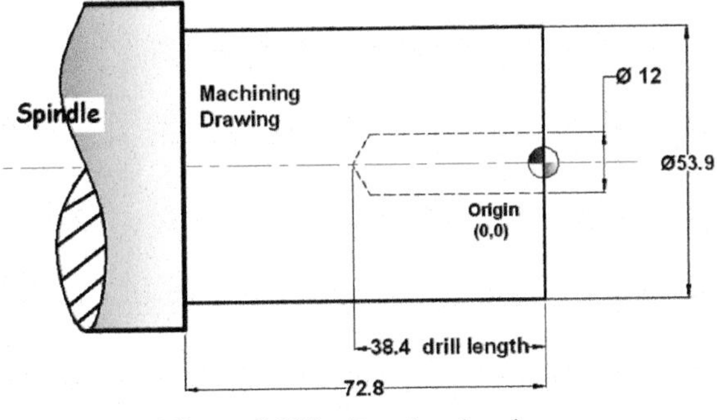

Figure 9.298 Practice drawing.

- **Drilling operation**

Raw Material – Mild steel (MS)
Cutting Tool –

- Turning tool
- Drilling tool

Note:

Turning tool can use for facing operation.

Drill Tool Ø12mm
Cutting Feed for Facing Operation – 0.12 mm/revolution
Cutting Feed for Drilling Operation – 0.050 mm/revolution
Spindle Speed for Facing Operation – 1500 rpm
Spindle Speed for Drilling Operation – 450 rpm

```
O83060;
G54;
------- S1500;
------- (Turning Tool);
G00 X200 Z300;
G00 X60.0 Z5.0;
G96------- S200;
-------;
G00 X60.0 Z0.0;
G01 ------- F0.12; Facing operation
G00 Z1.0;
M09;
-------;
```

```
G00 X200 Z300;
M00;
T0505 (Drilling Tool);
G00 ------- -------;
G00 X100.0 Z25.0;
------- M03 -------;
G00 ------- Z5.0;
M08;
G00 Z1.0;
------- Z-5.0 F0.05; Drilling operation start
------- Z5.0;
G00 Z-4.0;
G01 Z-10.0 F0.05;
G00 -------;
G00 Z-9.0;
G01 Z-15 -------;
G00 Z5.0;
G00 Z-14.0;
G01 ------- F0.05;
G00 Z5.0;
G00 Z-19.0;
------- Z-25.0 F0.05;
G00 Z5.0;
G00 -------;
G01 ------- F0.05;
-------Z5.0;
G00 -------;
G01 Z-35.0 -------;
G00 Z5.0;
G00 Z-34.0;
-------Z-38.4 F0.05;
G00 Z5.0;
M09;
-------;
G00 X200 Z300;
M30;
```

9.5.1.9 Figure 9.299 shows both the raw material and machining in one drawing

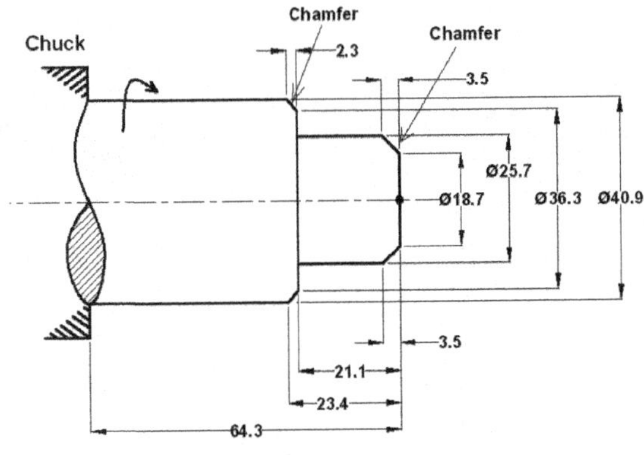

Figure 9.299 Practice drawing.

- Total length of work piece = 65.0 mm

Operation Planning of Profile Turning

Machining Operation – Facing, step turning and chamfering
Following three operations will be performed on the above job.

Facing Operation – with depth of cut **1.0 mm** (total material for facing 0.7 mm)

- **Step turning**
- **Chamfering**

Raw Material – Mild steel (MS)
Cutting Tool – Turning tool (rough step turning tool and finishing step turning tool with chamfer)

Note:

Turning tool can use for facing operation.

Depth of Cut for Roughing – Maximum1.5 mm one side/3.0 mm diametrical (both side)
Finishing Allowance in X-axis – 0.200 mm diametrical (both side)
Finishing Allowance in Z-axis – 0.100 mm
Cutting Feed for Roughing – 0.25 mm/revolution
Cutting Feed for Finishing – 0.12 mm/revolution
Spindle Speed – 1600 rpm

Note:

Finishing allowance means, during roughing operation CNC machine will leave some material on the work piece surface for final cut.

During following operations, we will use two cutting tools.

- Rough turning tool (T0101)
- Finish turning tool for Finishing cut (T0303)

O41046;

G50 -------;

T0101; (Rough Turning Tool)

G00 X100 Z150;

G00 X45.0 Z5;

------- M03 S200;

M08;

G00 X45 **Z0.100;** [0.100 mm material left for final cut]

G01 ------- F0.12; **Rough Facing** operation

G00 Z1.0;

G00 X37.9;

G01 Z-21.0 F0.25;

G01 X38.9 F0.25;

G01 X40.9 ------- F0.25;

G00 -------;

G00 -------;

G01 Z-21.0 F0.25;

G01 X36.9 F0.25;

G01 X40.9 Z-23.0 F0.25;

G00 Z1.0;

-------X31.9;

G01 Z-21.0 -------;

------- X37.0 F0.25;

G00 Z1.0;

G00 X28.9;

G01 Z-21.0 F0.25;

G01 ------- F0.25;

G00 Z1.0;

G00 X25.9;

```
G01 ------- F0.25;
G01 X37.0 F0.25;
G00 Z1.0;
G00 -------; Starting point of rough chamfer (SPRC)
G01 Z0.0 F0.12;
------- X25.9 Z-1.0 -------; Ending point of rough chamfer (EPRC)
G00 Z1.0;
G00 X21.9; SPRC
G01 ------- F0.12;
G01 ------- Z-2.0 F0.25; EPRC
G00 Z1.0;
G00 -------; SPRC
G01 Z0.0 F0.12;
-------X25.9 Z-3.0 F0.25; EPRC
G00 Z1.0;
G00 X18.9; SPRC
G01 ------- F0.25;
G01 X25.9 ------- F0.25; EPRC
-------Z1.0;
M09;
-------;
G00 X100 Z150;
-------;
T0303; (Final Turning Tool)
G00 X100 Z150;
G00 X30.0 Z5;
G96 ------- S230;
M08;
G00 X30 Z0.0; Final Facing operation
G01 ------- F0.12; Remaining material (0.100 mm) of the work piece face
                   will be remove.
-------Z1.0;
G00 X17.7;
G01 ------ Z0.0 F0.12; Starting point of finish chamfer
------- X25.7 Z-3.5 -------; Ending point of finish chamfer
G01 ------- F0.12;
G01 X36.3 F0.12;
G01 ------- Z-23.9 F0.12;
```

```
G00 Z1.0;
M09;
M05;
G00 X100 Z150;
-------
```

9.5.1.10 Figure 9.300 shows the raw material and machining drawing in one drawing

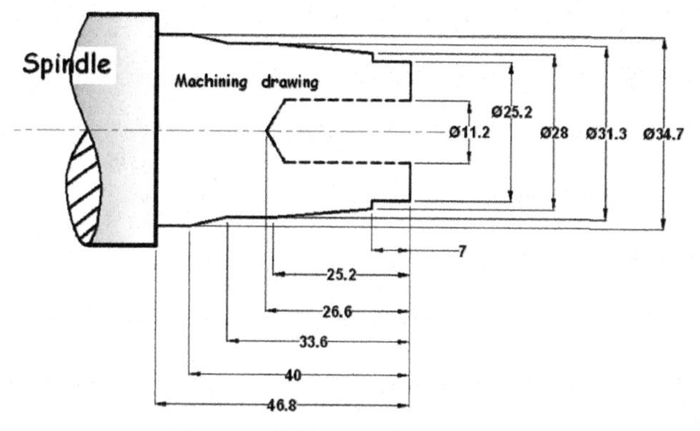

Figure 9.300 Practice drawing.

- Total length of work piece = 47.8 mm

Operation Planning of Profile Turning and Drilling Operation

Machining Operation – Facing and Drilling Operation
Following three operations will be performed on the above job.

- **Facing Operation** with depth of cut **1.0 mm** (total material for facing 1.0 mm)
- **Drilling Operation**
- **Profile Turning Operation**

Raw Material – Mild steel (MS)
Cutting Tool – Turning tool

- Rough turning tool – T0101
- Final turning tool – T0303

Drilling Tool – T0505
Note:

Turning tool can use for facing operation.

Drill Tool Ø11.2 mm
Depth of Cut for Roughing – Max.2.0 mm one side/4.0 mm diametrical (both side)
Cutting Feed for Turning Operation – 0.25 mm/revolution
Cutting Feed for Drilling Operation – 0.050 mm/revolution
Cutting Feed for Final Turning Operation – 0.12 mm/revolution
Finishing Allowance in X-axis – 0.200 mm diametrical (both side)
Finishing Allowance in Z-axis – 0.100 mm
Spindle Speed for Turning Operation – 1700 rpm
Spindle Speed for Drilling Operation – 450 rpm

Note:

Finishing allowance means, during roughing operation CNC machine will leave some material on the work piece surface for final cut.

```
O12126;
G54;
------- S1700;
T0101; (Rough Turning Tool)
G00 X200 Z300;
G00 X42.0 Z5.0;
------- M03 S200;
M08;
G00 X42.0-------; 0.100 mm material left for final turning operation.
-------X0.0 -------; Facing operation
G00 Z1.0;
------- X31.5;
------- Z-33.5 -------;
G01 ------- Z-39.9 F0.25;
G00 Z1.0;
G00 X28.2;
G01 ------- F0.25;
G01 ------- Z-25.1 F0.25;
G00 Z1.0;
------- X25.4;
------- Z-6.9 F0.25;
G01 ------ F0.25;
G00 Z1.0;
M09;
-------
```

```
G00 X200 Z300;
M00;
T0505; (Drilling Tool)
G00 X200 Z300;
G00 X100.0 Z25.0;
-------M03 S450;
G00 X0.0 Z5.0;
-------
G00 Z1.0;
------- Z-5.0 -------; Drilling operation starts
G00 -------;
-------Z-4.0;
------- Z-10.0 F0.05;
G00 Z5.0;
G00 -------;
G01 ------- F0.05;
G00 Z5.0;
G00 -------;
------- Z-20.0 -------;
G00 Z5.0;
G00 -------;
G01------- F0.05;
G00 Z5.0;
G00 Z-24.0;
G01 ------- F0.05;
G00 Z5.0;
M09;
M05;
------- X200 Z300;
M00;
---------; (Final Turning Tool)
G00 X200 Z300;
G00 X30.0 Z5.0;
G96 ------- S220;
M08;
G00 X30.0------;
------- X0.0 F0.12; Facing operation
G00 Z1.0;
G00 X25.2;
```

```
G01 ------ F0.12;
G01 X28.0 F0.12;
G01 ------ Z-25.2 F0.12;
G01 Z-33.6 F0.12;
------ X34.7 Z-40.0 F0.12;
G00 Z1.0;
M09;
-------
G00 X200 Z300;
M30; [3]
```

9.5.1.11 Figure 9.301 shows the raw material and machining drawing in one drawing

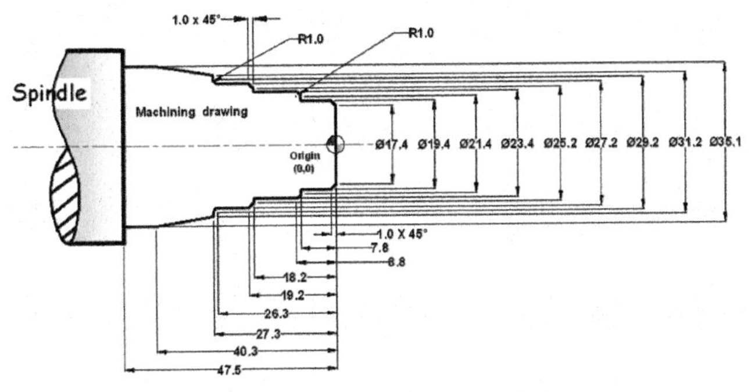

Figure 9.301 Practice drawing.

- Total length of work piece = 48.5 mm

Operation Planning of Profile Turning Operation

Machining Operation – Facing and Profile Turning operation
Following operation will be performed on the above job

i. **Facing operation** with depth of cut **1.0 mm** (total material for facing 1.0 mm)
ii. **Profile Turning Operation**

Raw Material – Mild Steel (MS)
Cutting Tool – Turning Tool

Note:

Turning tool can use for facing operation.

Depth of Cut for Roughing – Max.2.0 mm one side/4.0 mm diametrical (both side)
Cutting Feed for Turning Operation – 0.25 mm/revolution
Cutting Feed for Final Turning Operation – 0.12 mm/revolution
Finishing Allowance in X-axis – 0.200 mm diametrical (both side)
Finishing Allowance in Z-axis – 0.100 mm
Spindle Speed for Turning Operation – 1700 rpm

Note:

Finishing allowance means, during roughing operation CNC machine will leave some material on the work piece surface for final cut.

O55844;

```
-------;
G50 S1700;
T0101; (Rough Turning Tool)
G00 X100 Z150;
G00 X40.0 Z5.0;
------- M03 S200;
-------
G00 X40.0-------; 0.100 mm material left for final turning operation.
G01 X0.0 F0.12; Facing operation
G00 Z1.0;
G00 -------;
G01 Z-27.2-------;
G01 X35.3 ------- F0.25;
G00 Z1.0;
G00 X27.4;
G01 ------- F0.25;
G02------- Z-27.2 R1.0 F0.25;
G01 X35.5 F0.25;
G00 Z1.0;
G00 -------;
------- Z-18.1 F0.25;
G01 X25.4 F0.25;
G01 X27.4 ------- F0.25;
G00 Z1.0;
```

G00 -------;
G01 Z-7.7 F0.25;
G01 X21.6 F0.25;
------- X23.6 Z-8.7 ------- F0.25;
G00 Z1.0;
G00 X17.6;
G01 Z0.0 -------;
------- X19.6 Z-1.0 F0.25;
G00 Z1.0;
M09;

G00 X100 Z150;

T0303; (Final Turning Tool)
G00 X100 Z150;
G00 X25.0 Z5.0;
------- M03 S230;
M08;
G00 X25.0 **Z0.0;**
G01 ------- F0.12; **Facing** operation
G00 Z1.0;
G00 X16.4;
------- X17.4 ------ F0.12;
G01 ------- Z-1.0 F0.12;
G01 Z-7.8 F0.12;
G01 X21.4 F0.12;
------------- Z-8.8 **R1.0** F0.12;
G01 ------ F0.12;
------- X25.2 F0.12;
G01 ------- Z-19.2 F0.12;
G01 Z-26.3 F0.12;
-------X29.2 ------- ------- F0.12;
G01 -------F0.12;
G01 X35.1-------F0.12;
-------Z1.0;

M05;
M30;

References

[1] Drawings and sketches has been generated from Auto CAD 2014.
[2] CNC programming Manual of MTAB company: (Certificate course on CNC Turning, MTAB Technology Centre, MTAB, Chennai, India).
[3] Fanuc control system or similar control system.

10

Cutting Insert (Bit) and Cutting Tool Holders Nomenclatures

For becoming a good CNC programmer or Manufacturing engineer **you must know about the cutting tools and cutting tool bits nomenclatures**. It will help you to decide the best way of machining which is cost effective, time saving, good accuracy, good quality, with less rejection etc. You will see some important and generally used cutting tool holders and bits (inserts) nomenclatures in the following pages.

10.1 What are Cutting Inserts? What are the Benefits of the Right Cutting Insert?

Now a days, in the modern machining process we use cutting inserts with cutting tool holders/shanks. Cutting inserts are available in various types. It is also known as a bit. Using of cutting inserts depends on the type of machines. We can say bits are used in various type of machining (operation). Generally, cutting bits are used with tool holders/shank, where cutting bits are mounted or fitted with the help of screw and lever on the tool holder [1].

Benefits of the Right Cutting Insert:

- We get required surface finish as per drawing.
- It reduces material cutting load during metal removing.
- It increases machine life.
- It reduces chattering and material cutting noise during metal removing.
- It reduces over heating during metal removing.
- It increases insert's life.
- It increases production rate.
- It reduces rejection.
- It saves machining time.
- It will be very good for economically for the company.

10.2 When We Select an Insert/Bit for Machining. We Should Keep Following Important Points in Our Mind

Facts about: Insert has following characteristics

- Strength of insert (how much depth of cut insert can take or strength of insert).
- Power required for material cutting (how much spindle power required for cutting).
- Profiles cutting ability (can insert cut the profiles or how much insert is versatile).

10.2.1 Characteristics of the Inserts (Figure 10.1)

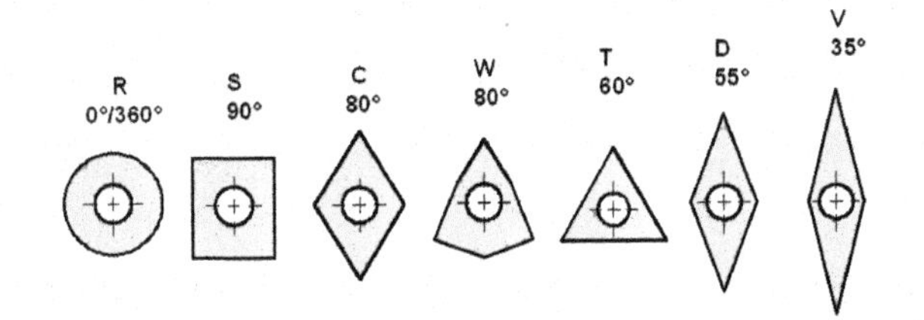

Vibration will be lower towards to V shape insert from R shape insert.

Cutting strength will increase towards to R shape insert from V shape insert.

Complex profile making ability will increase towards to V shape insert from R shape insert.

During machining, the cutting tool requires high spindle power towards to R shape insert from V shape insert.

Cutting tool can take a heavy cut towards to R shape insert from V shape insert.

Figure 10.1 Image is showing how the characteristics of inserts change with shape [2].

- Insert R has good strength respect of insert V.
- During complex profiling, insert R will be less useful but insert V will highly useful.
- During machining, insert V will require less spindle power but R shaped insert will require high power.
- During cutting operation vibration will lower for V-shape insert and higher for R.

Note: If you apply the above instructions you can save manufacturing time and manufacturing cost will be reduced. It means the company gets more profit.

10.3 What are Positive and Negative Inserts? Where and Which Type of Machining these are Used?

10.3.1 Negative Insert (Bit)

Figure 10.2 shows a negative insert.

Figure 10.3 shows the cutting edges of the insert.

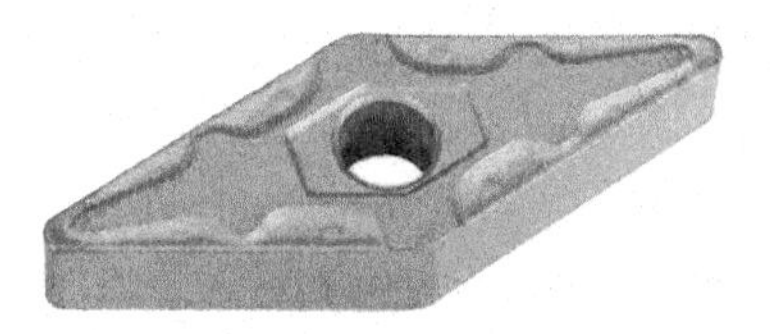

Figure 10.2 Negative insert, both side-cutting edges.

(Courtesy: VNMG Insert, Rawat Engg. Tech Pvt. Ltd., India)

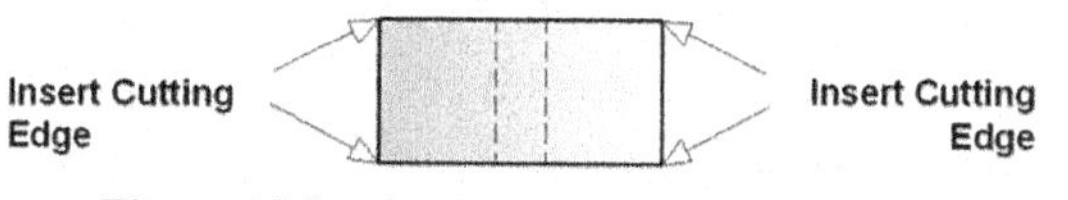

Figure 10.3 Cutting edges of the insert [3].

Following image is showing CNMG cutting insert, which is very popular in turning operation. See Figure 10.4.

Figure 10.4 CNMG cutting insert for turning operation, both side-cutting edges.

(Courtesy: CNMG Insert, Rawat Engg. Tech Pvt Ltd., India)

Figure 10.5 TNMG cutting insert for turning operation, both side cutting edges.

(Courtesy: TNMG Insert, Rawat Engg. Tech Pvt Ltd., India)

This type of insert is also used in turning operation, which is mounted on a suitable cutting tool holder/shank. See Figure 10.5.

10.3.1.1 Advantages of negative inserts

- Negative insert does not have any clearance angle but in tool holder, where the insert is fitted provides clearance angle.
- In this insert, the cutting forces more higher than positive inserts so it should not use where cutting forces less during cutting operation.
- The insert's wedge angles are in 90° therfore its cutting edges are stronger. Due to this benefit it is used in heavy rough and rigid cutting.
- Negative inserts cutting edges are avaible on both sides. It means you can use up the side and down the side of the insert.

Examples CNMG, TNMG, VNMG, etc. all these have four-sided cutting edges.

10.3.1.2 Names of some negative inserts

CNMG, DNMG, RNMG, SNMG, TNMG, VNMG, WNMG.

10.3.2 Positive Insert

Figure 10.6 shows the cutting edges of positive insert.

The VBMT insert is used in profile turning and grooving operations, which is mounted on the suitable cutting tool holder (Figure 10.7).

Figure 10.8 shows the cutting edges of the positive insert. The TCMT insert is used in medium machining.

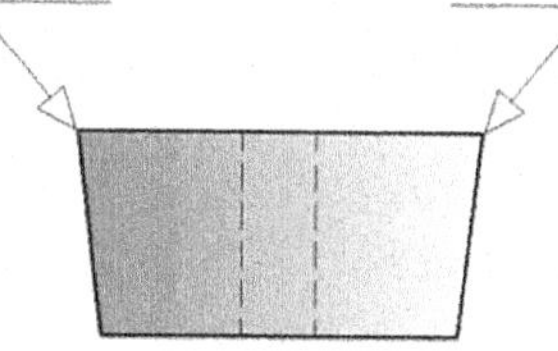

Figure 10.6 Cutting edges of the positive insert.

Figure 10.7 VBMT cutting insert, one-sided cutting edges.

(Courtesy of: VBMT Insert, Rawat Engg. Tech Pvt Ltd., India)

Figure 10.8 TCMT cutting insert, one-sided cutting edges.

(Courtesy: TCMT Insert, Rawat Engg. Tech Pvt Ltd., India)

10.3.2.1 Advantages of positive inserts

- Positive insert has clearance angle, the holder not provide clearance or rake angles.
- In this insert, the cutting edges are sharper than negative inserts. Positive inserts can bear less cutting force. It is generally used for finishing operation and low wall thickness surface.
- Insert's wedge angles are less than 90° therfore its cutting edges can not bear heavy and intruppted/rigid cutting. So it is used for fine and medium cutting.
- Due to clearance angle, positive insert's cutting edges are avaible only on one side (Figures 10.6 and 10.7).

Examples VBMT, TCMT, VCMT, etc. all these have two-sided cutting edges.

10.3.2.2 Names of some positive inserts

CCMT, DCMT, VCMT, SCMT, TPMR, TCMT, WCMT, TPGX, VCGX, CCGX.

10.4 Turning Tool Insert's Nomenclature [4]

Following images are showing cutting bits/inserts, which are mounted in different tool holders/shanks for internal and external turning operations (Figure 10.9).

Figure 10.9 Images of cutting inserts/bits.

(Courtesy: TCMT Insert, Rawat Engg. Tech Pvt Ltd., India)

Following chart is tells us turning cutting insert's nomenclature. It is tells us about insert's geometric parameters, which we need during material cutting operations. See Figure 10.10.

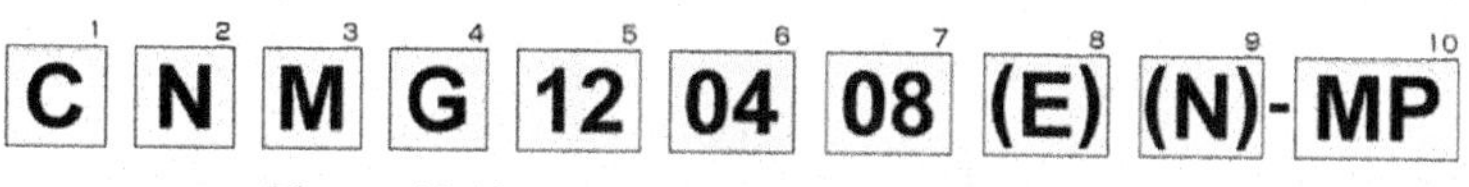

Figure 10.10 Cutting insert (Bit) nomenclature.

(Courtesy: MITSUBISHI MATERIALS CORPORATION METALWORKING SOLUTIONS COMPANY, Japan)

10.4.1 Insert Shape

Figure 10.11 shows the inserts/bits's shape.

Figure 10.12 shows the shapes of actual inserts.

1. Insert Shape		
Symbol	Insert Shape	
H	Hexagonal	
O	Octagonal	
P	Pentagonal	
S	Square	
T	Triangular	
C	Rhombic 80°	
D	Rhombic 55°	
E	Rhombic 75°	
F	Rhombic 50°	
M	Rhombic 86°	
V	Rhombic 35°	
W	Trigon	
L	Rectangular	
A	Parallelogram 85°	
B	Parallelogram 82°	
K	Parallelogram 55°	
R	Round	
X	Special Design	

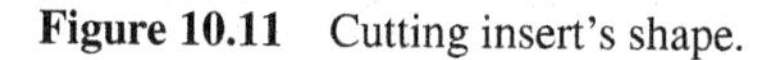

Figure 10.11 Cutting insert's shape.

(Courtesy: MITSUBISHI MATERIALS CORPORATION METALWORKING SOLUTIONS COMPANY, Japan)

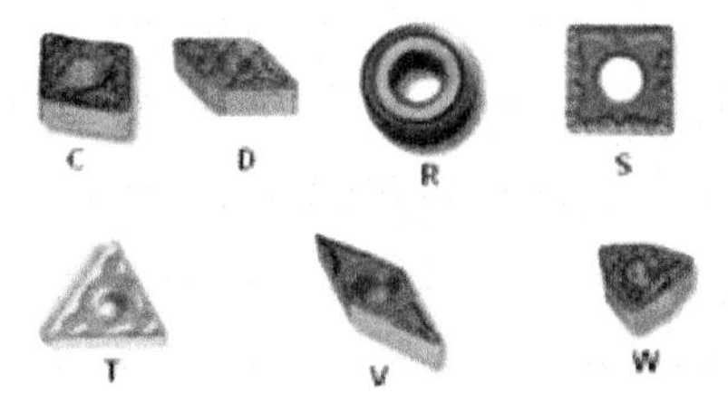

Figure 10.12 Insert's shape.

(Courtesy: MITSUBISHI MATERIALS CORPORATION METALWORKING SOLUTIONS COMPANY, Japan)

10.4.2 Insert Clearance Angle/Insert Relief Angle

Figure 10.13 shows the inserts' clearance angle (relief angle).

10.4.3 Insert Tolerance

Figure 10.14 shows the inserts'tolerance.

10.4.4 Types of Inserts

Figure 10.15 shows the inserts'clamping system.

2. Relief Angle	
Symbol	Normal Clearance
A	3°
B	5°
C	7°
D	15°
E	20°
F	25°
G	30°
N	0°
P	11°
O	Other Relief Angle
	Major Relief Angle

Figure 10.13 Insert clearance angle.

(Courtesy: MITSUBISHI MATERIALS CORPORATION METALWORKING SOLUTIONS COMPANY, Japan)

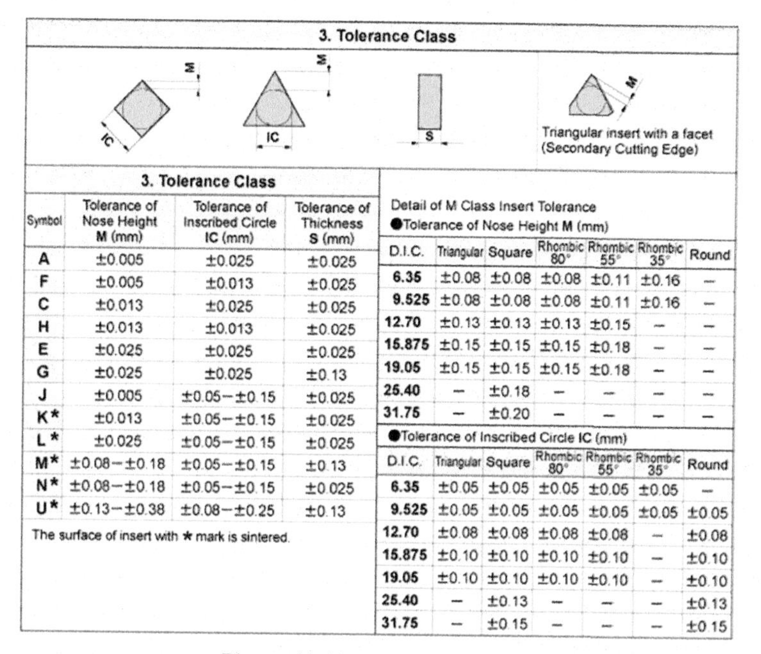

3. Tolerance Class

Symbol	Tolerance of Nose Height M (mm)	Tolerance of Inscribed Circle IC (mm)	Tolerance of Thickness S (mm)
A	±0.005	±0.025	±0.025
F	±0.005	±0.013	±0.025
C	±0.013	±0.025	±0.025
H	±0.013	±0.013	±0.025
E	±0.025	±0.025	±0.025
G	±0.025	±0.025	±0.13
J	±0.005	±0.05–±0.15	±0.025
K*	±0.013	±0.05–±0.15	±0.025
L*	±0.025	±0.05–±0.15	±0.025
M*	±0.08–±0.18	±0.05–±0.15	±0.13
N*	±0.08–±0.18	±0.05–±0.15	±0.025
U*	±0.13–±0.38	±0.08–±0.25	±0.13

The surface of insert with * mark is sintered.

Detail of M Class Insert Tolerance

●Tolerance of Nose Height M (mm)

D.I.C.	Triangular	Square	Rhombic 80°	Rhombic 55°	Rhombic 35°	Round
6.35	±0.08	±0.08	±0.08	±0.11	±0.16	–
9.525	±0.08	±0.08	±0.08	±0.11	±0.16	–
12.70	±0.13	±0.13	±0.13	±0.15	–	–
15.875	±0.15	±0.15	±0.15	±0.18	–	–
19.05	±0.15	±0.15	±0.15	±0.18	–	–
25.40	–	±0.18	–	–	–	–
31.75	–	±0.20	–	–	–	–

●Tolerance of Inscribed Circle IC (mm)

D.I.C.	Triangular	Square	Rhombic 80°	Rhombic 55°	Rhombic 35°	Round
6.35	±0.05	±0.05	±0.05	±0.05	±0.05	–
9.525	±0.05	±0.05	±0.05	±0.05	±0.05	±0.05
12.70	±0.08	±0.08	±0.08	±0.08	–	±0.08
15.875	±0.10	±0.10	±0.10	±0.10	–	±0.10
19.05	±0.10	±0.10	±0.10	±0.10	–	±0.10
25.40	–	±0.13	–	–	–	±0.13
31.75	–	±0.15	–	–	–	±0.15

Figure 10.14 Inserts tolerance.

(Courtesy: MITSUBISHI MATERIALS CORPORATION METALWORKING SOLUTIONS COMPANY, Japan)

4. Chipbreaker and Clamping System

Metric

Symbol	Hole	Hole Configuration	Chip Breaker	Figure	Symbol	Hole	Hole Configuration	Chip Breaker	Figure
W	With Hole	Cylindrical Hole +	No		A	With Hole	Cylindrical Hole	No	
T	With Hole	One Countersink (40–60°)	One Sided		M	With Hole	Cylindrical Hole	Single Sided	
Q	With Hole	Cylindrical Hole +	No		G	With Hole	Cylindrical Hole	Double Sided	
U	With Hole	Double Countersink (40–60°)	Double Sided		N	Without Hole	–	No	
B	With Hole	Cylindrical Hole +	No		R	Without Hole	–	Single Sided	
H	With Hole	One Countersink (70–90°)	One Sided		F	Without Hole	–	Double Sided	
C	With Hole	Cylindrical Hole +	No		X	–	–	–	Special Design
J	With Hole	Double Countersink (70–90°)	Double Sided						

Figure 10.15 Inserts' clamping system.

(Courtesy: MITSUBISHI MATERIALS CORPORATION METALWORKING SOLUTIONS COMPANY, Japan)

10.4.5 Insert Cutting Edge Length

Following table is showing inserts' cutting edge length. See Figure 10.16.

10.4.6 Insert Thickness

Following table is showing inserts' thickness. See Figure 10.17.

5. Insert Size							
Symbol							Diameter of Inscribed Circle (mm)
R	W	V	D	C	S	T	
	02		04	03	03	06	3.97
	L3	08	05	04	04	08	4.76
	03	09	06	05	05	09	5.56
06							6.00
	04	11	07	06	06	11	6.35
	05	13	09	08	07	13	7.94
08							8.00
09	06	16	11	09	09	16	9.525
10							10.00
12							12.00
12	08	22	15	12	12	22	12.70
15	10		19	16	15	27	15.875
16							16.00
19	13		23	19	19	33	19.05
20							20.00
			27	22	22	38	22.225
25							25.00
25			31	25	25	44	25.40
31			38	32	31	54	31.75
32							32.00

Figure 10.16 Insert's cutting edge length.

(Courtesy: MITSUBISHI MATERIALS CORPORATION METALWORKING SOLUTIONS COMPANY, Japan)

6. Insert Thickness

*Thickness is from the bottom of the insert to the top of the cutting edge.

Symbol	Thickness (mm)
S1	1.39
01	1.59
T0	1.79
02	2.38
T2	2.78
03	3.18
T3	3.97
04	4.76
06	6.35
07	7.94
09	9.52

Figure 10.17 Insert's thickness.

(Courtesy: MITSUBISHI MATERIALS CORPORATION METALWORKING SOLUTIONS COMPANY, Japan)

7. Insert Corner Configuration	
Symbol	Corner Radius (mm)
00	Sharp Nose
V3	0.03
V5	0.05
01	0.1
02	0.2
04	0.4
08	0.8
12	1.2
16	1.6
20	2.0
24	2.4
28	2.8
32	3.2
00 : Inch M0 : Metric	Round Insert

Figure 10.18 Sutting insert's nose radius.

(Courtesy: MITSUBISHI MATERIALS CORPORATION METALWORKING SOLUTIONS COMPANY, Japan)

10.4.7 Insert Corner Radius

Following table is showing inserts' nose radius. See Figure 10.18.

10.4.8 Insert Cutting Edge Condition

Figure 10.19 shows inserts' cutting edge condition.

10.4.9 Insert Cutting Direction/Hand of Tool

Figure 10.20 shows the inserts' cutting direction.

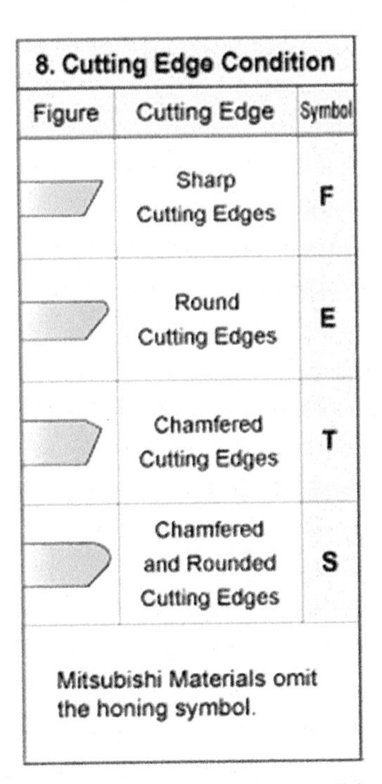

8. Cutting Edge Condition		
Figure	Cutting Edge	Symbol
	Sharp Cutting Edges	F
	Round Cutting Edges	E
	Chamfered Cutting Edges	T
	Chamfered and Rounded Cutting Edges	S
Mitsubishi Materials omit the honing symbol.		

Figure 10.19 Cutting edge condition.

(Courtesy: MITSUBISHI MATERIALS CORPORATION METALWORKING SOLUTIONS COMPANY, Japan)

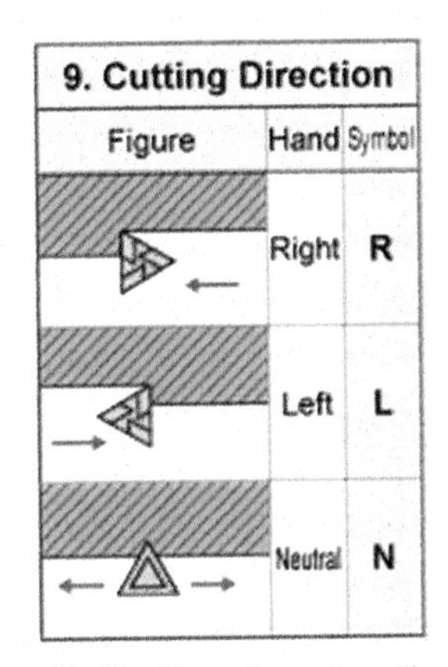

9. Cutting Direction		
Figure	Hand	Symbol
	Right	R
	Left	L
	Neutral	N

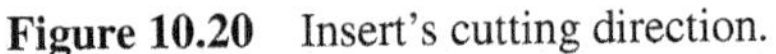

Figure 10.20 Insert's cutting direction.

(Courtesy: MITSUBISHI MATERIALS CORPORATION METALWORKING SOLUTIONS COMPANY, Japan)

10.4.10 Insert Chip Breaker

Figure 10.21 shows the inserts'chip breaking facility.

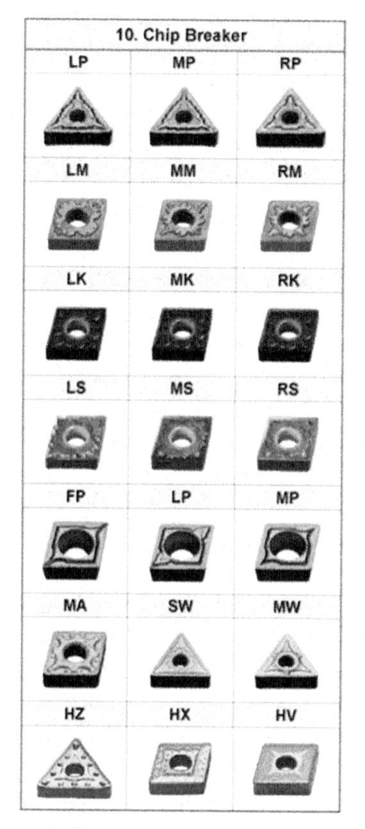

Figure 10.21 Inserts' chip breaker facility.

(Courtesy: MITSUBISHI MATERIALS CORPORATION METALWORKING SOLUTIONS COMPANY, Japan)

10.5 Turning Tool Holders (External Tool Holders): ISO Nomenclature

Turning tool holders are known by an ISO naming convention that describes their shapes and size. The example below is for a turning tool holder **P C L N R 25 25 M 12**.

Note: The tables below (Figure 10.22) may show only some of the options in each parameter. These keep evolving, and for a full list you must refer to the catalog of the cutting tool manufacturer.

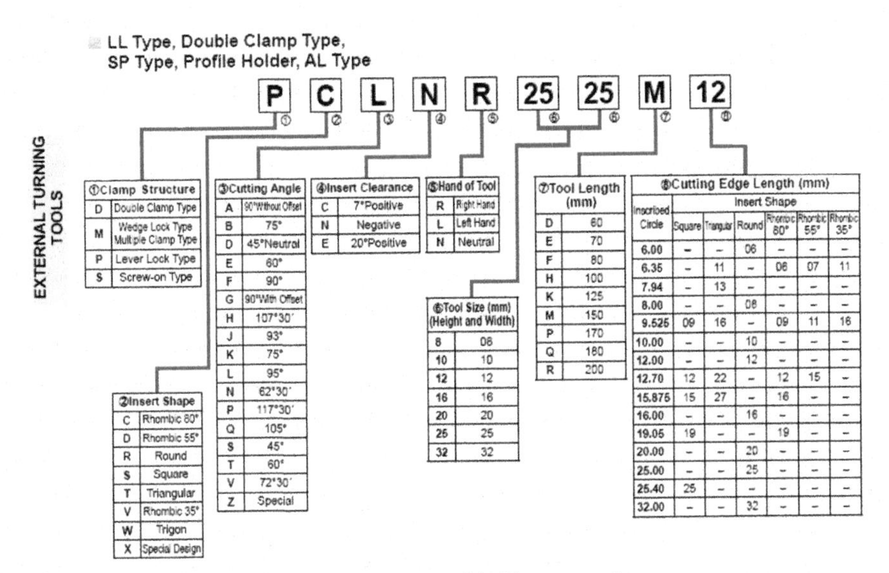

①Clamp Structure	
D	Double Clamp Type
M	Wedge Lock Type Multiple Clamp Type
P	Lever Lock Type
S	Screw-on Type

②Insert Shape	
C	Rhombic 80°
D	Rhombic 55°
R	Round
S	Square
T	Triangular
V	Rhombic 35°
W	Trigon
X	Special Design

③Cutting Angle	
A	90° Without Offset
B	75°
D	45° Neutral
E	60°
F	90°
G	90° With Offset
H	107°30′
J	93°
K	75°
L	95°
N	62°30′
P	117°30′
Q	105°
S	45°
T	60°
V	72°30′
Z	Special

④Insert Clearance	
C	7° Positive
N	Negative
E	20° Positive

⑤Hand of Tool	
R	Right Hand
L	Left Hand
N	Neutral

⑥Tool Size (mm) (Height and Width)	
8	08
10	10
12	12
16	16
20	20
25	25
32	32

⑦Tool Length (mm)	
D	60
E	70
F	80
H	100
K	125
M	150
P	170
Q	180
R	200

⑧Cutting Edge Length (mm)						
Inscribed Circle	Insert Shape					
	Square	Triangular	Round	Rhombic 80°	Rhombic 55°	Rhombic 35°
6.00	–	–	06	–	–	–
6.35	–	11	–	06	07	11
7.94	–	13	–	–	–	–
8.00	–	–	08	–	–	–
9.525	09	16	–	09	11	16
10.00	–	–	10	–	–	–
12.00	–	–	12	–	–	–
12.70	12	22	–	12	15	–
15.875	15	27	–	16	–	–
16.00	–	–	16	–	–	–
19.05	19	–	–	19	–	–
20.00	–	–	20	–	–	–
25.00	–	–	25	–	–	–
25.40	25	–	–	–	–	–
32.00	–	–	32	–	–	–

Figure 10.22 Turning tool holder nomenclatures.

(Courtesy: MITSUBISHI MATERIALS CORPORATION METALWORKING SOLUTIONS COMPANY, Japan)

10.6 Boring Tool Holders Nomenclature: ISO Codes Key

Figure 10.23 shows the boring tool holders' nomenclatures.

According to format, it is little similar to external cutting tool holder's nomenclature.

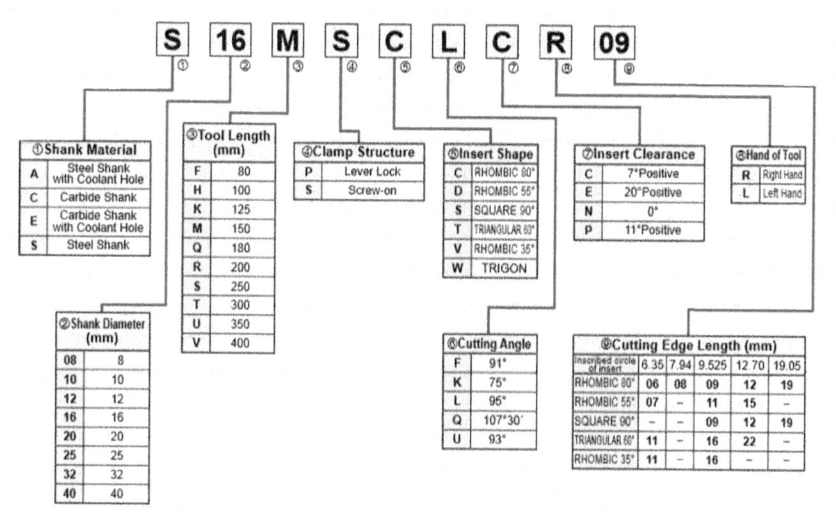

Figure 10.23 Turning tool holder nomenclatures.

(Courtesy: MITSUBISHI MATERIALS CORPORATION METALWORKING SOLUTIONS COMPANY, Japan)

10.7 What is Cutting Tool Entering Angle?

Entering angle/lead angle, is the cutting angle between cutting tool edge and the cutting tool feed direction. Whenever you select the cutting tool, always consider this angle. If we do not consider this angle it can damage the profile or shape. Large entering angle has the ability to turn shoulder of the workpiece and small entering angle can not turn 90° shoulder. See Figure 10.24.

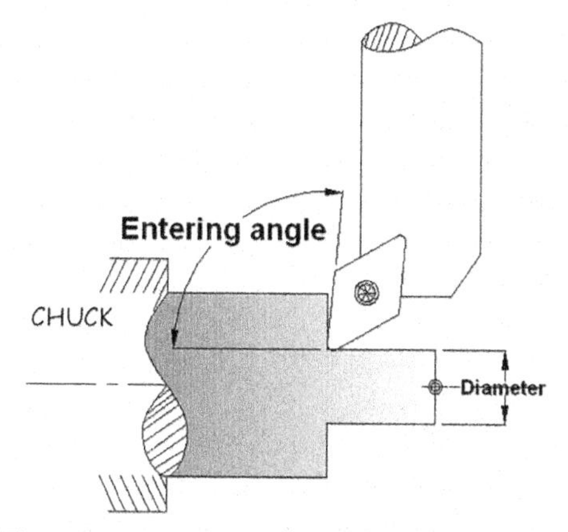

Figure 10.24 Entering angle of the cutting tool.

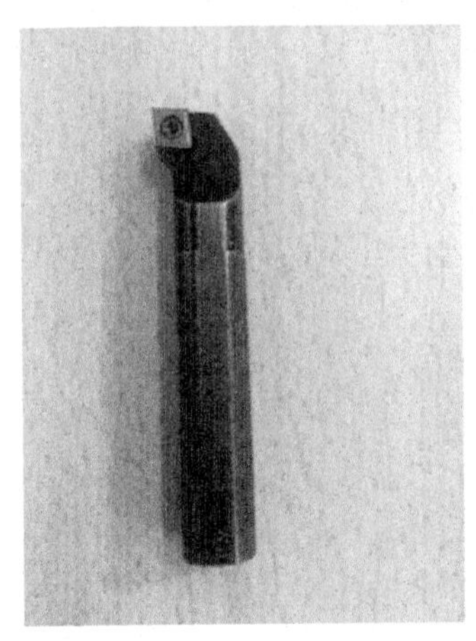

Figure 10.25 Right hand boring tool.

(Courtesy: Boring Tool, Sara Sae India Pvt Ltd, India)

10.8 What is Boring Bar or Boring Tool? Where it is Used?

Boring tool enlarges the hole and makes true the hole respectively of the central axis of the workpiece (spindle). The boring tool is used for machining the internal profile of the workpiece. See Figure 10.25.

10.9 Different Types of Boring Tools (Internal Machining Tools)

Figure 10.26 shows the boring tool holder with CNMG insert.

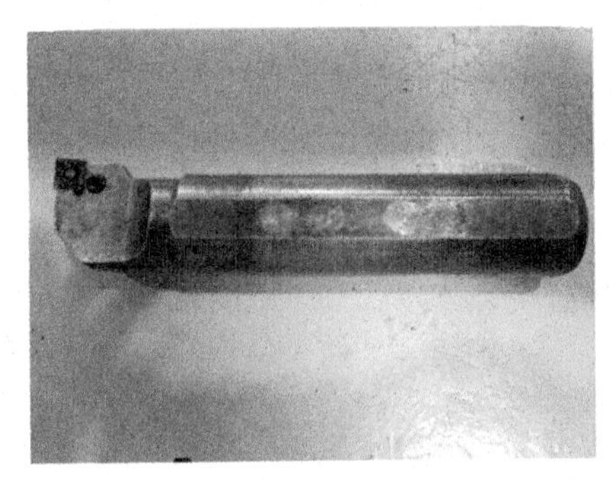

Figure 10.26 Boring tool holder for CNMG inserts.

(Courtesy: Boring Tool, Sara Sae India Pvt Ltd, India)

Figure 10.27 shows the different types of boring tool holders.
Figure 10.28 shows the boring tool holder with the sleeve.
Figure 10.29 shows the boring tool holder with TNMG insert.

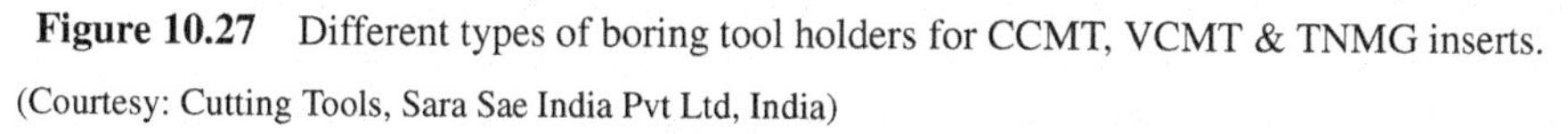

Figure 10.27 Different types of boring tool holders for CCMT, VCMT & TNMG inserts.
(Courtesy: Cutting Tools, Sara Sae India Pvt Ltd, India)

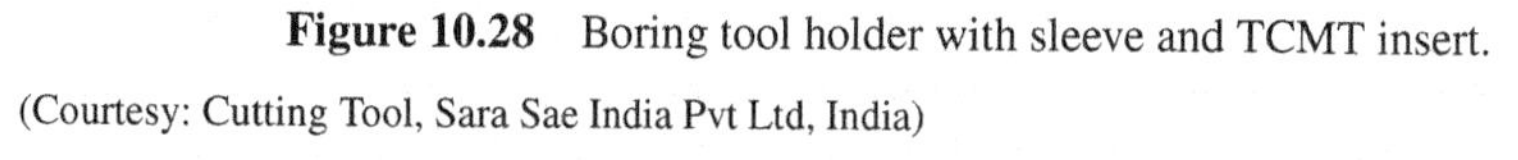

Figure 10.28 Boring tool holder with sleeve and TCMT insert.
(Courtesy: Cutting Tool, Sara Sae India Pvt Ltd, India)

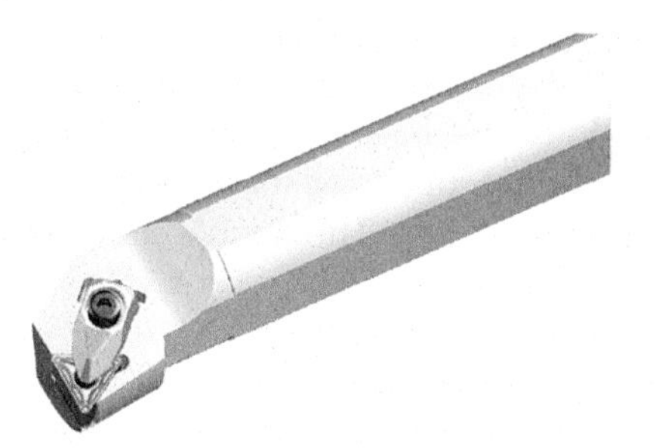

Figure 10.29 Boring tool holder with CNMG inserts.
(Courtesy: Cutting Tool, Sara Sae India Pvt Ltd, India)

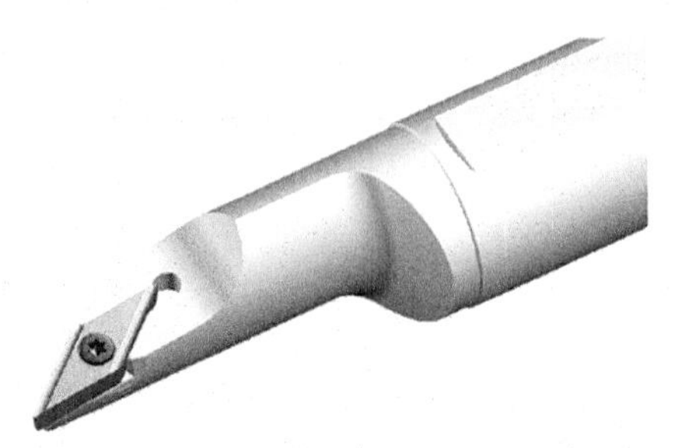

Figure 10.30 Boring tool holder with VBMT inserts.

(Courtesy: Cutting Tool, Sara Sae India Pvt Ltd, India)

Figure 10.30 shows boring tool holder with VBMT insert.

10.10 Different Types of Turning Tools (External Machining Tools)

10.10.1 What are Right Hand and Left Hand Cutting Tools?

From the ancient time humans are using cutting tools (ancient cutting tools made from stones, bones etc.) for cutting the material and these cuting tools have wedge shaped. Without wedge shaped we cannot cut the material in a proper way. Using of wedge shaped cutting tools we get less rejection/scrap, we save the cutting time, we get the perfect shape, finishing, quality and accuracy of the object.

We use the same idea/concept during metal cutting. Due to the above explained benefits. In lathe machine, we use wedge type cutting tools during workpiece rotation and these wedge type cutting tools are single point tools. When workpiece rotates, single point turning tool (SPTT) takes the depth of cut on the workpiece surface and move and removes the material towards to cutting direction.

When from, the modification has done on the lathe machine. We can cut the material from both directions for these both directions we use the left hand and right hand cutting tools. These tools called single point turning tool (SPTT).

10.10.1.1 Right hand cutting tool

When we put the cutting tool on the right hand palm. If, cutting tool head shows and match the right hand thumb direction. it means, right hand thumb direction represents the cutting tool direction. on the other hand we can say cutting tool edge should be on the right side direction. See Figure 10.31.

Figure 10.31 Right hand external tool with CNMG insert.

(Courtesy: Turning Tool, Sara Sae India Pvt Ltd, India)

For material removing, we apply clockwise circular motion of the spindle for right hand cutting tool. In CNC machine we use M03 code for clock wise circular motion of the spindle.

10.10.2 Left Hand Cutting Tool

When we put the cutting tool on the left hand palm. If, cutting tool head shows and match the left hand thumb direction. it means, left hand thumb direction represents the cutting tool direction. on the other hand we can say cutting tool edge should be on the left side direction. See Figure 10.32.

Figure 10.32 Left hand external tool with DCMT insert.

(Courtesy: Left Hand Tool, Sara Sae India Pvt Ltd, India)

For material removing, we apply counter clock wise circular motion of the spindle for left hand cutting tool. In CNC machine we use M04 code for the counter clock wise circular motion of the spindle.

Figure 10.33 **compares the left hand and the right hand tool.**

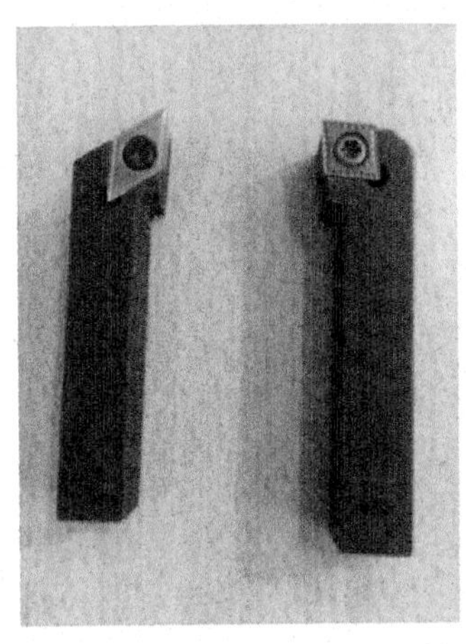

Figure 10.33 Comparison between right hand and left hand external tool.

(Courtesy: Cutting Tool, Sara Sae India Pvt Ltd, India)

10.10.3 Right Hand Cutting Tool Images

Figure 10.34 shows the right hand tool. It is used in medium material removing.

Figure 10.34 External turning tool holder with DNMG insert.

(Courtesy: Cutting Tool, Sara Sae India Pvt Ltd, India)

Figure 10.35 shows the right hand turning tool. It is used in heavy material removing.

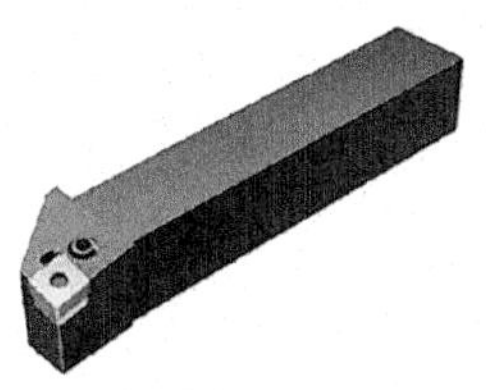

Figure 10.35 External turning tool holder with CNMG insert.
(Courtesy: Cutting Tool, Sara Sae India Pvt Ltd, India)

10.10.4 Left hand cutting tools images

Figure 10.36 shows the left hand turning tool. It is used in medium turning operation for profile turning operations.

Figure 10.37 shows left hand turning tool.

Figure 10.38 shows the left hand turning tool. It is used in complex profile machining.

Figure 10.36 External turning tool holder with VNMG insert.
(Courtesy: Cutting Tool, Sara Sae India Pvt Ltd, India)

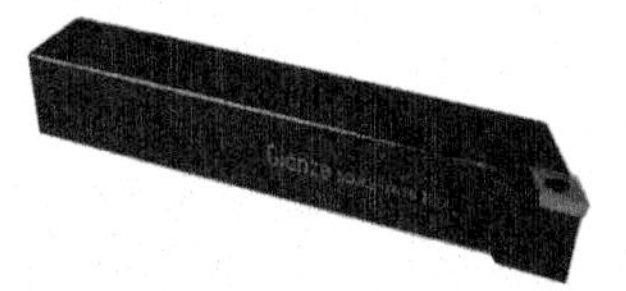

Figure 10.37 External turning tool holder with VCMT insert.
(Courtesy: Cutting Tool, Sara Sae India Pvt Ltd, India)

Figure 10.38 External turning tool holder with VCMT insert.
(Courtesy: Cutting Tool, Sara Sae India Pvt Ltd, India)

10.11 Neutral Cutting Tools

10.11.1 What is Neutral Tool? Where it is Used?

The neutral cutting tool has a straight shape from one end to another end. We can say neutral cutting tool does not have right side/left side cutting edge like ordinary cutting tools, its cutting head goes straight towards to shank direction or parallel to shank direction.

The neutral tool is used for internal and external grooving. It is used, where other cutting tools have entering angle problem. "Neutral cutting tool head has to be parallel to the shank axis". (Figure 10.39).

Figure 10.39 External neutral tool holder for profile turning and grooving.

(Courtesy: External Neutral tool holder, Sara Sae India Pvt Ltd, India)

Figure 10.40 CNC cutting tool rack.

(Courtesy: Cutting Tool Rack, Sara Sae India Pvt Ltd, India)

References

[1] Fundamentals of Metal Cutting and Machine Tools, B.L. Juneja, G.S. Sekhon and Nitin Seth, Revised Second Edition (2005), New Age International Publisher, India.
[2] CNC Technology & Programming, Tilak Taj, 2016, Dhanapat Rai Publishing Company, India.
[3] Drawings and sketches has been generated from Auto CAD 2014.
[4] Mitsubishi Materials Corporation, Advanced Material & Tools Company, Japan.

11

Drawings and CNC Programs

11.1 Drawings for CNC Programming: Exercise

1. Write the CNC program for the following step turning drawing (See Figure 11.1).
2. Write the CNC turning program for the following step turning and chamfering drawing. See Figure 11.2.
3. Write the CNC turning program for the following facing operation drawing. See Figure 11.3.
4. Write the CNC turning program for following corner radius drawing. See Figure 11.4.
5. Write the CNC turning program for following corner radius drawing. See Figure 11.5.
6. Write the CNC turning program for the following corner radius drawing. See Figure 11.6.
7. Write the CNC turning program for the following drawing of oval shape. See Figure 11.7.
8. Write the CNC program for the following drilling operation drawing. See Figure 11.8.
9. Write the CNC program for the following drawing of threading operation. See Figure 11.9.
10. Write the CNC program for the following taper turning drawing. See Figure 11.10.

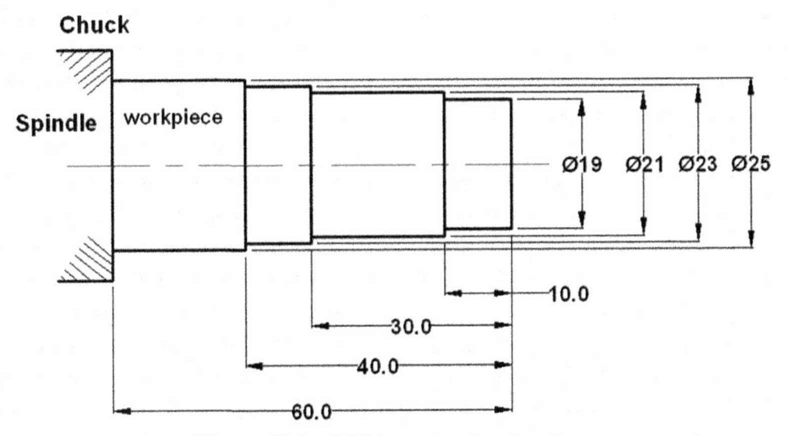

Figure 11.1 Multi-step turning drawing.

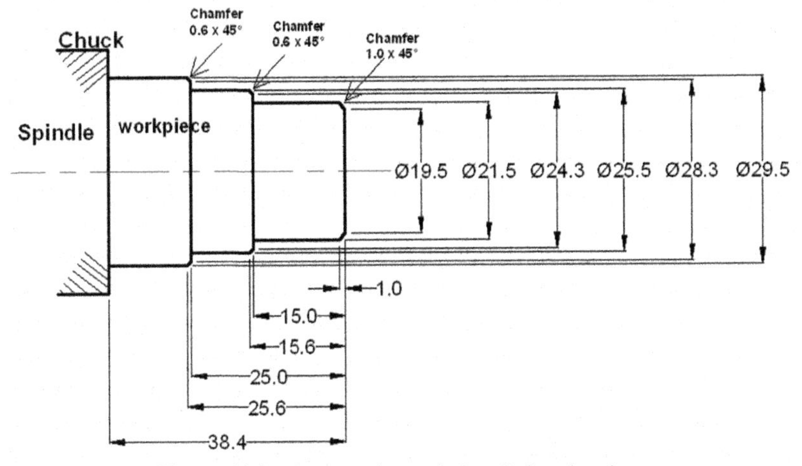

Figure 11.2 Step turning and chamfering drawing.

11. Write the CNC program for the following drawing of radius profile. See Figure 11.11.
12. Write the CNC program for following corner radius and taper turning drawing. See Figure 11.12.

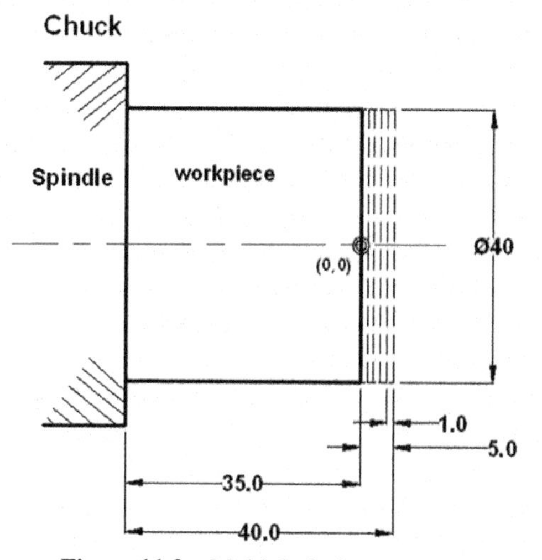

Figure 11.3 Multiple facing drawing.

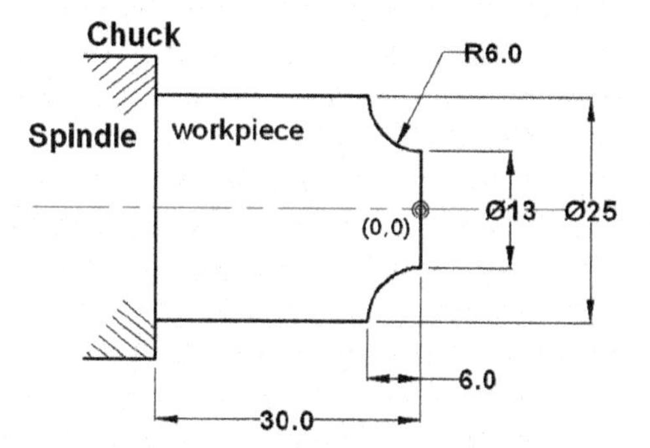

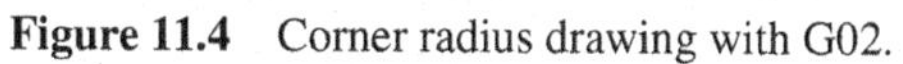

Figure 11.4 Corner radius drawing with G02.

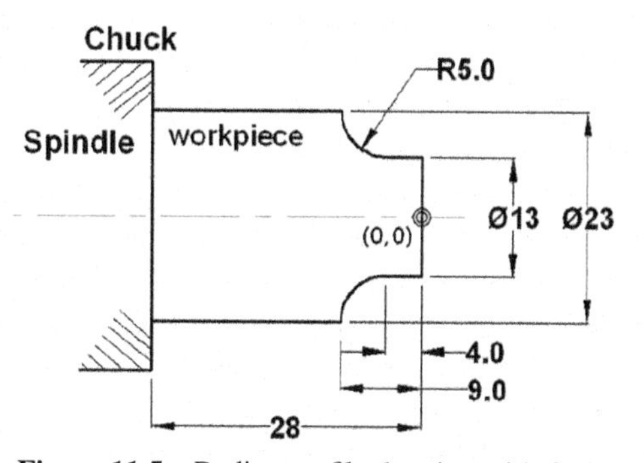

Figure 11.5 Radius profile drawing with G02.

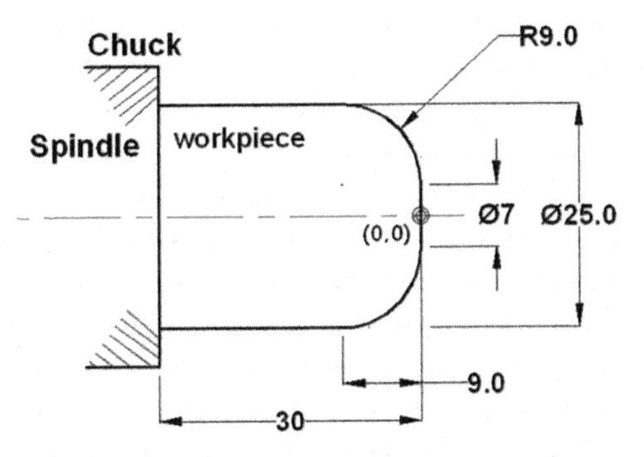

Figure 11.6 Corner radius drawing with G03.

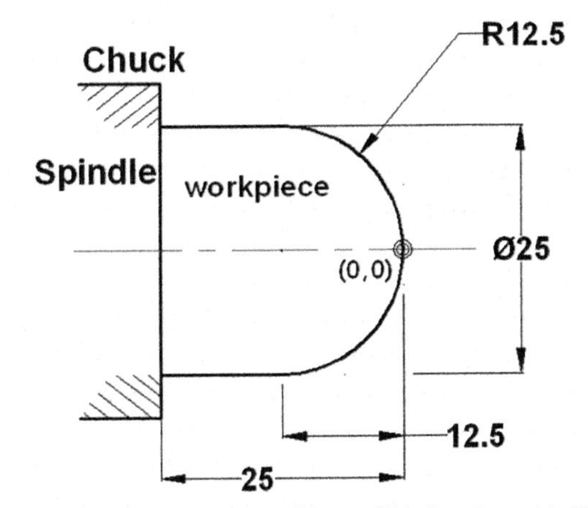

Figure 11.7 Oval shape (radius profile) drawing with G03.

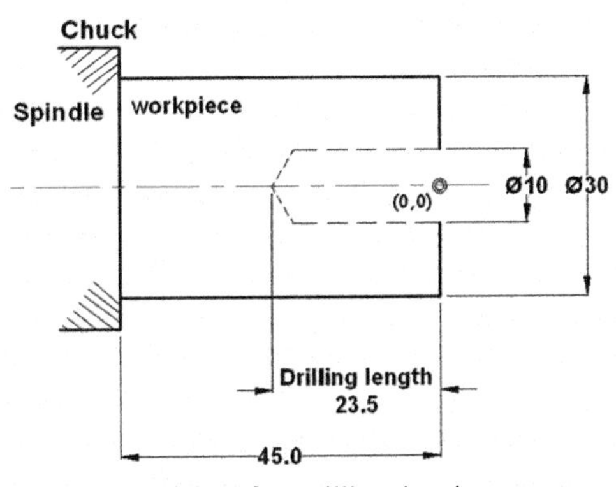

Figure 11.8 Drilling drawing.

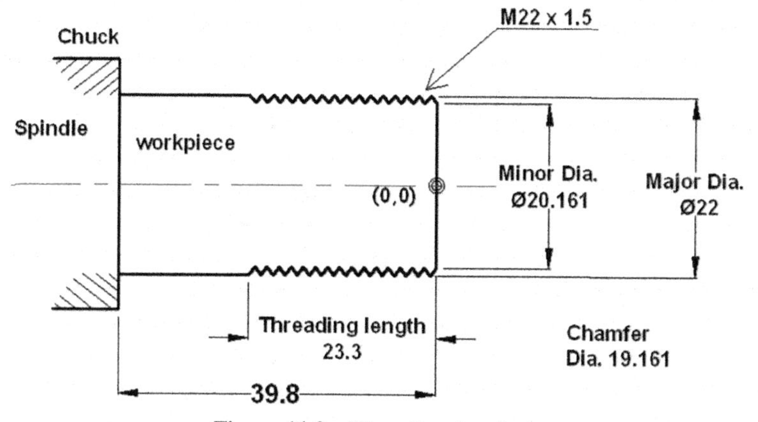

Figure 11.9 Threading drawing.

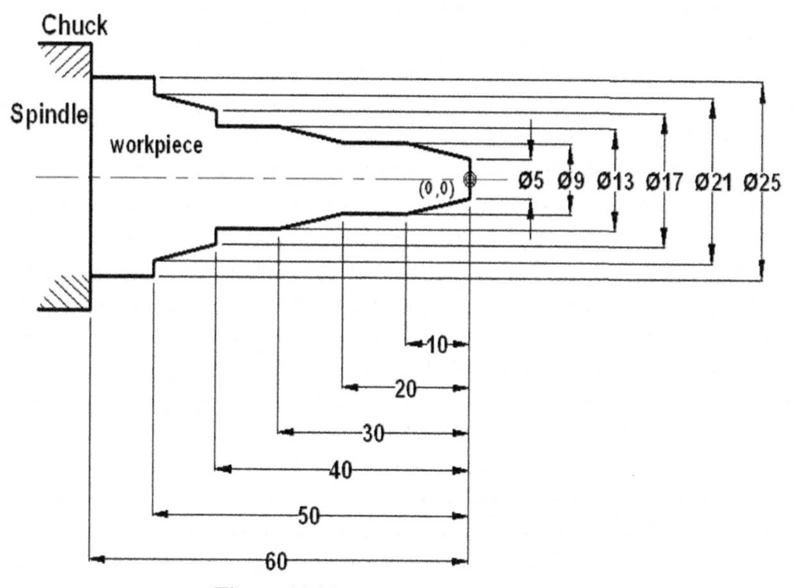

Figure 11.10 Taper profile drawing.

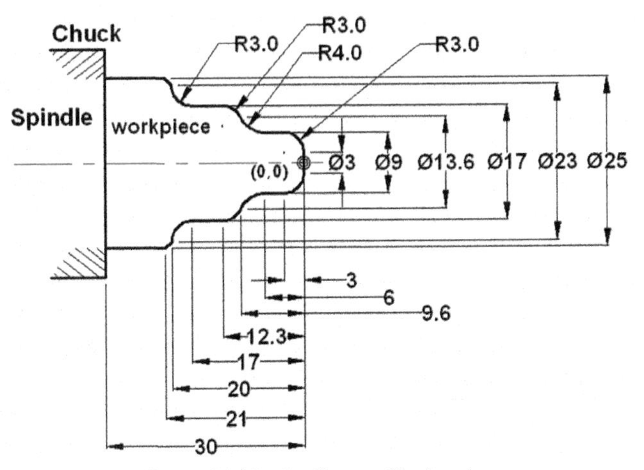

Figure 11.11 Radius profile drawing.

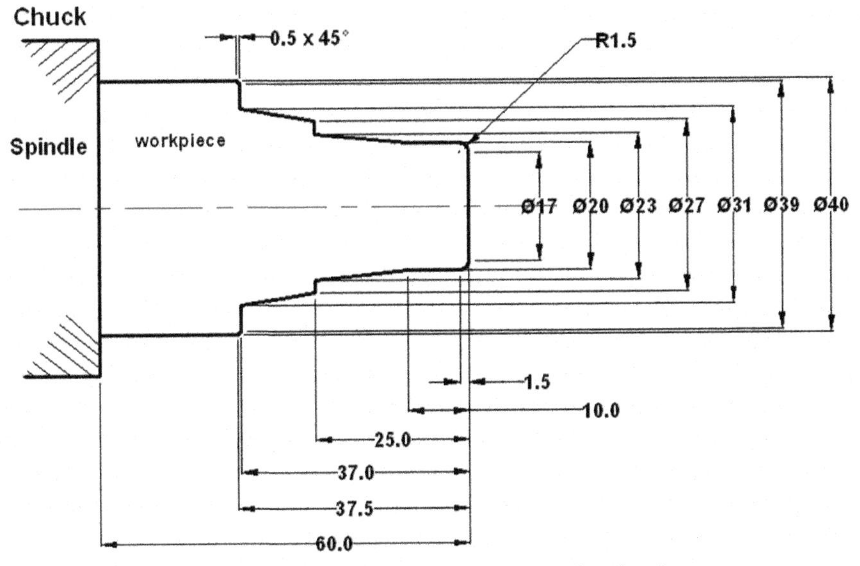

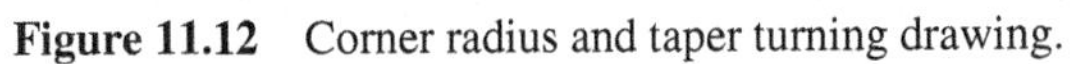

Figure 11.12 Corner radius and taper turning drawing.

11.2 Write Your Own CNC Programs by using Following Drawings and Match with Given Following CNC Programs

Note: Only for the demo program.

11.2.1 Following Drawings are Programmed with Finishing Cut Program

I. Raw material is the same for all drawings: mild steel
II. All dimensions are in millimeters.

CNC Program No. 1

O23666;
G54;
G50 S1800;
T0101 (Turning Tool – **Final cut program**);
G00 X100 Z100;
G00 X40.0 Z5;
G96 M03 S210;
M08;
G00 X40.0 Z0.0;
G01 X0.0 F0.12;

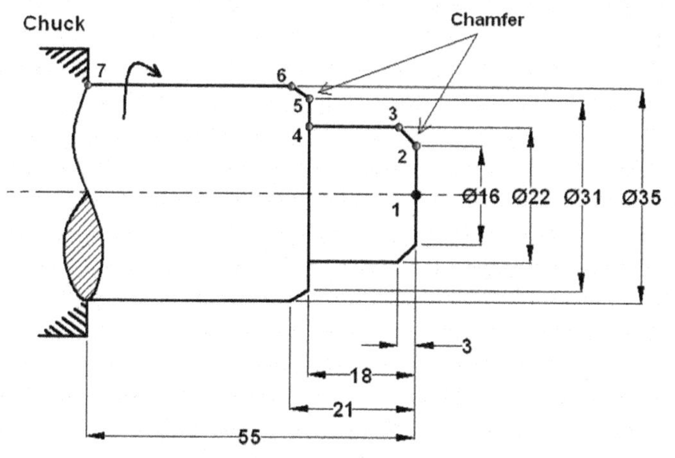

Figure 11.13 Practice drawing.

```
G00 Z1.0;
G01 X15.0 Z1.0;
G01 X16.0 Z0.0 F0.12;
G01 X22.0 Z-3.0 F0.12;
G01 Z-18.0 F0.12;
G01 X31.0 Z-18.0 F0.12;
G01 X35.0 Z-21.0 F0.12;
G00 X35.0 Z1.0;
M09;
M05;
G00 X100 Z100;
M30;
```

CNC Program No. 2

O20015;
G54;
G50 S1800;
T0101 (Turning Tool – **Final cut program**);
G00 X100 Z150;

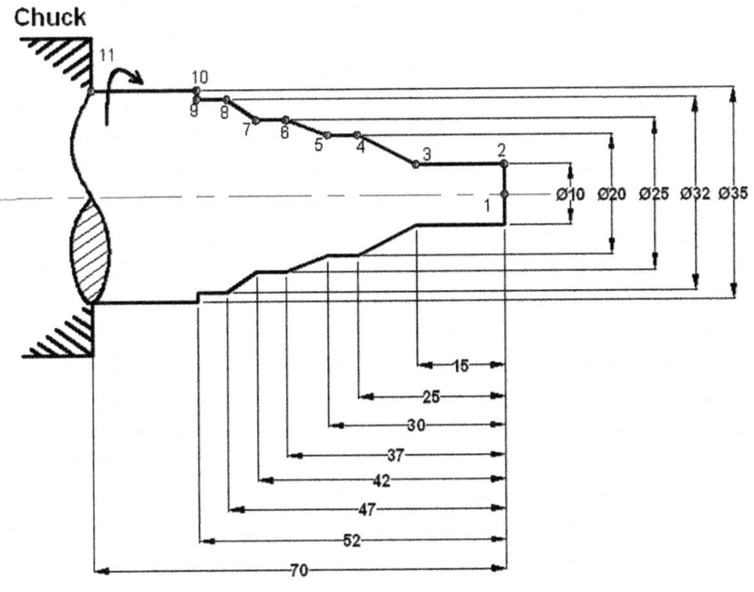

Figure 11.14 Practice drawing.

G00 X40.0 Z5;
G96 M03 S210;
M08;
G00 X40.0 Z0.0;
G01 X0.0 F0.12;
G00 Z1.0;
G00 X10.0 Z1.0;
G01 Z-15.0 F0.12;
G01 X20.0 Z-25.0 F0.12;
G01 Z-30.0 F0.12;
G01 X25.0 Z-37.0 F0.12;
G01 Z-42.0 F0.12;
G01 X32.0 Z-47.0 F0.12;
G01 Z-52.0 F0.12;
G01 X37.0 F0.12;
G00 X37.0 Z1.0;
M09;
M05;
G00 X100 Z150;
M30;

CNC Program No. 3

O97320;
G54;
G50 S1500;
T0101 (Turning Tool – **Final cut program**);
G00 X150 Z150;
G00 X67.0 Z5;
G96 M03 S220;
M08;
G00 X67.0 Z0.0;
G01 X0.0 F0.12;
G00 Z1.0;
G00 X13.0 Z1.0;
G01 X14.0 Z0.0 F0.12;
G01 X25.0 Z-17.0 F0.12;
G01 Z-25.0 F0.12;
G01 X28.0 Z-26.5 F0.12;

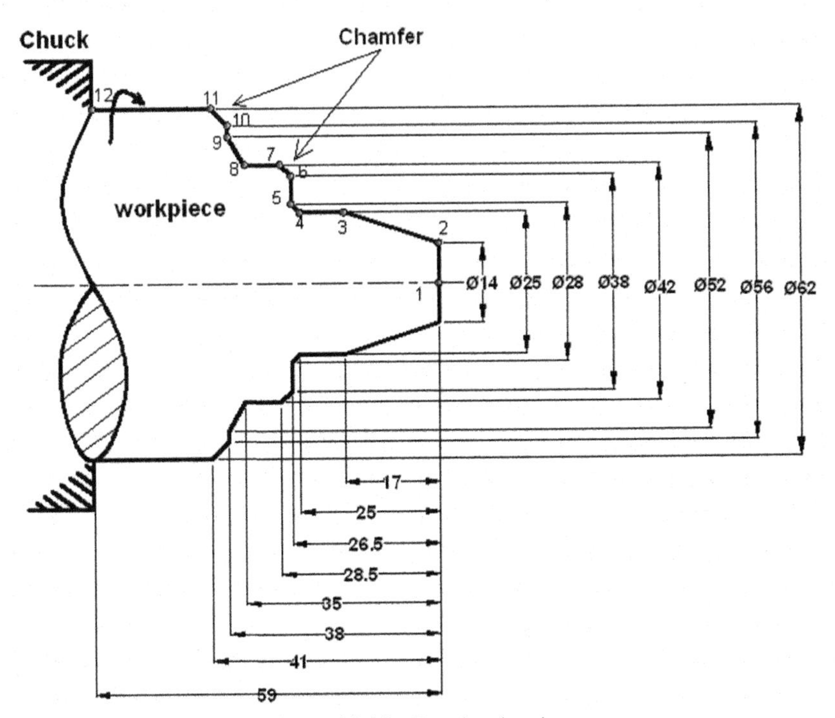

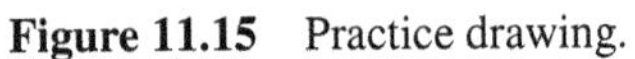

Figure 11.15 Practice drawing.

```
G01 X38.0 F0.12;
G01 X42.0 Z-28.5 F0.12;
G01 Z-35.0 F0.12;
G01 X52.0 Z-38.0 F0.12;
G01 X56.0 F0.12;
G01 X62.0 Z-41.0 F0.12;
G00 Z1.0;
M09;
M05;
G00 X150 Z150;
M30;
```

CNC Program No. 4

O55471;

```
G54;
G50 S1500;
```

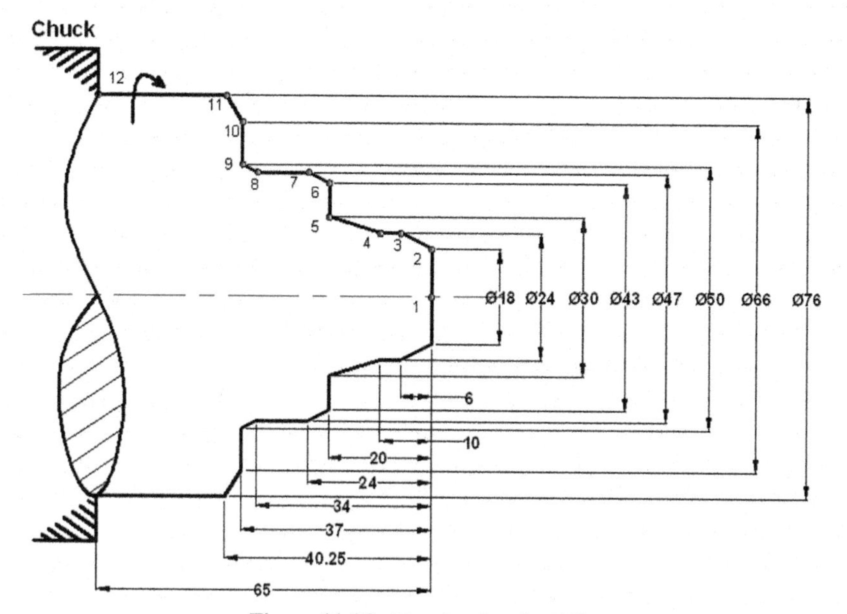

Figure 11.16 Practice drawing [1].

T0101 (Turning Tool – **Final cut program**);
G00 X150 Z150;
G00 X81.0 Z5;
G96 M03 S220;
M08;
G00 X81.0 Z0.0;
G01 X0.0 F0.12;
G00 Z1.0;
G00 X17.0 Z1.0;
G01 X18.0 Z0.0 F0.12;
G01 X24.0 Z-6.0 F0.12;
G01 Z-10.0 F0.12;
G01 X30.0 Z-20.0 F0.12;
G01 X43.0 F0.12;
G01 X47.0 Z-24.0 F0.12;
G01 Z-34.0 F0.12;
G01 X50.0 Z-37.0 F0.12;
G01 X66.0 F0.12;
G01 X76.0 Z-40.25 F0.12;
G00 Z1.0;

M09;
M05;
G00 X150 Z150;
M30;

CNC Program No. 5

O77710;
G54;
G50 S1700;
T0101 (Turning Tool – **Final cut program**);
G00 X150 Z150;
G00 X55.0 Z5;
G96 M03 S210;
M08;
G00 X55.0 Z0.0;
G01 X0.0 F0.12;
G00 Z1.0;
G00 X7.0 Z1.0;
G01 X8.0 Z0.0 F0.12;
G03 X20.0 Z-6.0 R6.0 F0.12;

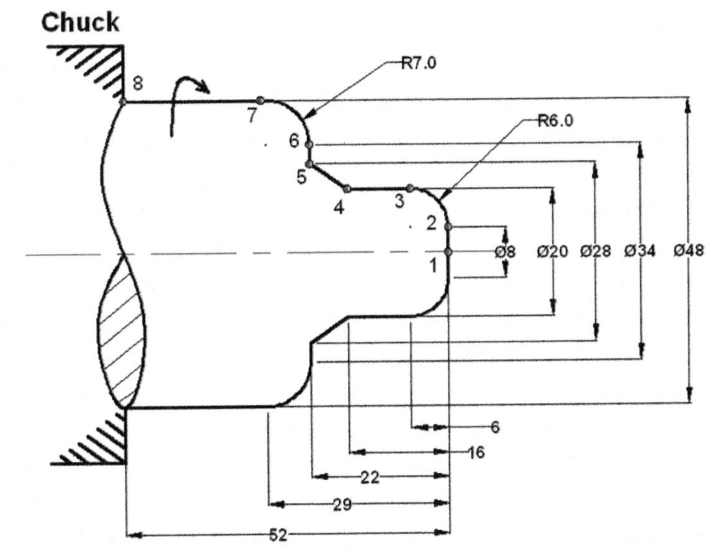

Figure 11.17 Practice drawing.

```
G01 Z-16.0 F0.12;
G01 X28.0 Z-22.0 F0.12;
G01 X34.0 F0.12;
G03 X48.0 Z-29.0 R7.0 F0.12;
G00 Z1.0;
M09;
M05;
G00 X150 Z150;
M30;
```

CNC Program No. 6

O92219;
G54;
G50 S1700;
T0101 (Turning Tool – **Final cut program**);
G00 X150 Z150;
G00 X50.0 Z5;

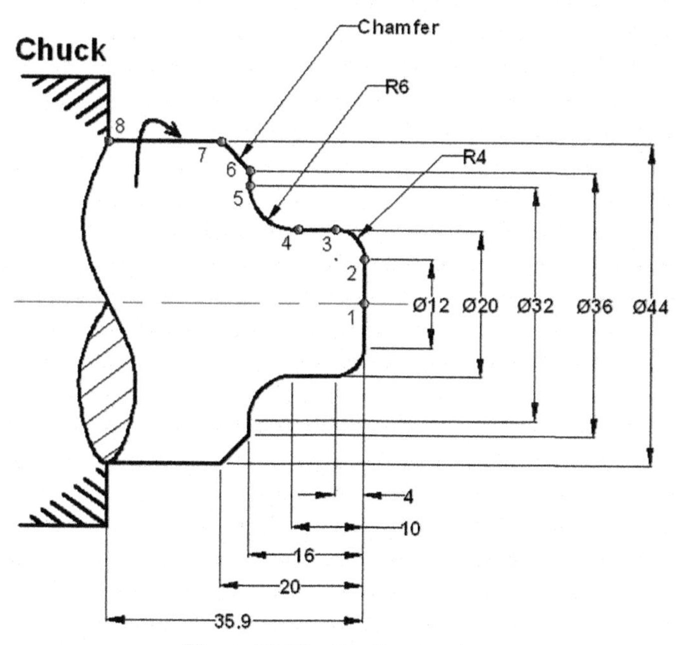

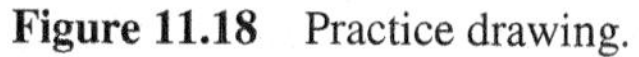

Figure 11.18 Practice drawing.

```
G96 M03 S210;
M08;
G00 X50.0 Z0.0;
G01 X0.0 F0.12;
G00 Z1.0;
G00 X11.0 Z1.0;
G01 X12.0 Z0.0 F0.12;
G03 X20.0 Z-4.0 R4.0 F0.12;
G01 Z-10.0 F0.12;
G02 X32.0 Z-16.0 R6.0 F0.12;
G01 X36.0 F0.12;
G01 X44.0 Z-20.0 F0.12;
G00 Z1.0;
M09;
M05;
G00 X150 Z150;
M30;
```

11.3 Important Tips Before and After Machining

11.3.1 Cutting Fluid (Coolant)

In CNC milling machine Cutting fluid is injecting on the work piece and Cutter. See Figure 11.19.

Different types (grade) of cutting oil (coolant) are used on different materials during machining (operation). Before using the cutting oils as a coolant we must consider the following instructions.

Figure 11.19 Water soluble cutting oil.
(Courtesy: Water Soluble Cutting Oil, Rawat Engg. Tech Pvt Ltd., India)

***About cutting fluid soluble cutting oil*:**

- Cutting oil (fluid) reduced the cutting temperature of cutting tool and work piece. It does not allow increasing the tool and work piece temperature above the maximum limit.
- The best cutting fluid maintains the temperature of the cutting tool and work piece during machining (operation). It means during machining, the work piece will not be more hot and cold due to this effect, work piece and cutting tool material properties can be changed.
- Cutting fluid reduces the cutting friction and increases the surface finish of the cutting material.
- Good cutting fluid increases the cutting tool life and machine life.
- Good cutting fluid saves the work piece and machine from corrosion and rust.
- Coolant saves the human body from the hot metal.
- Cutting fluid/lubricant made from petrochemical products, animal fat, plants, biochemical etc. [2].

Note: Valvoline Cool 103 SA is a high performance, long life metal working fluid for machining of ferrous and non-ferrous alloys and grinding. It is a versatile fluid that can be used for a wide variety of operations on multiple metals. Different - different mixing ratios are used for cutting fluid and water for preparing coolant, it depends on cutting oil grade like 1:19 (Courtesy: Valvoline Cool 103 SA (grade), Valvoline Cummins Private Limited, India) etc. where one liter is cutting oil (fluid) and 19 Ltr is water. Generally, it is called coolant.

11.3.2 Safety Precautions

- For CNC machine operating or CNC programming, you should have relevant experience or worked under expert's guidance previously.
- It is very important that always properly holds the work piece between jaws or vice with recommending pressure.
- Do not wear lose clothes, rings or hanging things like neck's chain/wrist chain.
- Do not talk/discuss while the CNC machine is working or CNC program is feeding.
- Do not enter inside the CNC machine without permission of immediate boss when CNC machine is stopped, because it is an electronic system. It can be run suddenly.

- Before working or operating the CNC machine always check the cutting tools, coolant condition, hydraulic and pneumatic pressure.
- Cutting tools have sharp cutting edges. Use gloves otherwise, without using hand gloves may cause of injuries.
- After cutting the material, cutting tools and work pieces become extremely hot, without using the hand gloves may cause burn.
- Do not open the machine door during the machining operation. Hot chips may be the cause of burn.
- When applying high rpm, machinery parts may be expelled due to centrifugal force. Before operating with machine please read the manuals and recommendations.

11.4 Interview Questions

11.4.1 The Following Questions Can Be Asked in Interview

- Can you use the Vernier calipers and micrometer or read them?
- What is the least count of the measuring instrument or what is the least count of Vernier calipers and micrometers.
- Can you describe and calculate the feed, cutting speeds, and depth of cut?
- Can you describe the routine maintenance; you have performed regularly on the CNC machine.
- Can you share an experience, where you identified a problem in the CNC program/CNC machine?
- Share an experience of finished work piece when you measured a work piece as per drawing.
- Describe your experience when you made the job.
- What are 3 jaws and 4 jaws chuck?
- Can you explain some G & M codes?
- What is canned cycle?
- Can you read the manufacturing drawing?
- Do you know about cutting tool holders and bits nomenclatures?
- What is cutting tool geometry in CNC machine?
- Which CNC control system(s) are used by you? Give the names of few CNC control systems.
- Do you know, how to write CNC program?
- Can you make a simple CNC program with different operations?
- Do you know CAD?

- Can you work in CAM?
- Do you know the geometric dimension, symbols and tolerances?
- Can you share CNC programming problem, which is solved by you on the shop floor.

11.5 Answers Keys

1. This answer key is for Figure 9.291 [Chapter No. 9]

O66971;
G54;
G50 S1800;
T0101 (Rough Turning Tool);
G00 X150 Z150;
G00 X32 Z5;
G96 M03 S200;
M08;
G00 X32 ***Z0.1***; [0.100 mm material left for final cut]
G01 X0.0 F0.12; **Rough Facing** operation
G00 Z1.0;
G00 X24.2;
G01 *Z-29.9* F0.25;
G01 X28.0 F0.25;
G00 Z1.0;
G00 X22.0;
G01 *Z-25.4* F0.25;
G01 X25.0 F0.25;
G00 Z1.0;
G00 X19.7;
G01 Z-14.1 *F0.25*;
G01 X23.0 F0.25;
G00 Z1.0;
G00 *X16.7*;
G01 Z-6.7 F0.25;
G01 X20.5 F0.25;
G00 Z1.0;
M09;
M05;
G00 X150 Z150;

M00;
T0303 (Final Turning Tool);
G00 X150 Z150;
G00 X32 Z5;
G96 *M03* S230;
M08;
G00 X32.0 **Z0.0; Final Facing** operation
G01 X0.0 F0.12; Remaining material (0.100 mm) on the work piece face will be remove.
G00 Z1.0;
G00 X16.5;
G01 *Z-6.8* F0.12;
G01 *X19.5* F0.12;
G01 Z-14.2 F0.12;
G01 X21.8 F0.12;
G01 Z-25.5 F0.12;
G01 *X24.0* F0.12;
G01 *Z-30.0* F0.12;
G01 X28.0 *F0.12*;
G00 Z1.0;
M09;
M05;
G00 X150 Z150;
M30;

2. This answer key is for Figure 9.292 [Chapter No. 9]

O50014;
G54;
G50 S1500;
T0101;
G00 X100 Z100;
G00 X30 Z5;
G96 M03 S180;
M08;
G00 X30.0 **Z-0.8**;
G01 *X0.0* F0.12;
G00 Z1.0;
G00 X30;
G00 ***Z-1.6***;

G01 X0.0 F0.12;
G00 Z1.0;
G00 X30;
G00 ***Z-2.4***;
G01 X0.0 F0.12;
G00 Z1.0;
M09;
M05;
G00 X150 Z150;
M30;

3. This answer key is for Figure 9.293 [Chapter No. 9]

O66971;
G54;
G50 *S1800*;
T0101;
G00 X150 Z150;
G00 X30 Z5;
G96 *M03* S200;
M08;
G00 X30.0 ***Z0.0***;
G01 X0.0 F0.12;
G00 Z1.0;
G00 X21.5;
G01 X22.5 Z0.0 F0.12; Starting point of chamfer
G01 X26.3 *Z-1.9* F0.12; Ending point of chamfer including clearance with 45° angle
G00 Z1.0;
M09;
M05;
G00 X150 Z150;
M30;

4. This answer key is for Figure 9.294 [Chapter No. 9]

O51020;
G54;
G50 S1500;
T0101 (Rough Turning Tool);

G00 X200 Z250;
G00 X70.0 Z25;
G00 X70.0 Z5.0;
G96 M03 S200;
M08;
G00 X70 ***Z0.20***; [0.200 mm material left for final cut]
G01 *X0.0* F0.12; **Rough Facing** operation
G00 Z1.0;
G00 *X62.8*;
G01 Z-63.3 F0.25;
G01 X67.0 F0.25;
G00 Z1.0;
G00 *X59.7*;
G01 *Z-63.3* F0.25;
G01 X67.0 F0.25;
G00 Z1.0;

G00
G01 Z-39.4 *F0.25*;
G01 X60.0 F0.25;
G00 Z1.0;
G00 X54.9;
G01 *Z-39.4* F0.25;
G01 X60.0 *F0.12*;
G00 Z1.0;
G00 X51.9;
G01 *Z-19.6* F0.25;
G01 X56.0 F0.25;
G00 Z1.0;
G00 X48.9;
G01 Z-19.6 F0.25;
G01 *X56.0* F0.25;
G00 Z1.0;
G00 X48.6;
G01 *Z-19.6* F0.25;
G01 X56.0 F0.25;
G00 *Z1.0*;
M09;
M05;
G00 X200 Z250;

M00;
T0303 (Final Turning Tool);
G00 X150 Z150;
G00 X55.0 Z5;
G96 M03 S230;
M08;
G00 X55 **Z0.0; Final Facing** operation
G01 X0.0 F0.12; Remaining material (0.200 mm) on the work piece face will be remove.
G00 Z1.0;
G00 *X48.4′*;
G01 *Z-19.7* F0.12;
G01 X54.7 F0.12;
G01 Z-39.5 F0.12;
G01 X59.5 F0.12;
G01 *Z-63.4* F0.12;
G01 X67.0 *F0.12*;
G00 Z1.0;
M09;
M05;
G00 X200 Z250;
M30;

5. **This answer key is for Figure 9.295 [Chapter No. 9]**

O58920;
G54;
G50 S2000;
T0101 (Rough Turning Tool);
G00 X100 Z150;
G96 M03 S200;
M08;
G00 X27.0 **Z0.0**;
G01 *X0.0* F0.12; **Facing** operation
G00 Z1.0;
G00 X20.8;
G01 X20.8 *Z0.0* F0.12; Rough cut, starting point of radius
G03 X22.8 *Z-1.0* **R1.0** F0.12 Ending point of radius
G00 Z1.0;
G00 *X18.4*;

G01 X19.4 Z0.0 *F0.12*; Finishing cut, starting point of radius
G03 X22.8 Z-1.7 ***R1.7*** F0.12; Ending point of radius
G00 Z1.0;
M09;
M05;
G00 X100 Z150;
M30;

6. **This answer key is for Figure 9.296 [Chapter No. 9]**

O51020;
G54;
G50 *S1500*;
T0101 (Turning Tool);
G00 X200 Z200;
G00 X45.0 Z25;
G96 M03 S200;
M08;
G00 X45.0 **Z0.0**;
G01 X0.0 *F0.12*; **Facing** operation
G00 Z1.0;
G00 X35.3;
G01 *X35.3* Z0.0 F0.15; Starting point of rough taper turning
G01 **X38.3 Z-43.0 F0.15**; Rough taper turning operation, ending point of rough taper turning
G00 Z1.0;
G00 X34.7;
G01 X33.7 Z0.0 *F0.12*; Starting point of final taper turning (SPT)
G01 X38.3 ***Z-44.7*** **F0.15**; Ending point of final taper turning (EPT)
G00 Z1.0;
M09;
M05;
G00 X200 Z250;
M30;

7. **This answer key is for Figure 9.297 [Chapter No. 9]**

O02876;
G54;
G50 S2000;
T0101 (Rough Turning Tool);
G00 X100 Z250;

G00 X27.0 Z5.0;
G96 *M03* S200;
M08;
G00 X27.0 ***Z0.0***;
G01 X0.0 F0.12; **Facing** operation
G00 Z1.0;
G00 X21.5;
G01 *X21.5* Z0.0 F0.12; Starting point of radius (SPR)
G03 X23.5 *Z-1.0* **R1.0** F0.12; Rough radius turning operation,
Ending point of radius (EPR)
G00 Z1.0;
G00 X19.5;
G01 *X19.5* Z0.0 F0.12; SPR
G03 X23.5 *Z-2.0* **R2.0** *F0.12*; Rough radius turning operation, EPR
G00 Z1.0;
G00 X17.0;
G01 *X17.0* Z0.0 F012; SPR
G03 X23.5 *Z-3.25* **R3.25** F0.12; Rough radius turning operation, EPR
G00 Z1.0;
G00 X15.5;
G01 X16.5 Z0.0 F0.12; **Finish radius** turning operation, SPR
G03 X23.5 Z-3.5 ***R3.5*** F0.12; EPR
G00 Z1.0;
M09;
M05;
G00 X100 Z250;
M30;

8. **This answer key is for Figure 9.298 [Chapter No. 9]**

O83060;
G54;
G50 S1500;
T0101 (Turning Tool);
G00 X200 Z300;
G00 X60.0 Z5.0;
G96 *M03* S200;
M08;
G00 X60.0 **Z0.0**;
G01 *X0.0* F0.12; **Facing** operation

G00 Z1.0;
M09;
M05;
G00 X200 Z300;
M00;
T0505 (Drilling Tool);
G00 *X200 Z300*;
G00 X100.0 Z25.0;
G97 M03 *S450*;
G00 *X0.0* Z5.0;
M08;
G00 Z1.0;
G01 Z-5.0 F0.05; Drilling operation starts
G00 Z5.0;
G00 Z-4.0;
G01 Z-10.0 F0.05;
G00 *Z5.0*;
G00 Z-9.0;
G01 Z-15 *F0.05*;
G00 Z5.0;
G00 Z-14.0;
G01 *Z-20.0* F0.05;
G00 Z5.0;
G00 Z-19.0;
G01 Z-25.0 F0.05;
G00 Z5.0;
G00 *Z-24.0*;
G01 *Z-30* F0.05;
G00 Z5.0;
G00 *Z-29.0*;
G01 Z-35.0 *F0.05*;
G00 Z5.0;
G00 Z-34.0;
G01 Z-38.4 F0.05;
G00 Z5.0;
M09;
M05;
G00 X200 Z300;
M30;

9. This answer key is for Figure 9.299 [Chapter No. 9]

O41046;
G54;
G50 *S1600*;
T0101 (Rough Turning Tool);
G00 X100 Z150;
G00 X45.0 Z5;
G96 M03 S200;
M08;
G00 X45 **Z0.100**; [0.100 mm material left for final cut]
G01 *X0.0* F0.12; **Rough Facing** operation
G00 Z1.0;
G00 X37.9;
G01 Z-21.0 F0.25;
G01 *X38.9* F0.25;
G01 X40.9 Z-22.0 F0.25;
G00 *Z1.0*;
G00 *X34.9*;
G01 Z-21.0 F0.25;
G01 X36.9 F0.25;
G01 X40.9 Z-23.0 F0.25;
G00 Z1.0;
G00 X31.9;
G01 Z-21.0 *F0.25*;
G01 X37.0 F0.25;
G00 Z1.0;
G00 X28.9;
G01 Z-21.0 F0.25;
G01 *X37.0* F0.25;
G00 Z1.0;

G00 X25.9;
G01 *Z-21.0* F0.25;
G01 X37.0 F0.25;
G00 Z1.0;
G00 *X23.9*; Starting point of rough chamfer (SPRC)
G01 Z0.0 F0.12;
G01 X25.9 Z-1.0 *F0.25*; Ending point of rough chamfer (EPRC)

G00 Z1.0;
G00 X21.9; SPRC
G01 *Z0.0* F0.12;
G01 *X25.9* Z-2.0 F0.25; EPRC
G00 Z1.0;
G00 *X19.9*; SPRC
G01 Z0.0 F0.12;
G01 X25.9 Z-3.0 F0.25; EPRC
G00 Z1.0;
G00 X18.9; SPRC
G01 *Z0.0* F0.25;
G01 X25.9 *Z-3.5* F0.25; EPRC
G00 Z1.0;
M09;
M05;
G00 X100 Z150;
M00;
T0303 (Final Turning Tool);
G00 X100 Z150;
G00 X30.0 Z5;
G96 *M03* S230;
M08;
G00 X30 **Z0.0; Final Facing** operation
G01 *X0.0* F0.12; Remaining material (0.100 mm) of the work piece face will be remove.
G00 Z1.0;
G00 *X17.7*;
G01 X18.7 Z0.0 F0.12; Starting point of finish chamfer
G01 X25.7 Z-3.5 *F0.12*; Ending point of finish chamfer
G01 *Z-21.1* F0.12;
G01 X36.3 F0.12;
G01 *X41.9* Z-23.9 F0.12;
G00 Z1.0;
M09;
M05;
G00 X100 Z150;
M30;

10. This answer key is for Figure 9.300 [Chapter No. 9]

O12126;
G54;
G50 S1700;
T0101 (Rough Turning Tool);
G00 X200 Z300;
G00 X42.0 Z5.0;
G96 M03 S200;
M08;
G00 X42.0 ***Z0.1***; 0.100 mm material left for final turning operation.
G01 X0.0 *F0.12*; **Facing** operation
G00 Z1.0;
G00 X31.5;
G01 Z-33.5 *F0.25*;
G01 *X34.7* Z-39.9 F0.25;
G00 Z1.0;
G00 X28.2;
G01 *Z-6.9* F0.25;
G01 *X31.5* Z-25.1 F0.25;
G00 Z1.0;
G00 X25.4;
G01 Z-6.9 F0.25;
G01 *X28.5* F0.25;
G00 Z1.0;
M09;
M05;
G00 X200 Z300;
M00;
T0505; (Drilling Tool)
G00 X200 Z300;
G00 X100.0 Z25.0;
G97 M03 S450;
G00 X0.0 Z5.0;
M08;
G00 Z1.0;
G01 Z-5.0 *F0.05*; Drilling operation starts
G00 *Z5.0*;

```
G00 Z-4.0;
G01 Z-10.0 F0.05;
G00 Z5.0;
G00 Z-9.0;
G01 Z-15 F0.05;
G00 Z5.0;
G00 Z-14.0;
G01 Z-20.0 F0.05;
G00 Z5.0;
G00 Z-19.0;
G01 Z-25.0 F0.05;
G00 Z5.0;
G00 Z-24.0;
G01 Z-26.6 F0.05;
G00 Z5.0;
M09;
M05;
G00 X200 Z300;
M00;
T0303 (Final Turning Tool);
G00 X200 Z300;
G00 X30.0 Z5.0;
G96 M03 S220;
M08;
G00 X30.0 Z0.0;
G01 X0.0 F0.12; Facing operation
G00 Z1.0;
G00 X25.2;
G01 Z-7.0 F0.12;
G01 X28.0 F0.12;
G01 X31.3 Z-25.2 F0.12;
G01 Z-33.6 F0.12;
G01 X34.7 Z-40.0 F0.12;
G00 Z1.0;
M09;
M05;
G00 X200 Z300;
M30;
```

11. This answer key is for Figure 9.301 [Chapter No. 9]

O55844;
G54;
G50 S1700;
T0101; (Rough Turning Tool)
G00 X100 Z150;
G00 X40.0 Z5.0;
G96 M03 S200;
M08;
G00 X40.0 ***Z0.1***; 0.100 mm material left for final turning operation.
G01 X0.0 F0.12; **Facing** operation
G00 Z1.0;
G00 *X31.4*;
G01 Z-27.2 *F0.25*;
G01 X35.3 *Z-40.2* F0.25;
G00 Z1.0;
G00 X27.4;
G01 *Z-26.2* F0.25;
G02 *X29.4 Z-27.2* R1.0 F0.25;
G01 X35.5 F0.25;
G00 Z1.0;
G00 *X23.6*;
G01 Z-18.1 F0.25;
G01 X25.4 F0.25;
G01 X27.4 *Z-19.1* F0.25;
G00 Z1.0;
G00 *X19.6*;
G01 Z-7.7 F0.25;
G01 X21.6 F0.25;
G03 X23.6 Z-8.7 *R1.0* F0.25;
G00 Z1.0;
G00 X17.6;
G01 Z0.0 *F0.12*;
G01 X19.6 Z-1.0 F0.25;
G00 Z1.0;
M09;
M05;
G00 X100 Z150;

M00;
T0303; (Final Turning Tool)
G00 X100 Z150;
G00 X25.0 Z5.0;
G96 M03 S230;
M08;
G00 X25.0 **Z0.0**;
G01 *X0.0* F0.12; **Facing** operation
G00 Z1.0;
G00 X16.4;
G01 X17.4 *Z0.0* F0.12;
G01 *X19.4* Z-1.0 F0.12;
G01 Z-7.8 F0.12;
G01 X21.4 F0.12;
G03 *X23.4* Z-8.8 **R1.0** F0.12;
G01 *Z-18.2* F0.12;
G01 X25.2 F0.12;
G01 *X27.2* Z-19.2 F0.12;
G01 Z-26.3 F0.12;
G02 X29.2 *Z-27.3* ***R1.0*** F0.12;
G01 *X31.2* F0.12;
G01 X35.1 *Z-40.3* F0.12;
G00 Z1.0;
M09;
M05;
M30;

11.6 Formulas of Cutting Speed, Spindle Speed, Feed, Feed Per Tooth and Cutting Time

Cutting speed $Vc = \frac{\pi \times D \times n}{1000}$

Spindle speed $n = \frac{Vc \times 1000}{D \times \pi}$

Feed $Vf = n \times fz \times Z$

Feed per tooth $Fz = \frac{Vf}{n \times Z}$

where,

$\mathbf{V_c}$ = Cutting speed (mm/min)
π = 3.14
$\mathbf{D}$ = Diameter (mm)
$\mathbf{n}$ = Spindle speed (rev./min)
$\mathbf{V_f}$ = Feed (mm/min)
$\mathbf{f_z}$ = Feed per tooth (mm/tooth)
$\mathbf{Z}$ = Number of flutes

11.6.1 Cutting Time

$$t = \frac{L}{f \times n}$$

where,

$\mathbf{n}$ = Spindle speed (revolutions/min)
$\mathbf{f}$ = Feed (mm/min)
$\mathbf{L}$ = Cutting length

11.7 Important Notes

- **Machining:** Machining is the manufacturing process in which extra material is removed by the cutting tool during turning, milling, drilling, taping, grinding operations etc.
- **Symbol Ø** is called diameter. It expresses the work piece has round shape.
- During the process of the material removing in CNC machine, when the cutting tool comes near the work piece to take position before machining, cutting tool travels and cover rapidly (G00) this distance in few segments. Due to the safety of cutting tool, machine, work piece and operator.
- Generally metric (mm) system is used in mechanical manufacturing drawings but in some cases inches system is also used in mechanical manufacturing drawings.

References

[1] Drawings and sketches has been generated from Auto CAD 2014.
[2] Fundamentals of metal cutting and machine tools, B. L. Juneja, G. S. Sekhon and Nitin Seth, Revised Second Edition (2005), New Age International Publisher, India.

Index

About the Authors

Pawan Kumar Negi received his Mechanical Engineering degree from The Institution Of Mechanical Engineers (India), Mumbai, India, in 2015. Previously, he did his Diploma in I & C Engineering from Government Institute, Kanpur, India. Also he obtained various short term professional engineering courses in CNC Milling Machine Operating and Programming, Computer Aided Manufacturing (CAM) and CAD Design (CAD) from Government of India between 2001 and 2009. Presently, he is serving as a CNC programmer and trainer for more than six years in Graphic Era (Deemed to be University), India. He has taught CNC Technology, Automation and CNC programming. Before that he served as a CNC programmer in a reputed Oil and Gas industry, Weatherford Oil Tool Middle East Limited in United Arab Emirates and in EuroTechnology Group, in Saudi Arabia. In addition, he served in Sara Services & Engineers Pvt. Ltd., India, a reputed oil field company, as a CNC programmer cum Shop supervisor. Mr. Pawan has 11 years industrial and more than 6 years technical educational experience in the field of Mechanical engineering.

Dr. Mangey Ram received the Ph.D. degree major in Mathematics and minor in Computer Science from G. B. Pant University of Agriculture and Technology, Pantnagar, India, in 2008. He has been a Faculty Member for around ten years and has taught several core courses in pure and applied mathematics at undergraduate, postgraduate, and doctorate levels. He is currently a Professor at Graphic Era (Deemed to be University), Dehradun, India. Before joining the Graphic Era, he was a Deputy Manager (Probationary Officer) with Syndicate Bank for a short period. He is Editor-in-Chief of *International Journal of Mathematical, Engineering and Management Sciences* and the Guest Editor & Member of the editorial board of various journals. He is a regular Reviewer for international journals, including IEEE, Elsevier, Springer, Emerald, John Wiley, Taylor & Francis and many other publishers. He has published 125 research publications in IEEE, Taylor & Francis, Springer, Elsevier, Emerald, World Scientific and many other national and international journals of repute and also presented his works at national and international conferences. His fields of research are reliability theory and applied mathematics. Dr. Ram is a Senior Member of the IEEE, life member of Operational Research Society of India, Society for Reliability Engineering, Quality and Operations Management in India, Indian Society of Industrial and Applied Mathematics, member of International Association of Engineers in Hong Kong, and Emerald Literati Network in the U.K. He has been a member of the organizing committee of a number of international and national conferences, seminars, and workshops. He has been conferred with "Young Scientist Award" by the Uttarakhand State Council for Science and Technology, Dehradun, in 2009. He has been awarded the "Best Faculty Award" in 2011 and recently Research Excellence Award in 2015 for his significant contribution in academics and research at Graphic Era.

Om Prakash Yadav is a Professor of Industrial and Manufacturing Engineering (IME) at the North Dakota State University. He has also served as Department chair for 3 years, where he is responsible for teaching, research, and outreach activities for IME Department. He is the founding director of the Center for Quality, Reliability, and Maintainability Engineering (CQRME) located in IME Department at North Dakota State University. The CQRME is sponsored and fully funded by 10 member companies and has undertaken some interesting research projects. His research interests lie in the area of quality and reliability engineering (reliability-based design, robust design, design optimization, and failure analysis), product and operations management (logistics and supply chain, inventory modeling, and manufacturing systems analysis), and quantitative modeling using optimization techniques, statistical modeling, data mining, fuzzy logic and neural network approaches. His diverse educational and academic background combined with industrial experience allows him to undertake challenging and integrative type of research problems. He has been collaborating with researchers across the world and published several high quality research papers resulting through his collaboration with colleagues from other universities and departments. He has extensive experience in failure analysis and reliability analysis of complex mechanical, electrical/electronic, and hybrid systems including complex physical systems. He has developed an active research group with external funding from several sources including National Science Foundation (NSF), Department of Energy (DoE), NASA, Department of Defense supported Manufacturing Institute (NextFlex), ND State, and industry. Establishing CQRME with the support of local member companies is a great achievement and honor to be able to have first center in the history of Engineering College approved by ND State Board of Higher Education. He has over 60 journal papers, 3 book chapter, 80 conference proceedings. He has been invited to give around 20 talks at national and international meetings, and serves as a reviewer and referee in his field on a regular basis. His involvement in professional societies and profession is very intense including taking up leadership roles. He has served as president of Quality Control and Reliability Engineering (QCRE) division of IISE, Board of Directors of QCRE, and member of Management Committee of Reliability and maintainability Symposium (RAMS) since 2010.

Contact for Lab Manuals and Practical

Contact Details

E-mail id: pawancncprogindia@gmail.com

Author

Pawan Kumar Negi
India